AIR CONDITIONING SYSTEMS

AIR CONDITIONING SYSTEMS

PRINCIPLES, EQUIPMENT, AND SERVICE

Air Conditioning and
Refrigeration Institute

Joseph Moravek
Lee College
Baytown, Texas

Prentice
Hall

Upper Saddle River, New Jersey
Columbus, Ohio

Library of Congress Cataloging-in-Publication Data

Air conditioning systems: principles, equipment, and service / Air Conditioning and
Refrigeration Institute and Joseph Moravek.
     p. cm.
   Includes index.
   ISBN 0-13-517921-1
    1. Air conditioning. I. Moravek, Joseph. II. Air-Conditioning and Refrigeration
Institute.

TH7687.A4833 2001
697.9'3—dc21

00-038487

Vice President and Publisher: Dave Garza
Editor in Chief: Stephen Helba
Executive Editor: Ed Francis
Production Editor: Christine M. Buckendahl
Production Coordination: Lisa Garboski, bookworks
Design Coordinator: Robin G. Chukes
Production Manager: Pat Tonneman
Marketing Manager: Jamie Van Voorhis

This book was set in Meridien by The Clarinda Company, and was printed and bound by Courier Kendallville, Inc. The
cover was printed by Phoenix Color Corp.

10 9 8 7 6 5 4 3 2 1

ISBN: 0-13-517921-1

To my wife Martha
and our sons Zachary and Luke

PREFACE

Much of the information in this textbook is based on the Air Conditioning and Refrigeration Institute (ARI) Curriculum Guide. The curriculum guide was developed by leaders of the air-conditioning industry to act as a standard of what entry-level air-conditioning technicians should know when they enter the field. The ARI Curriculum Guide is a compilation of many labor-hours of serious discussion and thought about what to expect of an entry-level technician. This book addresses many of these important points.

The title, *Air Conditioning Systems: Principles, Equipment, and Service* prepares the reader for the book's content. The focus on the book is on air-conditioning systems and the components that make up these systems. The text discusses how the system and system components operate as a cooling unit. Another emphasis of the book is to provide information that will be useful to the service technician and suggest ways to maintain, service, and troubleshoot air-conditioning equipment. There are many practical and useful ideas in the chapters that follow.

Our goal is to provide theoretical and practical information to the entry-level technician. The practical information could be used to design laboratory activities. Generally, educators and the air-conditioning industry are advocates of competency-based education. Competency-based education means that students exhibit their learned skills in several different ways. The student should be able to write or draw an explanation of how something in an air-conditioning system operates. For example, the stu-

dent should be able to draw the basic refrigeration cycle and list all the entering and leaving conditions of each of the four major components. In addition, he or she should be able to exhibit his or her practical knowledge of an air-conditioning system. Practical knowledge is tested by having the student identify the four major components of an active air-conditioning system and measure the pressures and temperatures entering and leaving each of these components. The critical part of competency-based education is being able to exhibit to the instructor that the student can do basic field-related work. Competency needs to be displayed on an ongoing basis throughout a student's coursework. Most students can do a lab activity correctly one time. The problem with doing it one time is that the behavior is not repeated, so it is quickly forgotten. For example, when describing charging procedures, the text begins with the basic pressure-temperature method of charging. It builds on this charging procedure by adding other methods, such as checking superheat, subcooling, and amperage. Each time a new charging procedure is introduced, students use previously learned charging methods along with the newly learned charging skill. This builds on the charging procedures and reinforces the charging methods previously learned. Students will retain this knowledge when they demonstrate repetitive skilled behavior. In any event, repeating the exercises correctly several times during a course will commit the behavior to memory and it will become second nature.

This book was written to supplement and advance skills of the entry-level student. It is not intended to be a textbook for first-semester students unless they are enrolled in other basic coursework. The user of this textbook should have a basic knowledge of the refrigeration cycle, basic skills in electrical troubleshooting, and an understanding of the uses of the multimeter. Students should have already taken an air-conditioning course that included basic laboratory activities. The sooner students place this information into practice, the sooner they will develop habits that are important for a professional career in air conditioning.

Chapters include extensive coverage of charging air-conditioning systems and refrigerant and electrical troubleshooting procedures. The troubleshooting section includes both component diagnostics and system diagnostics. The book includes important chapters on safety and the refrigeration cycle. Preventive maintenance is a big part of equipment longevity. The chapter on preventive maintenance is extensive and can be used as a guide for servicing. The chapter on industry certification explains the different options available to the entry-level technician and the experienced professional.

The student's success is the most important objective of the writer and reviewers. Information must be placed into practice for it to be useful and retained. This is a useful source of technician knowledge in the classroom and in the air-conditioning field. Enjoy your learning journey. As long as you are in air-conditioning technology, you will always be learning something new.

ACKNOWLEDGMENTS

I would like to thank all those who were instrumental in facilitating the completing of this book, starting with my wife, Martha, who has always been supportive of my many ventures over the many years of our marriage. Leslie Sandler of ARI encouraged me to write this book and set up the contacts to begin this journey.

In the 1970s, my start in air conditioning began with Walter Hunt at the Gulf Coast Community Services, Houston, Texas, energy conservation program. Next, my knowledge and experience expanded while working under the supervision of Barby Barbara at the Texas Energy Extension Services at the University of Houston. My career diversified with employment at the City of Houston, Texas. While working for the City of Houston, I was exposed to commercial energy conservation, HVAC/R maintenance, and air-conditioning inspections. Guy Ellyson, Chief HVAC Inspector for the City of Houston, motivated me to learn more about the air-conditioning trade. Next, Larry Giroux, chief trainer of AES Houston, became a driving force behind my knowledge of the trade. For the past eighteen years I have learned from Larry and bothered him with problems and concerns I have had regarding air-conditioning systems and instructional delivery methods used to train technicians. He is a mentor to me.

My current employer, Lee College, has been a supporter of this project by allowing me the opportunity to write and learn during the summer. Allowing me free rein of an air-conditioning training program has taught me much. Lee College encourages writing as a way of growth and professional development. I must thank all my students who keep me challenged. Many of their mistakes have been shared in this book so as not be repeated by new students. I have learned much from my students. They are creative in ways of getting things done while maintaining a professional sense about their work in the classroom and in the laboratory.

I would be remiss if I did not thank Prentice Hall editor Ed Francis for his guidance on this long project. I must give a special thanks for the final editing done by Linda Thompson, and the project management skills of Lisa Garboski of bookworks. They really "cleansed" the material and molded it into the professional document that you see. These women know their air conditioning grammar, or at least they have me fooled. I must also thank Larry Jeffus of Eastfield College, a prolific author himself, for sharing his experiences and advice as an author.

Finally, I must thank all those organizations, companies, and corporations that contributed and allowed me to use their material in this book. This was one of the most difficult tasks on the road to developing this textbook. Gathering appropriate materials and gaining approval for their use was a monumental effort. The completion of this book would have been impossible without their help. The list of supporters includes:

A-1 Components
Alco Controls
Air-Conditioning Refrigeration Institute
Alnor Instruments
Amprobe Promax
Armstrong Air Conditioning
Bard Manufacturing
Copeland Corporation
Domestic Manufacturing
Fasco Motors and Blowers
General Electric Company
Heatcraft Refrigeration Products
Henry Valve Company
Honeywell Corporation

ICM Corporation
Intercity Products
J/B Industries
Lennox Industries
MSA Instrument Division
Marley Cooling Towers
Mars Corporation
North American Technicians Excellence, Incorporated
Neutronics Incorporated
Prentice-Hall Publishers
Rheem/Ruud Air Conditioning Company
Ritchie Manufacturing Company, Yellow Jacket Division

Robinair, SPX Corporation
Rotorex Company, Incorporated, Fedders Corporation
Sears, Roebuck and Company
Sentech Corporation
Shortridge Instruments, Incorporated
Spectronics Corporation
Sporlan Valve Company

TIF Instruments, Incorporated
Tecumseh Compressor Company
Thermal Engineering Company
Toxalert, Incorporated
Trane Company
Worker's Compensation Board of British Columbia
York International Corporation

All these organizations are supporters of education and providing information to get the job done professionally.

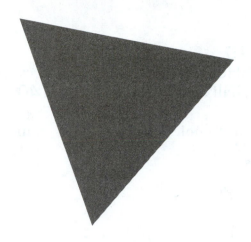

CONTENTS

AIR CONDITIONING SYSTEMS

CHAPTER ONE

SAFETY ON THE JOB

AFTER STUDYING THIS CHAPTER, THE STUDENT WILL BE ABLE TO:

•Describe recommended protective clothing and equipment for the HVAC/R technician.
•Perform service activities conforming with approved practices in personal safety.
•Select the correct fire extinguisher for classes of fire.
•Demonstrate the use of safe shop practices.
•Recognize unsafe conditions and list rules for shop and personal safety.

•Demonstrate the use of safe electrical practices.
•State the effect of different levels of amperes on the human body.
•Discuss rescue procedures.
•Explain the procedure for lockout and tagout system.
•List specific safety rules pertaining to the refrigeration trade.

KNOWLEDGE EXTENSION

Landmarks in the development of refrigerants are as follows.

1834 The first practical refrigerating machine (Jacob Perkins') used ether in a vapor compression cycle.

1849 Ferdinand Carre used ammonia and water in his refrigeration machine.

1850 The first absorption machine (Edmond Carre's) used water and sulfuric acid.

1866 Vapor compression systems used a mixture of petrol ether and naphtha (patented as chemogene). The same year carbon dioxide was introduced as a refrigerant.

1873 Ammonia was first used in vapor compression systems.

1875 Sulfur dioxide and methyl ether were introduced as refrigerants.

1878 Methyl chloride emerged as a refrigerant, soon followed by dichloroethene (dilene) used by Willis Carrier in his first centrifugal compressor.
All these were either flammable, toxic, or both. Accidents were common.

1926 Carrier switched to methylene chloride for his centrifugal compressors.

1930 R-12, nontoxic and nonflammable, was synthesized. It was the first of the chlorofluorocarbon (CFC) refrigerants. R-12, R-11, R-114, and R-113 were produced from 1931 to 1934.

1936 The first hydrochlorofluorocarbon (HCFC) refrigerant, R-22 was produced.

By 1963 the above five refrigerants constituted 98 percent of the total production of a thriving organic fluorine industry.

PERSONAL SAFETY

In all trades, safety is a major concern. Accidents are caused by carelessness, as well as lack of awareness of proper safety procedures. This chapter deals with some of the safety tips and procedures the refrigeration technician should follow—whether on the job site or on related locations where hazards could exist.

The following are some of the important areas in practicing personal safety that, if followed, will help in preventing accidents:

1. *Protective clothing and equipment,* appropriate to activity being performed.
2. Safety procedures for *handling harmful substances.*
3. *Safe work practices,* including proper care and use of tools.
4. Necessary precautions for *preventing electrical shocks.*
5. *Avoiding refrigerant contacts* with any part of your body. Always *keep the pressure of a confined gas within safe limits.*

PROTECTIVE CLOTHING AND EQUIPMENT

The following clothing and equipment should be used by refrigeration technicians:

HEAD PROTECTION

An approved hard hat (Figure 1–1) or cap should be worn whenever there is a danger of things dropping on the head or where the head may be bumped. On a construction site, proper head gear for protection is a *must.*

Figure 1–1 Hard hat for head protection (showing ear protectors). (*Use granted by Worker's Compensation Board of British Columbia*).

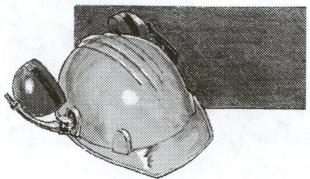

Confine long hair and loose clothing before operating rotating equipment.

EAR PROTECTION

Hearing protection devices (Figure 1–2) must be worn whenever there is exposure to high noise levels of any duration. These devices are of two types: (1) ear plugs, which are inserted in the ear, and (2) ear muffs, which cover the ear. Either one must be properly selected on the basis of how much protection is required.

EYE AND FACE PROTECTION

Approved eye or face protectors (Figure 1–3) must be worn whenever there is a danger of objects striking the eyes or face. Eye and face protectors have various shapes and sizes, some of them very specialized. If prescription eye glasses are worn, they must have side shields.

Special eye protectors must be worn when arc welding, gas welding, and burning to cut out harmful radiation. These come in various shades, which filter out the harmful emissions. Take time to identify the right one for the job. For example, *never wear oxyacetylene welding goggles when arc welding.* Always wear safety glasses when brazing, soldering and handling refrigerant and high-pressure gases.

Figure 1–2 Types of hearing protection equipment. (*Use granted by Worker's Compensation Board of British Columbia.*)

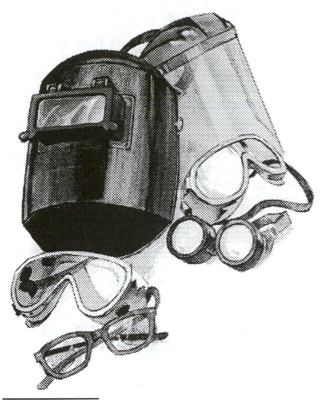

Figure 1–3   Eye and face protection equipment. *(Use granted by Worker's Compensation Board of British Columbia.)*

RESPIRATORY PROTECTION

There are two main types of respirators as shown in Figure 1–4: (1) those that purify the air by filtering out harmful dusts, mists, metals, fumes, gases and vapors; and (2) those that supply clean breathing air from a compressed air source. The second type should always be worn when working in a confined space where concentrations of harmful substances are very high or where the concentration is unknown. Remember that most refrigerants are odorless, tasteless, and invisible, and can cause asphyxiation in a very short time.

Respirators must fit tightly against the skin so that there is no leakage from the outside into the face. Workers who are required to use respirators at any time must be instructed in their use, care, maintenance, and limitations. A self-contained breathing apparatus (SCBA) is required in areas heavily contaminated with refrigerant.

HAND AND FOOT PROTECTION

HANDS

There are many different kinds of gloves (Figure 1–5). Some are made for special usages, such as gloves of steel mesh or Kevlar to protect against cuts and puncture wounds. Different glove materials are needed to protect against a variety of different

Figure 1–4   Various types of respirators. *(Use granted by Worker's Compensation Board of British Columbia.)*

Figure 1–5   Hand and foot protection equipment. *(Use granted by Worker's Compensation Board of British Columbia.)*

chemicals. The EPA recommends butyl lined rubber gloves when handling refrigerants. Choose the right kind from a dependable supplier who can supply this information. Discard the damaged ones.

FEET

When choosing foot protection (Figure 1–5) use the following guidelines:

1. All footwear must be well constructed to support the foot and to provide secure footing.
2. Where there is danger of injury to the toes, top of the foot, or from electrical shock, the proper shoe or boot must have Construction Safety Approval (CSA) indicated.
3. Where there is danger of injury to the ankle, footwear must cover the ankle and have a built-in protective element/support.
4. If there is danger of harmful liquids dropping on the foot, the top of the shoe must be completely covered with an impervious material or treated to keep the dripped substance from contacting the skin.

Figure 1–6   Safety belts and harnesses. *(Use granted by Worker's Compensation Board of British Columbia.)*

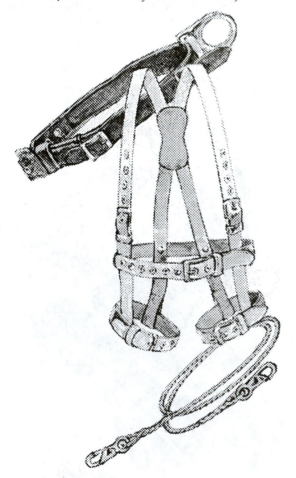

FALL PROTECTION

Two methods of preventing injury from falling are: (1) fall prevention equipment and (2) fall-arresting equipment. Either of these methods are required when working at heights over 10 ft above grade when no other means have been provided for preventing falls. Figure 1–6 illustrates safety belts and harnesses.

In *fall prevention,* workers are prevented from getting into a situation where they can fall. For example, a safety belt attached to a securely anchored lanyard will limit the distance a worker can move.

In *fall arresting* the worker must wear a safety harness attached to a securely anchored lanyard that will limit the fall to a safe distance above impact. The harness helps prevent the worker from suffering internal damage. Belts should not be used to arrest a fall. Where a fall-arresting system is not practicable, a safety net should be suspended below the work activity. The worker should be secured separately from the tools and equipment.

HARMFUL SUBSTANCES

Workers in the mechanical trades can be exposed to a variety of harmful substances, such as dust, asbestos, carbon monoxide, refrigerants, resins, adhesives, and solvents.

All *dust* can be harmful. Where dust cannot be controlled by engineering methods, an approved respirator designed to filter out specific dust must be worn.

When *asbestos*-containing material (insulation) is being cut or shaped, the particles must be removed by a ventilation system which discharges them through a high efficiency particulate air (HEPA) filter. All waste materials that contain asbestos must be placed in impervious bags for transfer to an approved disposal site. These fibers, when inhaled, are considered carcinogenic.

Mobile equipment operating in an enclosed area can produce dangerous levels of *carbon monoxide* (CO). Oil or gas-fired space heaters without suitable vents can also produce carbon monoxide. Areas must be well ventilated while being heated with these devices.

Some *refrigerants* are more dangerous than others. All refrigerants are dangerous if they are allowed to replace the oxygen in the air. Even the so-called safe refrigerants can produce a poisonous phosgene gas when heated to high temperatures. Phosgene gas is formed when refrigerant comes in

ON THE JOB

How safe are refrigerants? All have concerns associated with toxicity, flammability, and physical hazards. Now, more so than in the past, these can be used safely if one recognizes there is not now, or likely to be discovered in the future, the "ideal refrigerant," free from the above concerns.

The new, alternative refrigerants can be used with comparable or higher safety than those they replace when applied following recommended selection, handling, installation, and operating practices in equipment conforming to recognized safety standards.

Inherent risks to refrigerants are:

1. *Quantity* or *concentration levels. Acute toxicity* refers to the impacts of single exposures, often at high concentrations. *Chronic toxicity* refers to the effects of repeated or sustained exposures over a long time. The Program for Alternative Fluorocarbon Toxicity (PAFT) Testing is a cooperative effort to reduce both risks. Sponsored by the major CFC producers from nine countries, this program has initiated intensive testing for R-123, R-134a, R-142b, R-124, R-125, R-225ca, R-225cb, and R-32 refrigerants, and is ongoing. These studies investigate acute, subchronic and chronic toxicity as well as genetic effects on man and other living organisms in the environment. Another program simply investigates effects on tumors, benign or malignant.

2. *Accidents, failures* or *incidents of short duration.*
   a. Direct exposure to refrigerants at low temperatures can cause frostbite. Use appropriate eye protection and gloves.
   b. Suffocation from leaks and spills is also a risk. Install leak sensors and use proper installation, handling, and storage procedures.
   c. Flammability increases with temperature and pressure, especially in the presence of combustible lubricants. Accidents have occurred even with completely nonflammable refrigerants. Never use R-22 mixed with air for leak testing; use dry nitrogen or carbon dioxide gas instead.
   d. Other preventive measures: Properly identify container and storage vessels. Install one or more pressure-relief devices on all refrigerating systems in case of fire or other abnormal conditions. Limit machinery room access to authorized personnel. Provide at least one self-contained breathing apparatus (SCBA) for emergency use.

contact with an open flame such as a torch, gas furnace, or boiler burners. Hydrochloric or hydrofluoric acid can form in the refrigerant oil and damage the skin during installation of refrigerant hoses. Refrigerants sprayed on any part of the body can quickly freeze tissue. The safe handling of refrigerants is discussed in detail in a later part of this chapter.

Resins, adhesives, and *solvents* can be dangerous if not properly handled. Ensure that the workspace is continuously ventilated with large amounts of fresh air.

Never use *carbon tetrachloride* for any purpose, because it is extremely toxic, either inhaled or on the skin. Even slight encounters with it can cause chronic problems. Consult a physician if exposed to it.

SAFE WORK PRACTICES

The refrigeration and air conditioning technician works in many areas: in the shop, in various types of buildings, in equipment rooms, on rooftops, and on the ground outside buildings. Each location requires different activities where safe performance is essential.

In addition, the worker deals with many potentially dangerous conditions, such as handling pressurized liquids and gases, moving equipment and machines, working with electricity and chemicals, and exposure to heat and cold. It is important, therefore, that the technician practice good safety procedures wherever the work is being done or whatever part of work in which he or she is engaged.

HAND TOOLS

1. Keep all hand tools sharp, clean, and in safe working order.
2. Repair or replace defective tools.
3. Use correct, proper-fitting wrenches for nuts, bolts, and objects to be turned or held.
4. Do not work in the dark; use plenty of light.
5. Do not leave tools on the floor.

POWER TOOLS

1. Use only power tools that are properly grounded.
2. Stand on dry nonconductive surfaces when using electrical tools.
3. Use only properly sized electrical cords in good condition.
4. Turn on the power only after checking to see that there is no obstruction to proper operation.
5. Disconnect the power from an electrical tool (or motor) before performing the maintenance task of oiling or cleaning.
6. Disconnect the power supply when equipment is not in use.
7. Disconnect the power supply when changing blades, bits, or accessories.

SHOP SAFETY

1. Keep the shop or laboratory floor clear of scraps, litter, and spilled liquid.
2. Store oily shop towels or oily waste in metal containers in an open, airy place.
3. Clean the chips from a machine with a brush; do not use a towel, bare hands, or compressed air.
4. Keep safety glasses and gloves in a prominent location adjacent to machinery used for grinding, buffing, or hammering and where material with sharp edges is handled.

5. Establish cleaning periods. Make sure everyone is clear when using compressed air to clean.

STEPS IN MAINTAINING AN ORDERLY SHOP

1. Arrange machinery and equipment to permit safe, efficient work practices and ease in cleaning.
2. Materials, supplies, tools, and accessories should be safely stored in cabinets, on racks, or in other readily available locations.
3. Working areas and work benches should be clear. Floors should be clean. Keep aisles, traffic areas, and exits free of materials and obstructions.
4. Combustible materials should be properly disposed of or stored in approved containers.
5. Drinking fountain and wash facilities should be clean and in good working order at all times.

FIRE EXTINGUISHERS

The danger of fire is always present. Oil, grease or paint soaked rags can ignite spontaneously. Keep them in metal containers.

Sparks, open flames, and hot metal can ignite many materials. Always have a fire extinguisher close at hand when welding or burning.

Extreme caution should be taken with highly flammable and volatile solvents. Due to its low flash point (the temperature at which vapors will ignite), gasoline should never be used as a cleaning solvent.

Fire extinguishers should be readily accessible, properly maintained, regularly inspected, and promptly refilled after use. They are classified according to their capacity for handling specific types of fires (Figure 1–7).

Class A Extinguishers These are used for fires involving ordinary combustible materials such as wood, paper, and textiles, where a quenching, cooling action is required.

PREVENTIVE MAINTENANCE

Proper maintenance and care is required to extend the life and reliability of all tools and test equipment. This includes cleaning tools of grease and oils regularly upon job completion. Gauges should be hung up or securely fastened and recalibrated monthly. Test instruments should always be turned off when work is complete and reinstalled in carrying cases. Low batteries should be replaced immediately and discarded in appropriate containers. Vacuum pumps should be drained of oil and refilled with fresh oil when work is complete. Oxygen-acetylene welding units should have caps on containers and be secured when moving. All welding regulators should be back sealed and bled free of any residual gas pressures.

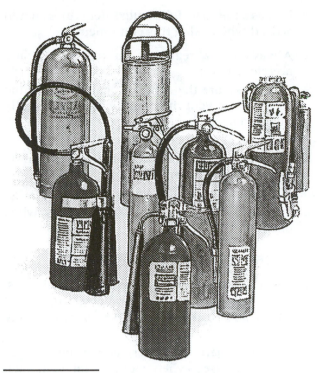

Figure 1—7 Types of fire extinguishers. *(Use granted by Worker's Compensation Board of British Columbia.)*

Class B Extinguishers These are for flammable liquid and gas fires involving oil, gas, paint, and grease, where oxygen exclusion or flame interruption is essential.

Class C Extinguishers These are for fires involving electrical wiring and equipment where non-conductivity of the extinguishing agent is critical. This type of extinguisher should be present whenever functional testing and system energizing take place.

MATERIAL HANDLING

Use mechanical lifting devices whenever possible. Use a hoist line when lifting tools or equipment to a roof. If you are required to lift a heavy object, get help. In order not to strain your back, the following procedures should be observed when lifting heavy objects. (Figure 1–8):

1. Bend your knees and pick up the object, keeping your back straight up.
2. Gradually lift the weight using your leg muscles, continuing to keep your back straight up.
3. You may not be able to lift as much weight with proper lifting procedures. Get help when lifting heavy or bulky objects.

ACCESS EQUIPMENT

Access equipment refers to ladders and scaffolds that are used to reach locations not accessible by

Figure 1–8 Properly lifting heavy loads. *(Use granted by Worker's Compensation Board of British Columbia.)*

other means. The following precautions should be practiced in the use of *ladders* (Figure 1–9):

1. Only use CSA- or ANSI-approved ladders. Maintain ladders in good condition. Inspect ladders before each use and discard ladders needing frequent repairs or showing signs of deterioration.

Figure 1—9 Proper use of a ladder. *(Use granted by Worker's Compensation Board of British Columbia.)*

2. All portable ladders must have nonslip feet.

3. Place ladders on a firm footing, no further out from the wall than ¼ the length of the ladder.

4. Ladders must be tied, blocked, or otherwise secured to prevent them from slipping sideways.

5. Never overload a ladder. Follow the maximum carrying capacity of the ladder, including the person and equipment. The American National Standards Institute (ANSI) sets the standard for ladders.

6. Only one person should be on a ladder, unless the ladder is designed to carry more people. Follow maximum load rating.

7. Never use a broken ladder or a ladder on top of scaffolding.

8. Always face the ladder and use both hands when climbing or descending a ladder.

9. Use fiber glass or wood ladders when doing any work around electrical lines.

10. Ladders should be long enough so you can perform the work comfortably, without learning or having to go beyond the two rungs below the top rung safety barrier.

11. Step ladders should only be used in their fully open positions.

The following recommendations apply to *scaffolds:*

1. Scaffolds must be supported by solid footings.

2. A scaffold having a height exceeding 3 times its base dimension must be secured to the structure.

3. Wheels of rolling scaffolds must be locked when used by workers and/or their materials.

4. No worker is to remain on the scaffold while it is being moved.

5. Access to the work platform must be a fixed vertical ladder or other approved means.

Welding and Cutting

Welding and cutting is a specialized skill and requires special training. Many refrigeration and air conditioning technicians acquire this training due to the need to perform some of these operations as part of their work. It must be recognized, strictly from a safety standpoint, that this work should not be attempted without adequate knowledge and instruction.

Air-Acetylene and Oxyacetylene Torches

Torches are used by HVAC/R technicians to produce sufficient heat for silver (hard) soldering and brazing. These torches use a mixture of air or oxygen and acetylene as a fuel. The following safety rules should be practiced when using this equipment:

1. Always use a regulator on the oxygen and acetylene tanks.

2. Always secure the cylinder against something solid to prevent it from being accidentally knocked over.

3. Wear the proper safety glasses.

4. Open the valve on the acetylene cylinder only one-quarter of a turn.

5. Light the torch with a striker.

6. Do not stand in front of the gauges when opening.

Commercial Job Site Safety

1. Always follow contractors safety guidelines. Most safety guidelines parallel Occupational Safety and Health Act (OSHA) or state safety rules.

2. Hard hat, hard-toe shoes, and safety glasses should be used when required for your own safety.

3. Every person who goes to work in the morning has a right to return home without injury after his or her work is done.

First Aid

Refrigeration and air-conditioning workers are advised to enroll in an approved first-aid course. Prompt and correct treatment of injuries not only reduce pain but could also save lives. A classification of accidents that occur to HVAC/R personnel, related to the hazards described, includes the following:

1. Injuries due to mechanical causes

2. Injuries due to electrical shocks

3. Injuries due to high pressure

4. Injuries due to burns and scalds. Injuries due to explosions

5. Injuries due to breathing toxic gases

Steps to Be Followed in Case of an Accident

1. All accidents, injuries, and illnesses should be reported to whomever is in charge, no matter how minor injuries may seem.

2. First aid should be administered if needed, only by those qualified to do so. Posted emergency procedures should be followed, as applicable. For example: If liquid refrigerant is sprayed on the skin or in the eyes, flush the area with cold water and get treatment.

3. The victim may be sent or taken to receive medical services.

4. An accident report form should be filled out by the person in charge.

5. The area should be cleaned up to remove any contaminants causing the injury before permitting the area to be used again.

6. An investigation of the accident should be conducted to determine the cause of the accident and to determine ways to prevent similar incidents.

SAFETY WHEN WORKING WITH ELECTRICITY

All possible precautions must be practiced to prevent electrical shock—current passing through the body. Very few realize the damage that can be done even by a small amount of current.

The following information applies to low-voltage circuits where current is measured in milliamps. One amp is equal to 1000 milliamps.

The tabulation and the illustration in Figure 1–10 indicate the effect on the body when various

Electric Current	Body Sensation
Less than 0.5 mA	No sensation
1 to 2 mA	Muscular contraction
5 to 25 mA	Painful shock, inability to let go
More than 25 mA	Violent muscular contractions
50 to 500 mA	Heat convulsions, death
More than 100 mA	Paralysis of breathing, burns

amounts of current pass through the body—all 100 mA or less.

ELECTRICAL SAFETY RULES

1. Check all circuits for voltage before doing any service work. Tag and lock all electrical disconnects when working on live circuits.

2. Stand on dry nonconductive surfaces when working on live circuits.

3. Work on live circuits only when absolutely necessary.

4. Never bypass an electrical protective device.

5. Properly fuse all electrical lines.

6. Properly insulate all electrical wiring.

7. Ground out component before you reach in to touch it. Finally, backhand a component before touching it.

Figure 1–10 Amperage ratings of electric current creating various shock effects. *(Use granted by Worker's Compensation Board of British Columbia.)*

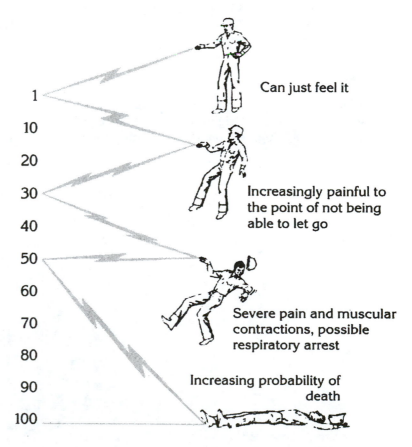

Can just feel it

Increasingly painful to the point of not being able to let go

Severe pain and muscular contractions, possible respiratory arrest

Increasing probability of death

PRACTICING PROFESSIONAL SERVICES

Lockout/tagout is a safety procedure which has been implemented nationwide to establish an easy-to-understand guideline for locking and tagging the control of equipment and sources of energy to prevent unanticipated actuation while work repairs and maintenance are being performed.

When there is no place to physically attach a lock, a tag is substituted and attached to the machines, control switch, energy master switch, or main power panel is there to warn other workers that someone is working on a piece of equipment and could be injured if the equipment is actuated.

Machines or equipment must be stopped and isolated from all sources of energy and the controls must be locked before employees begin to perform service, replacement, or maintenance because unexpected start-up of the equipment could cause injury.

▼ REFRIGERATION SAFETY

The hazards associated with refrigeration service are principally associated with the proper use of refrigerants and their storage in closed containers and systems. A large improvement was made when the industry started using the so-called safe refrigerants (Class I or fluorocarbons) which were nontoxic and nonflammable. Dangers now relate to the use of pressurized gas or liquid and the fact that these chemicals when released accidentally can replace oxygen in a confined space without sensory detection.

The following rules can help decrease the hazards of refrigeration service.

USE OF REFRIGERANTS

1. Good ventilation is essential where work is being done with refrigerants, and whenever welding, brazing, or using a cutting torch.

Don't use torches in high concentrations of refrigerant.
2. Wear safety goggles and gloves when working with refrigerants. Liquid refrigerant can cause "frostbite" when in contact with eyes and skin.
3. Wrap cloth around hose fittings before removing them from a pressurized system or cylinder. Inspect all fittings before attaching hoses or working on them. Use low-loss fittings to reduce refrigerant loss or frostbite (Figure 1–11).
4. Ventilate a room containing refrigerants, pressurized gas, or liquid before entering to work. Breathing a mixture of air and refrigerant can cause unconsciousness because the refrigerant contains no oxygen.
5. Install oxygen monitors and alarm systems. These are required in machinery rooms with refrigeration equipment. Never enter an area where refrigerant is above exposure limits without proper breathing apparatus.

Figure 1–11   Quick-coupling, low-loss fitting, used to prevent loss of refrigerant when attaching and detaching refrigeration hoses. *(Courtesy of Ritchie Engineering Company, Inc., Yellow Jacket Division)*

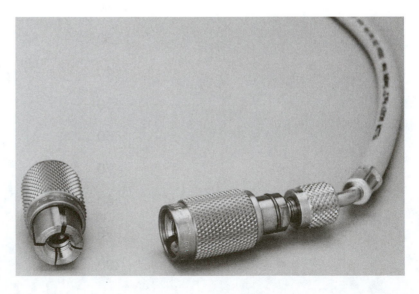

6. Wear gloves when servicing a system where a compressor has burned out. Refrigerant oil contained in the system can be very acidic. It should never be allowed to touch the skin.

7. Shut off and tag valves before working on refrigerant, steam and water lines.

8. Follow all codes when making modifications or repairing any system.

9. Never chip ice or frost from refrigeration line, coils, or sight glass. Use an alcohol spray to melt the ice on a sight glass.

10. Read material safety data sheets (MSDSs).

11. Follow warning and caution signs.

12. Following refrigerant recovery, ensure the system is open to the atmosphere before brazing.

HANDLING REFRIGERANT CYLINDERS

1. Do not fill a cylinder with liquid refrigerant to more than 80% of its volume. Heat can expand the refrigerant and create a rupture pressure. Space must be available inside the cylinder for proper expansion to take place. In recovering refrigerants, this is particularly important. Special cylinders have been designed for recovery that have an automatic volume limiting device.

2. In using a cylinder or transporting it, the cylinder must be secured with a chain or a rope in an upright position. Do not drop a cylinder.

3. Mixing refrigerants is dangerous. Cylinders are color coded to help identify each refrigerant. Each system has an identifying label. Do not mix refrigerants. Maintain the identification system.

4. Never apply a torch to a system containing refrigerant. If heat is needed to vaporize refrigerant, use hot water at a temperature not to exceed 125°F.

5. Do not refill disposable refrigerant cylinders.

6. Replace the cylinder cap when not using a cylinder. The cap protects the valve. Do not lift or carry a cylinder by the valve.

SYSTEM SAFETY

1. Never use compressed air oxygen or acetylene to pressurize a system. Use dry nitrogen or carbon dioxide from a tank properly fitted with a pressure regulator.

2. When isolating a section of piping or component of a system, exercise caution to prevent damage and potential hazard from liquid expansion.

3. Always charge refrigerant *vapor* into the low side of the system. *Liquid* refrigerant entering the compressor could damage the compressor or cause it to burst.

4. Never service a refrigeration system where an open flame is present. The flame must be enclosed and vented outdoors. If a fluorocarbon refrigerant comes in contact with intense heat, it can produce poisonous phosgene gas.

5. Always prevent moisture (water) from entering the refrigeration system. It can cause considerable damage. All parts must be kept dry. Containers of oil must be sealed to prevent contamination from absorption of moisture.

SAFETY EQUIPMENT
Minimum safety equipment should include

> Hard hats
> Safety glasses
> Safety shoes

ON THE JOB

Material safety data sheets (MSDSs) provide detailed information about various chemicals MSDSs are divided into eight sections:

1. Product identification
2. Hazardous ingredients
3. Physical data
4. Fire and explosion information
5. Health and hazard data
6. Reactivity data
7. Spill or leak procedures
8. Safe handling and use

A master file of MSDSs for hazardous chemicals to which employees may be exposed is kept in the MSDS folder, available for review as required.

Gloves

Fire extinguisher

First-aid kit

List of phone numbers (hospital, doctor, fire department, police, etc.)

SUMMARY

- In summary safety is important. This section recommends clothing and equipment:
  - Safety glasses
  - Gloves for refrigerant handling
  - Hardhats for construction sites
  - Work shoes with rubber soles and safety toes
  - Long-sleeve work shirt when brazing or soldering
  - Ear (hearing) protection in machinery rooms
  - Face shields where flying particles may be present
  - Safety lines and belts to prevent falling
  - Respiratory protection in confined spaces or areas of large refrigerant leaks
- Fire extinguisher for classes of fire:
  - Class A—ordinary combustion material (paper, wood, etc.)
  - Class B—flammable liquids
  - Class C—electrical
- Rules for shop and personal safety:
  - Keep shop and lab areas clean.
  - Store oily shop towels in approved metal containers.
  - Clean metal chips from machines with a brush.
  - Wear protective clothing and use proper equipment.
  - Keep work areas organized and aisles, traffic areas free of material and obstacles.
  - Dispose of hazardous waste properly.
  - Keep fire extinguishers present.
  - Use only CSA- or ANSI-approved ladders.
  - Follow recommended safety procedures.
- Material safety data sheets:
  - Provide safety information on all substances.
  - Labs and shops should display and update this material for students and workers.
- Everyone should review electrical safety rules:
  - Check for voltages present before servicing.
  - Use the lockout-tagout procedure to disable electrical circuits when servicing.
  - Stand on a dry surface.
  - Use a high-voltage insulating pad when required.
  - Work on live circuits only when necessary.
  - Never bypass protective devices.
  - Properly fuse or circuit breaker protect.
  - Use one-handed "Hop-Skip" techniques when probing for voltages.
  - Ground out the circuit before touching it.
- Effects of current on the human body range from no sensation at 0.5 mA to heat convulsions and death at 50 to 500 mA. Refer to the section on safety when working with electricity for details.
- Rescue procedures:
  - Turn off the power.
  - Use a nonconductive object to push the victim away.
  - Call for emergency assistance.
  - Perform CPR and/or rescue breathing until help arrives.
- Refrigerant safety:
  - Provide good ventilation.
  - Wear safety goggles and butyl-lined gloves.
  - Wrap cloth around hose fittings before removing them from pressurized system or cylinder or use low-loss fittings.
  - Ventilate room containing refrigerants before entering and/or wear self-contained breathing apparatus.
  - Install refrigerant alarms and monitors.
  - Wear gloves when working with compressor burnouts.
  - Do not fill a cylinder to more than 80% of its volume.
  - Transport refrigerant in the upright, secured position.
  - Do not pressurize refrigeration system with air or oxygen; use only dry nitrogen, carbon dioxide, or a mixture of nitrogen and R-22.
  - Do not mix refrigerants.
  - Never apply a torch to a pressurized vessel.
  - Do not refill a disposable cylinder.
  - Replace cylinder caps.
  - When isolating sections of refrigeration system, exercise caution to prevent liquid expansion.
  - Use only approved refrigeration service techniques.
  - Never service where an open flame is present.

- Prevent moisture from entering the sealed system.
- Environmental safety principles:
  - Become properly trained and certified.
  - Isolate and/or use recovery techniques before opening the sealed system.
  - Never purge refrigerant charge to the atmosphere.
  - Use approved equipment.
  - Use approved service techniques.

PROBLEMS AND QUESTIONS

1. What safety clothing and equipment are recommended for the air conditioning and refrigeration technician?

2. How do you clear your gauge hoses of air?

3. When is it acceptable to release a minimum amount of refrigerant during service or normal operation?

4. What type of fire extinguisher is required for fuel-oil fires?

5. What type of fire extinguisher should be used for electrical fires?

6. What is the purpose of the lockout/tagout system? Describe the procedure.

7. Is it OK to use two hands when probing an electrical circuit for voltages with a multimeter? Why or why not?

8. When should you use electrical rubber gloves?

9. How many amperes of electrical current could kill a person?

10. Describe a safe procedure for using a multimeter to check for voltage in an energized electrical circuit.

11. If one of your fellow workers is being shocked by electricity, how can you safely rescue the person? Discuss the rescue procedure.

12. What safety rules are required when using refrigerants?

13. When charging a refrigeration/air conditioning system, what safety equipment should you use? Why?

14. When working on refrigeration systems, when is a self-contained breathing apparatus required?

15. Describe the procedures for assuring that no voltage is applied to a component that is to be changed out.

CHAPTER TWO

REFRIGERATION CYCLE: HOW THE SYSTEM COOLS

AFTER STUDYING THIS CHAPTER, THE STUDENT WILL BE ABLE TO:

- Describe the four basic components in the refrigeration cycle.
- Describe the operation of the refrigeration cycle.
- Explain the operation of the compressor, condenser, metering device, and evaporator.
- Identify system components.
- List the three basic metering devices.

- Identify the piping found on the compressor.
- State safety practices around air-conditioning systems.
- Determine the condensing and evaporating temperature.
- Define superheat and subcooling.

INTRODUCTION

An air-conditioning system simply removes heat from an indoor area and dumps that heat outside. Whenever heat is removed from the air, the air is cooled. The air in a building is pulled through an air filter, return-air-duct system; and evaporator coil (where the heat and moisture are removed from the air) and redistributed to rooms through a supply-duct system.

The process of cooling the air also removes moisture, thereby lowering the humidity in the air. The air in a building is recirculated over and over again as it is filtered, cooled, and dehumidified by this process called air conditioning . . . conditioning the air.

The four basic parts of any air conditioner are the

- Compressor
- Condenser
- Flow control
- Evaporator

These four parts make up the circuit through which the refrigerant travels. Copper tubing carries the refrigerant to each of the major components. Now let us see how each of these components operates and how each is interrelated with the others. (Figure 2–1)

The transfer of heat in a refrigeration system is performed by a refrigerant operating in a closed system. The refrigeration process has its application in both refrigerated systems and air-conditioning systems. Refrigerated systems are chiefly concerned with cooling products, whereas air-conditioning systems cool (or heat) people. Air-conditioning systems use refrigeration to provide comfort cooling and dehumidification of air.

One of the very useful properties of the refrigerant is the pressure-temperature relationship of the saturated vapor. A refrigerant vapor is said

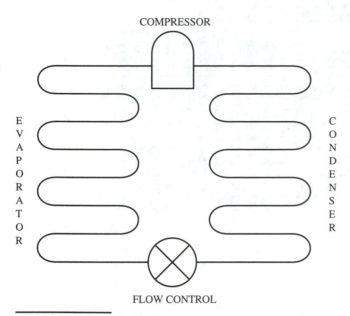

COMPRESSOR

E V A P O R A T O R

C O N D E N S E R

FLOW CONTROL

Figure 2–1   Refrigeration cycle.

to be saturated whenever both liquid and vapor are present in the same container, in stable equilibrium.

Under these conditions a fixed relationship exists between the temperature of the refrigerant in the container and its pressure. A typical temperature-pressure table is shown in Figure 2–2.

COMPRESSOR

The first component of an air-conditioning system is the compressor.

PURPOSE

The compressor is a refrigerant vapor pump. It creates a pressure difference and starts the circulation of refrigerant in the system.

LOCATION

The condenser is in the outdoor condensing unit.

PHYSICAL DESCRIPTION

It is a black, cylinderlike vessel with a domed top (Figure 2–3). Copper or steel refrigerant lines and electrical wiring are connected to this component.

OPERATION

It compresses refrigerant gas (vapor). Compression creates a hot, high-pressure superheated gas.

Think of the compressor as a gas or vapor pump. The compressor pushes the refrigerant through the basic components and interconnecting piping.

ON THE JOB

When you go to a job site for the first time, locate the four major components of the refrigeration cycle—the compressor, evaporator, condenser, and metering device. Think about the basics you learned in this chapter. Locate the discharge line, suction line, and process tube on the compressor. Identify the type of metering device at the evaporator.

Figure 2-2 Pressure-temperature table.

Temperature °F.	Refrigerant—Code				
−60	19.0	12.0	17.0	7.2	18.6
−55	17.3	9.2	15.0	3.8	16.6
−50	15.4	6.2	12.8	0.2	14.3
−45	13.3	2.7	10.4	1.9	11.7
−40	11.0	0.5	7.6	4.1	8.7
−35	8.4	2.6	4.6	6.5	5.4
−30	5.5	4.9	1.2	9.2	1.6
−25	2.3	7.4	1.2	12.1	1.3
−20	0.6	10.1	3.2	15.3	3.6
−18	1.3	11.3	4.1	16.7	4.6
−16	2.0	12.5	5.0	18.1	5.6
−14	2.8	13.8	5.9	19.5	6.7
−12	3.6	15.1	6.8	21.0	7.9
−10	4.5	16.5	7.8	22.6	9.0
−8	5.4	17.9	8.8	24.2	10.3
−6	6.3	19.3	9.9	25.8	11.6
−4	7.2	20.8	11.0	27.5	12.9
−2	8.2	22.4	12.1	29.3	14.3
0	9.2	24.0	13.3	31.1	15.7
1	9.7	24.8	13.9	32.0	16.5
2	10.2	25.6	14.5	32.9	17.2
3	10.7	26.4	15.1	33.9	18.0
4	11.2	27.3	15.7	34.9	18.8
5	11.8	28.2	16.4	35.8	19.6
6	12.3	29.1	17.0	36.8	20.4
7	12.9	30.0	17.7	37.9	21.2
8	13.5	30.9	18.4	38.9	22.1
9	14.0	31.8	19.0	39.9	22.9
10	14.6	32.8	19.7	41.0	23.8
11	15.2	33.7	20.4	42.1	24.7
12	15.8	34.7	21.2	43.2	25.6
13	16.4	35.7	21.9	44.3	26.5
14	17.1	36.7	22.6	45.4	27.5
15	17.7	37.7	23.4	46.5	28.4
16	18.4	38.7	24.1	47.7	29.4
17	19.0	39.8	24.9	48.8	30.4
18	19.7	40.8	25.7	50.0	31.4
19	20.4	41.9	26.5	51.2	32.5
20	21.0	43.0	27.3	52.4	33.5
21	21.7	44.1	28.1	53.7	34.6
22	22.4	45.3	28.9	54.9	35.7
23	23.2	46.4	29.8	56.2	36.8
24	23.9	47.6	30.6	57.5	37.9
25	24.6	48.8	31.5	58.8	39.0
26	25.4	49.9	32.4	60.1	40.2
27	26.1	51.2	33.2	61.5	41.4
28	26.9	52.4	34.2	62.8	42.6
29	27.7	53.6	35.1	64.2	43.8
30	28.4	54.9	36.0	65.6	45.0
31	29.2	56.2	36.9	67.0	46.3
32	30.1	57.5	37.9	68.4	47.6
33	30.9	58.8	38.9	69.9	48.9
34	31.7	60.1	39.9	71.3	50.2
35	32.6	61.5	40.9	72.8	51.6
36	33.4	62.8	41.9	74.3	52.9
37	34.3	64.2	42.9	75.8	54.3
38	35.2	65.6	43.9	77.4	55.7
39	36.1	67.1	45.0	79.0	57.2
40	37.0	68.5	46.1	80.5	58.6
41	37.9	70.0	47.1	82.1	60.1
42	38.8	71.4	48.2	83.8	61.6
43	39.8	73.0	49.4	85.4	63.1
44	40.7	74.5	50.5	87.0	64.7
45	41.7	76.0	51.6	88.7	66.3
46	42.6	77.6	52.8	90.4	67.9
47	43.6	79.2	54.0	92.1	69.5
48	44.6	80.8	55.1	93.9	71.1
49	45.7	82.4	56.3	95.6	72.8
50	46.7	84.0	57.6	97.4	74.5
55	52.0	92.6	63.9	106.6	83.4
60	57.7	101.6	70.6	116.4	92.9
65	63.8	111.2	77.8	126.7	103.1
70	70.2	121.4	85.4	137.6	114.1
75	77.0	132.2	93.5	149.1	125.8
80	84.2	143.6	102.0	161.2	138.3
85	91.8	155.7	111.0	174.0	151.7
90	99.8	168.4	120.6	187.4	165.9
95	108.2	181.8	130.6	201.4	181.1
100	117.2	195.9	141.2	216.2	197.2
105	126.6	210.8	152.4	231.7	214.2
110	136.4	226.4	164.1	247.9	232.3
115	146.8	242.7	176.5	264.9	251.5
120	157.6	259.9	189.4	282.7	271.7
125	169.1	277.9	203.0	301.4	293.1
130	181.0	296.8	217.2	320.8	—
135	193.5	316.6	232.1	341.2	—
140	206.6	337.2	247.7	362.6	—
145	220.3	358.9	264.0	385.0	—
150	234.6	381.5	281.1	408.4	—
155	249.5	405.1	298.9	432.9	—

Italic figures represent vacuum-inches of mercury. **Bold figures** represent pressure-pounds per square inch. *(Courtesy of Sporlan Valve Company).*

16

Figure 2–3   Small welded hermetic compressor using HFC-134a refrigerant. *(Courtesy of Copeland Corporation)*

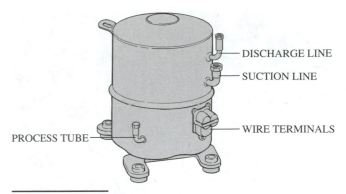

Figure 2–4   Compressor piping identification.

The most common type of compressor looks like a black domelike structure located in the outdoor unit. The outdoor unit is referred to as the condensing unit. Compressors on large systems may be found in a mechanical room. Other types and styles of compressors are shown in Figure 2–6.

The compressor has three wires that supply power to operate a motor inside the compressor. The compressor has two copper lines running to its steel shell. Some small refrigeration appliances have steel lines connecting the compressor to the refrigerant circuit. A third, short copper line may be noticed on the compressor housing. This copper line may protrude from the compressor housing a few inches with nothing connected to it. This copper line is called the *process tube*. The process tube can be used by the manufacturer or technician to charge the unit. The manufacturer reseals the process tube after it is used. This process tube contains suction (low) pressure. The process tube can also be used to field-service the suction side of the system.

The two copper refrigerant lines running to the compressor are called the *suction line* and *discharge line* (Figure 2–4). The suction line is identified as the larger, cooler line when the unit is in operation. Compressor lines may vary, depending on manufac-

turer. Do not use these labeled lines as common connection points. Check the compressor.

The discharge line is the smaller of the two lines and can be very hot when the unit is in operation. The discharge line could burn your skin. Touch the lines quickly if you are trying to identify the difference between the discharge and suction lines. Cool, gaseous refrigerant flows from the suction side (large copper pipe) into the compressor. The refrigerant is compressed inside the compressor. When the gas is compressed, both the temperature and pressure rise dramatically. The hot gas is also known as discharge superheat. This compressed gas is forced out of the discharge side of the compressor. The discharge pressure may be as high as 400 psig with a temperature as high as 250°F.

Safety First   Never stand in front of the compressor terminals when starting a condensing unit. On rare occasions the compressor terminals have blown out, shooting a stream of high-pressure refrigerant, oil, and fire that could seriously injure the HVAC technician. The black plastic compressor terminal cover and clip may not be able to contain this powerful eruption.

The compressor has three power wires going to electrical terminals on the compressor. These compressor terminals can be inspected. It is recommended that the wires carefully be removed from the terminals to prevent terminal leakage or blowout. Remove wiring only if necessary or to tighten the connection. The wire connectors should be snug on the compressor terminals. Loose connections will cause high amperage draw and wire burnout. Terminal leakage or blowout is extremely rare, but as a technician you should take caution when working with this part of the compressor to prevent accidental damage to you or the equipment (Figure 2–5).

If you do inspect these terminals, remember to disconnect the power to the unit and verify that the power is not present anywhere in the outdoor condensing unit. Also, turn off the thermostat, because

Figure 2–5   Compressor terminals.

the thermostat controls another voltage source to the condensing unit that may not be disconnected unless you disconnect the power to the blower/furnace unit. This control voltage source is usually 24-V. The 24-V control wiring is usually covered in a dark brown insulated sheath. In most installations the thermostat wiring *will not* be in protective conduit. The 24-V circuit starts and stops the condensing unit through the thermostat. In most cases the source for this 24-V control wiring is in the furnace/blower unit. Window units and package units may use 120 or 240 V as control voltage.

The compressor is a mechanical device for pumping refrigerant vapor from a low-pressure area (the evaporator) to a high-pressure area (the condenser). Because pressure, temperature, and volume of a gas are related, a change in pressure from low to high causes an increase in temperature and a decrease in volume or a compression of the vapor. The main types of compressors are reciprocating (piston), rotary, centrifugal, screw, and scroll. See Figure 2–6.

Compressors derive their names from the mechanical parts found in the component. In the reciprocating compressor, a piston travels back and forth in a cylinder. Refrigerant pressure is increased as the piston compresses the refrigerant in the cylinder.

The rotary compressor has a vane that rotates within a cylinder decreasing the size of the cylinder, which increases the refrigerant pressure.

The centrifugal compressor has a very high-speed centrifugal impeller with multiple blades. The impeller rotates within a housing, increasing the pressure on the refrigerant by centrifugal force.

The screw compressor uses rotating screws within a tapered housing. The screws are intermeshed to squeeze the refrigerant which increases the pressure.

The scroll compressor has a stationary and an orbiting scroll. The orbiting scroll moves within the stationary scroll. The compartments between the orbiting and stationary scroll reduce in size, thus building up refrigerant pressure that is discharged into the condenser.

▼ CONDENSER

PURPOSE

The condenser converts hot, high-pressure, superheated refrigerant gas to a warm, high-pressure subcooled liquid refrigerant. It removes heat that was absorbed by the refrigerant and removes heat developed by the heat of compression and motor operation.

LOCATION

The condenser is located in an outdoor condensing unit or in a water-cooled condenser.

ALTERNATIVE NAME

It is also called an outdoor coil or condenser coil.

PHYSICAL DESCRIPTION

The condenser is made of copper or aluminum tubing that is surrounded by aluminum fins or aluminum spines. The aluminum fins or spines aid in heat rejection. Most aluminum spines have aluminum tubing for the passage of refrigerant (Figure 2–7).

OPERATION

Outside air is circulated by a fan forcing air over or through the outdoor condensing fins and coil. This forced-air circulation removes heat from the superheated gas flowing through the condenser piping, causing it to condense into a warm liquid. There will be little pressure drop as the refrigerant travels

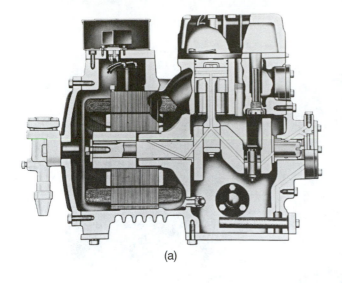

(a)

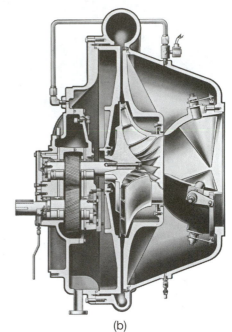

(b)

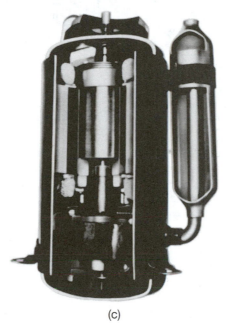

(c)

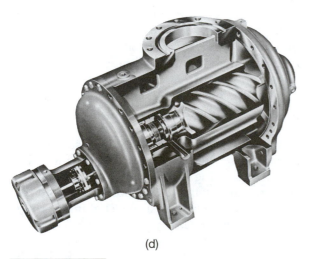

(d)

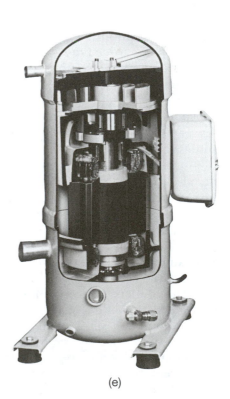

(e)

Figure 2–6 Types of compressors. *(a, Courtesy of Copeland Corporation; b, d, courtesy of York International Corporation; c, courtesy of Rotorex Company, Inc.; e, The Trane Company)*

19

ALUMINUM FINS   ALUMINUM SPINES

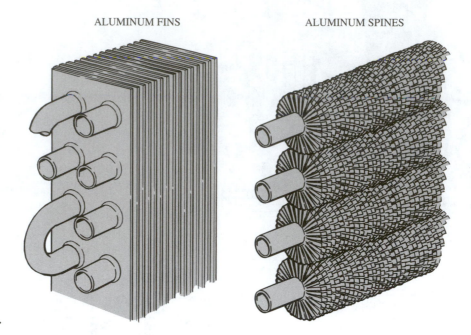

Figure 2–7   Condenser coil types.

through the condenser. A pressure drop of 1 to 3 psig may be noticed. This small pressure drop is insignificant and is undetectable by your refrigeration gauges. A pressure drop greater than 3 psi indicates a restriction in the condenser tubing.

Some equipment uses water instead of air to cool and condense the refrigerant. This operation is called a *water-cooled condenser.* Water is circulated through tubing. Refrigerant, which is contained in the condenser shell, transfers heat into the cooler water, thus beginning the condensing process.

There are three types of condensers: air-cooled, water-cooled, and evaporative (Figure 2–8). The air-cooled condenser uses air as the condensing medium, the water-cooled condenser uses water as the condensing medium, and the evaporative condenser uses both air and water to remove heat and condense the refrigerant.

Air-cooled condensers consist of two types: forced-air condensers and natural-draft (static) condensers. Air-cooled condensers can be further classified by their construction: (1) fin and tube, (2) bare pipe, and (3) plate. There are four types of water-cooled condensers: (1) double pipe, (2) open vertical shell and tube, (3) horizontal shell and tube, and (4) shell and coil. See Figure 2–8.

The condenser receives hot, high-pressure, superheated vapor or gas refrigerant from the compressor. The outdoor air is circulated around the condenser tubing, thereby removing heat from the hot refrigerant. The condenser is a heat exchanger. Heat exchangers transfer heat from warmer to cooler areas. The outdoor air may be 100°F, but the refrigerant temperature is much hotter. The hot, gaseous refrigerant gives up heat to the forced air circulating through the condenser coil. When the hot, gaseous refrigerant gives up enough heat, it will begin to condense into a liquid refrigerant. Removal of heat from the hot, superheated gas is also known as desuperheating.

The high-pressure gas that enters will be approximately the same pressure as the liquid refrigerant that leaves. For example, if the gas pressure entering the condenser is 250 psig, then the liquid pressure leaving the condenser is approximately 250 psig (Figure 2–9).

There is little or no pressure drop as the refrigerant travels through the condenser coil. A pressure drop greater than 3 psi may indicate a restriction in the condenser tubing. In most cases you will not have an access valve to check this pressure drop across the condenser; however, one can be installed to check the pressure drop across a condenser.

Another purpose of the condenser is to further cool the liquid refrigerant after it has condensed. The condensing temperature can be determined by examining the high-pressure-gauge reading of the refrigeration gauge set. This pressure has a correlating temperature on the face of the high-pressure gauge. The outer ring on the high-side gauge reads pressure. The inner rings read refrigerant temperature.

Most air-conditioning units use the refrigerant R-22. In that case, if you are reading a high-side pressure of 250 psig, then the corresponding condensing temperature will be 117°F. These readings are obtained from your refrigeration gauge set or from a pressure-temperature chart.

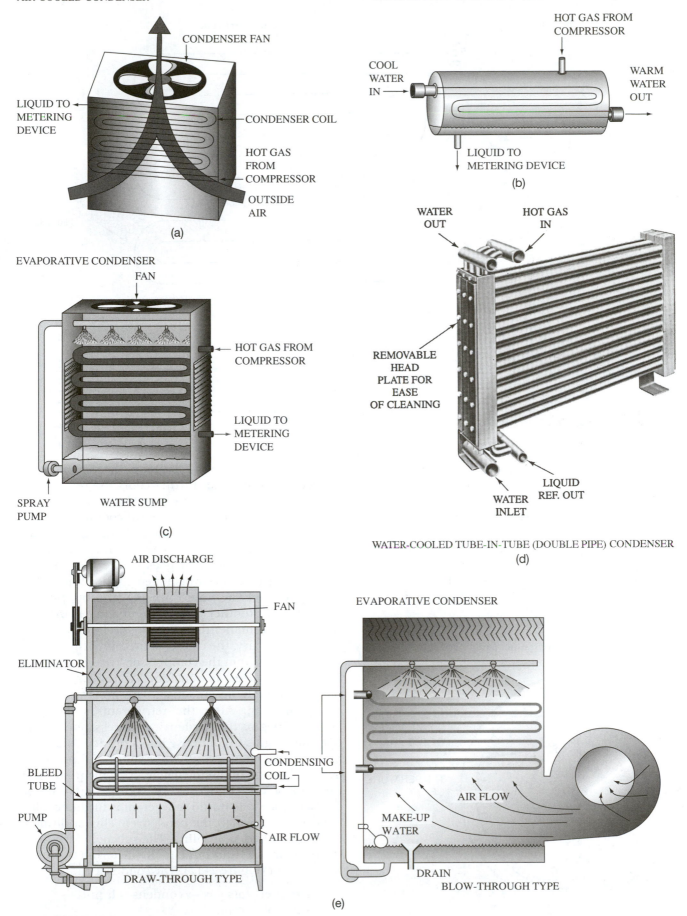

AIR-COOLED CONDENSER

CONDENSER FAN

LIQUID TO
METERING
DEVICE

CONDENSER COIL

HOT GAS
FROM
COMPRESSOR

OUTSIDE
AIR

(a)

WATER-COOLED CONDENSER (SHELL AND TUBE)

HOT GAS FROM
COMPRESSOR

COOL
WATER
IN

WARM
WATER
OUT

LIQUID TO
METERING DEVICE

(b)

EVAPORATIVE CONDENSER

FAN

HOT GAS FROM
COMPRESSOR

LIQUID TO
METERING
DEVICE

SPRAY
PUMP

WATER SUMP

(c)

WATER
OUT

HOT GAS
IN

REMOVABLE
HEAD
PLATE FOR
EASE
OF CLEANING

WATER
INLET

LIQUID
REF. OUT

WATER-COOLED TUBE-IN-TUBE (DOUBLE PIPE) CONDENSER

(d)

AIR DISCHARGE

FAN

ELIMINATOR

BLEED
TUBE

CONDENSING
COIL

PUMP

AIR FLOW

DRAW-THROUGH TYPE

EVAPORATIVE CONDENSER

AIR FLOW

MAKE-UP
WATER

DRAIN

BLOW-THROUGH TYPE

(e)

Figure 2–8   Types of condensers. (d, Courtesy of Heatcraft Refrigeration Products)

Condenser Pressure

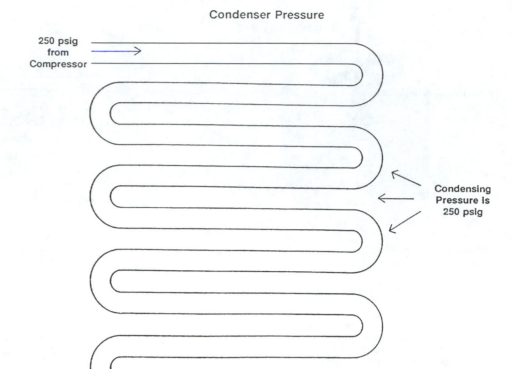

250 psig
from
Compressor

Condensing
Pressure Is
250 psig

Figure 2–9   Condenser operation.

250 psig to **Flow Control**

Check your unit's nameplate for the refrigeration type (Figure 2–10). With the advent of many new refrigerants, do not take chances by guessing that the equipment uses R-22 or any other common refrigerant.

You can also obtain the condensing temperature by noting the high-side pressure and using a pressure-temperature chart (Figure 2–11).

Figure 2–10   High-pressure gauge. *(Amprobe/Promax Instruments)*

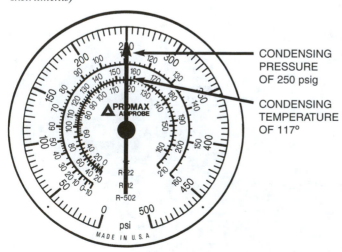

CONDENSING
PRESSURE
OF 250 psig

CONDENSING
TEMPERATURE
OF 117°

After the refrigerant is desuperheated and condensed, the condenser further cools the refrigerant. This cooling of the liquid refrigerant below its condensing temperature is called subcooling. We use this information when we cover charging methods. The high-pressure, subcooled liquid refrigerant leaves the condenser and goes to the flow-control device by way of an uninsulated copper line called the liquid line. The liquid line is warm but not hot to the touch. Another identifying feature is that the liquid line is smaller than the suction line.

In summary, the condenser receives the hot superheated gas from the compressor. Heat is removed from this gas. Heat is transferred to the air or water passing through the condenser. This process is called desuperheating. After the refrigerant is desuperheated, it begins the condensing process, and the refrigerant begins to condense into a liquid. This point provides the condensing temperature and pressure, as found on the high-side gauge. At some point near the end of the condenser, the refrigerant will be 100% liquid. This liquid refrigerant is further cooled as it passes through the condenser. This additional cooling is called subcooling. The subcooled liquid is fed through the liquid line to the metering device or flow control (Figure 2–12). The condenser:

Desuperheats gas → condenses liquid →
subcools liquid

SPORLAN TEMPERATURE PRESSURE CHART

Vacuum-Inches of Mercury – Italic Figures

Pressure-Pounds Per Square Inch Bold Figures

TEMPERATURE °F	12-F	22-V	500-D	502-R	717-A
-60	19.0	12.0	17.0	7.2	18.6
-55	17.3	9.2	15.0	3.8	16.6
-50	15.4	6.2	12.8	0.2	14.3
-45	13.3	2.7	10.4	1.9	11.7
-40	11.0	0.5	7.6	4.1	8.7
-35	8.4	2.6	4.6	6.5	5.4
-30	5.5	4.9	1.2	9.2	1.6
-25	2.3	7.4	1.2	12.1	1.3
-20	0.6	10.1	3.2	15.3	3.6
-18	1.3	11.3	4.1	16.7	4.6
-16	2.0	12.5	5.0	18.1	5.6
-14	2.8	13.8	5.9	19.5	6.7
-12	3.6	15.1	6.8	21.0	7.9
-10	4.5	16.5	7.8	22.6	9.0
-8	5.4	17.9	8.8	24.2	10.3
-6	6.3	19.3	9.9	25.8	11.6
-4	7.2	20.8	11.0	27.5	12.9
-2	8.2	22.4	12.1	29.3	14.3
0	9.2	24.0	13.3	31.1	15.7
1	9.7	24.8	13.9	32.0	16.5
2	10.2	25.6	14.5	32.9	17.2
3	10.7	26.4	15.1	33.9	18.0
4	11.2	27.3	15.7	34.9	18.8
5	11.8	28.2	16.4	35.8	19.6
6	12.3	29.1	17.0	36.8	20.4
7	12.9	30.0	17.7	37.9	21.2
8	13.5	30.9	18.4	38.9	22.1
9	14.0	31.8	19.0	39.9	22.9
10	14.6	32.8	19.7	41.0	23.8
11	15.2	33.7	20.4	42.1	24.7
12	15.8	34.7	21.2	43.2	25.6
13	16.4	35.7	21.9	44.3	26.5
14	17.1	36.7	22.6	45.4	27.5
15	17.7	37.7	23.4	46.5	28.4
16	18.4	38.7	24.1	47.7	29.4
17	19.0	39.8	24.9	48.8	30.4
18	19.7	40.8	25.7	50.0	31.4
19	20.4	41.9	26.5	51.2	32.5
20	21.0	43.0	27.3	52.4	33.5
21	21.7	44.1	28.1	53.7	34.6
22	22.4	45.3	28.9	54.9	35.7
23	23.2	46.4	29.8	56.2	36.8
24	23.9	47.6	30.6	57.5	37.9
25	24.6	48.8	31.5	58.8	39.0
26	25.4	49.9	32.4	60.1	40.2
27	26.1	51.2	33.2	61.5	41.4
28	26.9	52.4	34.2	62.8	42.6
29	27.7	53.6	35.1	64.2	43.8
30	28.4	54.9	36.0	65.6	45.0
31	29.2	56.2	36.9	67.0	46.3
32	30.1	57.5	37.9	68.4	47.6
33	30.9	58.8	38.9	69.9	48.9
34	31.7	60.1	39.9	71.3	50.2
35	32.6	61.5	40.9	72.8	51.6
36	33.4	62.8	41.9	74.3	52.9
37	34.3	64.2	42.9	75.8	54.3
38	35.2	65.6	43.9	77.4	55.7
39	36.1	67.1	45.0	79.0	57.2
40	37.0	68.5	46.1	80.5	58.6
41	37.9	70.0	47.1	82.1	60.1
42	38.8	71.4	48.2	83.8	61.6
43	39.8	73.0	49.4	85.4	63.1
44	40.7	74.5	50.5	87.0	64.7
45	41.7	76.0	51.6	88.7	66.3
46	42.6	77.6	52.8	90.4	67.9
47	43.6	79.2	54.0	92.1	69.5
48	44.6	80.8	55.1	93.9	71.1
49	45.7	82.4	56.3	95.6	72.8
50	46.7	84.0	57.6	97.4	74.5
55	52.0	92.6	63.9	106.6	83.4
60	57.7	101.6	70.6	116.4	92.9
65	63.8	111.2	77.8	126.7	103.1
70	70.2	121.4	85.4	137.6	114.1
75	77.0	132.2	93.5	149.1	125.8
80	84.2	143.6	102.0	161.2	138.3
85	91.8	155.7	111.0	174.0	151.7
90	99.8	168.4	120.6	187.4	165.9
95	108.2	181.8	130.6	201.4	181.1
100	117.2	195.9	141.2	216.2	197.2
105	126.6	210.8	152.4	231.7	214.2
110	136.4	226.4	164.1	247.9	232.3
115	146.8	242.7	176.5	264.9	251.5
120	157.6	259.9	189.4	282.7	271.7
125	169.1	277.9	203.0	301.4	293.1
130	181.0	296.8	217.2	320.8	—
135	193.5	316.6	232.1	341.2	—
140	206.6	337.2	247.7	362.6	—
145	220.3	358.9	264.0	385.0	—
150	234.6	381.5	281.1	408.4	—
155	249.5	405.1	298.9	432.9	—

Figure 2–11   Pressure-temperature chart. (*Courtesy of Sporlan Valve Company*)

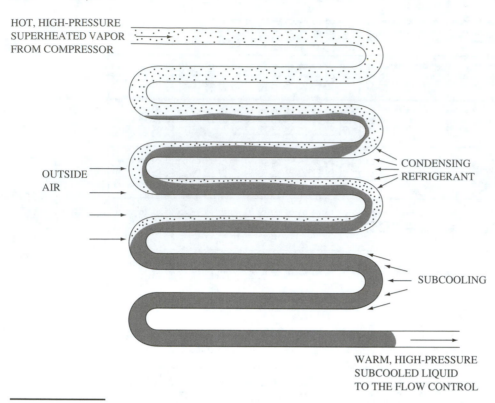

HOT, HIGH-PRESSURE
SUPERHEATED VAPOR
FROM COMPRESSOR

OUTSIDE
AIR

CONDENSING
REFRIGERANT

SUBCOOLING

WARM, HIGH-PRESSURE
SUBCOOLED LIQUID
TO THE FLOW CONTROL

Figure 2–12   Condenser operation.

▼ METERING DEVICE

PURPOSE

The metering device controls flow of refrigerant to the evaporator. It is a restrictive device that reduces the pressure and temperature of the liquid refrigerant.

LOCATION

The metering device is located inside the evaporator case or immediately outside the evaporator case.

ALTERNATIVE NAME

It is also called a flow-control, refrigerant-control, or restrictor device.

PHYSICAL DESCRIPTION

There are numerous types of flow controls; therefore, the physical description varies. The most common types are

- Thermostatic expansion valve (TXV or TEV)
- Capillary tube (cap tube)
- Fixed-orifice device (piston, accurator, or flow rater)

The automatic expansion valve, electronic expansion valve, and high-side and low-side floats are also used to control refrigerant flow. These are not commonly found in small air-conditioning systems. The low-side float is used in larger low-pressure chilled-water systems.

OPERATION

The metering device controls the refrigerant flow to the evaporator or chilled-water system. It receives warm, high-pressure, subcooled refrigerant from the condenser and converts it into a cold, low-pressure, liquid-vapor mixture. This liquid-vapor mixture is also known as a saturated mixture. There is about 80% liquid and 20% vapor. Think of the flow control as a device that automatically adjusts the amount of liquid refrigerant going into the evaporator. The liquid refrigerant in the evaporator absorbs heat from the air in a building or the fluid passing through it. Heat transfers from hot to cold; therefore, outdoor heat transfers into a building at a greater rate as the day warms. Lighting, occupants, and equipment generate indoor heat. The evaporator is designed to pick up the additional heat load added to a building. When the temperature is mild, the inside of the building will not need as much air conditioning; therefore, the flow control will reduce the amount of liquid refrigerant entering the evaporator. The metering device will feed more refrigerant to the evaporator as the heat load on the building increases.

The flow control feeds the maximum amount of liquid refrigerant to the evaporator on the hottest days of the summer and lesser amounts of refrigerant on cooler days and evening hours when the load is less.

A flow-control, or metering, device is a restrictive device that drops the pressure and temperature of the high-pressure, subcooled liquid refrigerant it receives from the condenser. The subcooled liquid refrigerant entering the flow control is warm (above 90°F) and is at a high pressure (above 175 psig). The restrictive nature of the flow control drops the pressure of the entering liquid refrigerant, causing a corresponding drop in refrigerant temperature.

Remember that at saturation the temperature and pressure of refrigerant travel in the same direction. If the pressure increases, so does the temperature. When the pressure decreases, the temperature of the refrigerant drops.

In summary, a metering device is a refrigerant-control device. A metering device restricts the refrigerant flow and drops the pressure and temperature of the refrigerant entering the evaporator in order for a cold, low-pressure, liquid-vapor refrigerant mixture to enter the evaporator coil (Figure 2–13). There are two pressure changes in the refrigeration cycle. The compressor increases the refrigerant pressure and creates the flow circulation to the condenser. The metering device reduces the refrigerant pressure with a corresponding drop in temperature, thus feeding the evaporator with a medium to absorb heat from the air or water circulating around it.

DETERMINING THE TYPE OF METERING DEVICE

To determine the type of metering device on the unit you are servicing, first locate the liquid line. The flow-control, or metering, device is located inside the evaporator case or inside the chiller barrel. In a few cases the metering device is located just outside the evaporator case or chiller.

Follow the liquid line from the condensing until the metering device is located. The liquid line is the smaller, uninsulated copper line coming from the outdoor condensing unit. On window units or package units, the flow-control device will be right at the evaporator coil. On these units the liquid line is short.

It is important to know the type of flow-control device you have on the unit being checked (Figure 2–14). This knowledge will assist you in charging the unit correctly. If the flow control is located inside the evaporator housing, it will be a challenge to find the flow-control device.

In some instances, the evaporator housing will have an inspection plate that will allow removal for easy flow-control identification. In most cases it will not be this simple to access.

Here are some pointers on locating the flow-control device: Follow the liquid line to the evaporator case. This is probably the side where the flow control is located. On many horizontal attic units, there is ductwork between the evaporator and furnace. This ductwork is called a transition. Try to remove a side of this transition ductwork nearest the liquid line entering the evaporator. If the ductwork is difficult to remove, penetrate this ductwork with sheet metal shears. Care must be taken with this action. You do not want to cut into the evaporator or furnace; cutting an opening may damage these components. However, this access opening will allow you to locate the flow-control device.

If the flow control is not located on this side of the evaporator, go through the same routine on the opposite side of the evaporator. Even if the flow control is not on this side of the evaporator, at least you will be able to inspect the condition of the evaporator coil.

If the evaporator coil has lint and dirt buildup, clean its surface with a stiff brush. Short, downward

Figure 2–13   Flow-control operation.

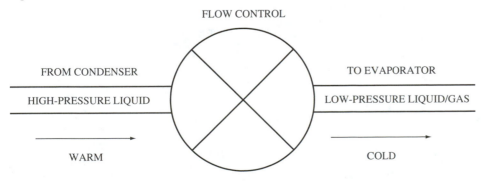

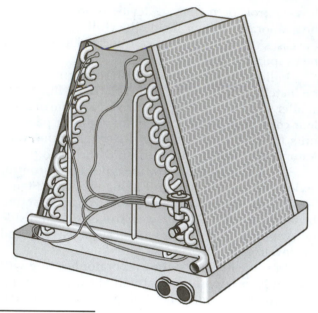

Figure 2–14 Evaporator case: flow control (TXV) located outside the case. *(Courtesy of Lennox Industries)*

excess lint and dirt on the surface of the coil. Then use the sprayer. Spray downward, trying not to force dirt deeper into the coil. Remove rust and algae from the drain pan. Pour a cup of bleach into the evaporator drain pan to temporarily kill algae growth. It is not advisable to mix cleaning agents such as bleach and coil cleaners. The evaporator must be clean in order to achieve a correct charge and correct operating standards. Algae tablets can be placed in the drain pan to reduce the possibility of drain blockage. Notify the owner of the unit that you are going to use chemicals to clean the coil. They may have allergies to bleach or other chemicals used in coil cleaners.

Close the inspection opening with a piece of sheet metal if it is a sheet metal duct plenum or ductboard if it is fiberglass ductboard plenum. Use silver duct tape for better sealing results. The best action at this point is to install a new insulated sheet metal or ductboard inspection plate over the existing opening. Screw the new inspection plate over the inspection opening and tape the joints around the plate to prevent air leakage. Have a number of inspection plates available for this inspection and sealing procedure, because they will be needed on other jobs.

When searching for the flow control on the supply air side of the evaporator, you can remove ductwork hanging on the supply plenum box. This will allow an inspection opening on this side of the evaporator. Use an inspection mirror and flashlight when looking through the supply plenum box opening.

strokes should remove the lint. Care should be taken to remove the trapped lint or dirt without pushing these substances deeper into the coil or dropping them into the evaporator drain pan. An evaporator coil cleaner can be used to complete the job.

A garden sprayer with hot water can also be used to clean the evaporator coil. First, remove all the

Figure 2–15 Attic installation.

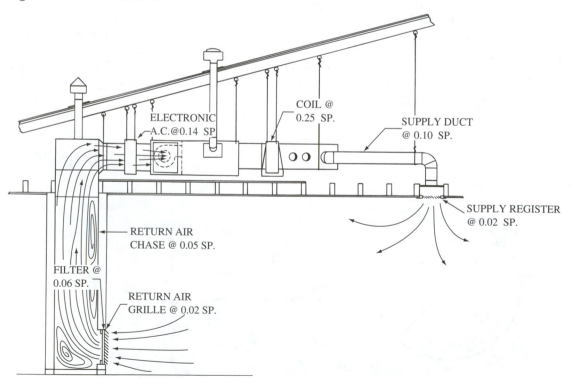

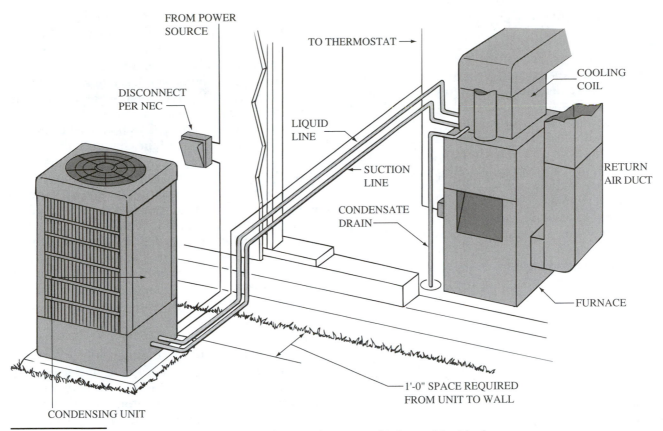

FROM POWER SOURCE

TO THERMOSTAT →

DISCONNECT PER NEC

COOLING COIL

LIQUID LINE

SUCTION LINE

RETURN AIR DUCT

CONDENSATE DRAIN

FURNACE

1'-0" SPACE REQUIRED FROM UNIT TO WALL

CONDENSING UNIT

Figure 2–16   Closet, garage, or basement installation. Flow control is located inside the cooling-coil case at the liquid line as it enters the case. This is a split system.

When the inspection is completed, secure the duct to the plenum box. Strapping or screwing the duct to the plenum duct collar is recommended. Duct tape will not hold long and the duct will slip off the duct collar. Flexible duct can be secured to the duct collar using a nylon tie on the inner duct liner as well as on the outer flexible duct vapor barrier. Plumber strap can also be used. The important point here is to make sure that the duct is tight on the duct collar. Sheet metal ductwork should be secured with screw. The Hardcast system of cloth tape and adhesive makes an excellent sealant over the vapor barrier and duct collar.

Insulate all exposed metal on the duct system. Exposed metal may sweat and cause condensation to form and drip onto the insulation or drywall. Operate the blower and check for air leaks.

Examine Figures 2–15 and 2–16 for evaporator location and possible inspection points.

In summary, there are many types of flow-control devices used to control the flow of refrigerant to the evaporator. The flow control is also called a *metering device*. There are three basic types of flow controls, or metering devices, found in residential air-conditioning equipment:

- Thermostatic expansion valve (TXV or TEV) (Figure 2–17)

- Capillary tube (CAP tube) (Figure 2–20)
- Fixed orifice (flow rater or piston type) (Figure 2–21)

Next we discuss the operation of various types of flow-control devices. You will need to know the type of metering device in the system prior to checking the charge.

Figure 2–17   Thermostatic expansion valve (TXV). *(Courtesy of Sporlan Valve Company)*

THERMOSTATIC EXPANSION VALVE

The thermostatic expansion valve, also known as the TEV, TXV, or expansion valve (Figures 2–17 and 2–18) opens and partially closes to regulate the flow of a liquid/gas refrigerant into the evaporator. The valve is preadjusted to maintain the optimum amount of refrigerant flow into the evaporator under varying indoor and outdoor temperatures.

The thermostatic expansion valve attempts to maintain an evaporator with adequate refrigerant flow to keep a building cool. The TXV tries to maintain a constant superheat in the evaporator by modulating refrigerant flow. The valve opens and closes around a narrow range of superheat temperatures.

PHYSICAL DESCRIPTION

A thermostatic expansion valve is shown in Figure 2–19. The TEV looks like a mushroom with a tentacle. The TEV has a liquid line input from the con-

Figure 2–18 Thermostatic expansion valve. *(Courtesy of Alco Controls)*

denser and an output, which feeds a mixture of cold liquid and gas refrigerant to the evaporator.

The mushroomlike top is called the power assembly, power cap, or thermostatic element. This mushroom top, or power cap, has a long, small tube that terminates in a refrigerant-charged sensing bulb. The refrigerant-charged sensing bulb is usually the same refrigerant type as that used in the installed system. The refrigerant charge in the power cap and sensing bulb is separated from the charge in the system by a diaphragm in the power cap.

The sensing bulb is about the size of a thumb and should be securely strapped to the output of the evaporator. This sensing bulb should be insulated to reduce the influences of the air temperature around the bulb. The sensing bulb should be strapped to the side or top of the suction line, never the bottom. The bottom side of a horizontal suction line has oil pockets that will act as insulation between the sensing bulb and the suction-line temperature it is trying to sense.

Additionally, most thermostatic expansion valves have a quarter-inch copper line extending from the side of the expansion valve body and attached to the output of evaporator. The quarter-inch copper line is called an external equalizer line. The external equalizer connection gives a pressure reading back to the TEV, which controls the output of refrigerant through the valve. This line should not be pinched or capped for proper valve operation. Some valves do not have an external equalizer. When purchasing a replacement valve the technician may need to install an equalizer line. Most new valves require the equalizer line for improved performance.

CAPILLARY TUBE AND FIXED-ORIFICE DEVICE

The capillary tube (or cap tube) in Figure 2–20 and the fixed orifice (Figure 2–21) are physically different from thermostatic expansion valve flow-control device. The refrigerant flow through these devices depends upon the condenser pressure, the inside diameter of the cap tube or piston orifice, and the length of the capillary tubing. The condenser pressure regulates the flow to the evaporator.

The condenser pressure is high on a hot day and low on a mild day. When it is high, more liquid refrigerant is forced through the cap tube or fixed orifice and into the evaporator. More liquid refrigerant in the evaporator is required to remove the heat

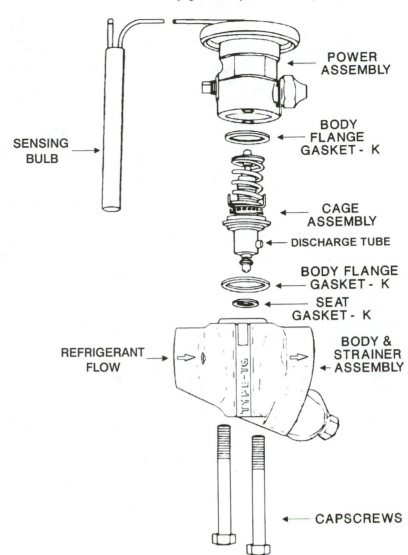

SENSING BULB

POWER ASSEMBLY

BODY FLANGE GASKET - K

CAGE ASSEMBLY

DISCHARGE TUBE

BODY FLANGE GASKET - K

SEAT GASKET - K

REFRIGERANT FLOW

BODY & STRAINER ASSEMBLY

CAPSCREWS

Figure 2–19 Breakdown of TXV. *(Courtesy of Sporlan Valve Company)*

Figure 2–20 Capillary Tube. *(Courtesy A-1 Components, Inc.)*

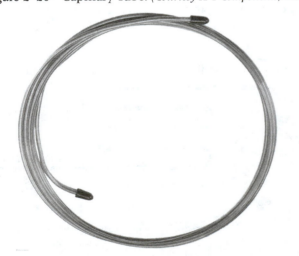

transferred into a building during the hottest part of the day.

As the day cools, the building needs less cooling capacity. The outdoor air entering the condenser drops in temperature, thereby dropping the condenser pressure. This drop in condenser pressure reduces the amount of liquid pushed through the cap tube or fixed-orifice device.

Less heat is transferred into a structure in the evening hours or on cloudy days. Less refrigerant is needed to cool the structure. The drop in condensing pressure reduces the amount of cooling capacity (liquid refrigerant) delivered to the evaporator. This drop in condenser pressure correspondingly reduces the amount of electric power (watts) drawn by an air conditioner during milder periods of air conditioning. Figure 2–22 shows the capillary tube installed at the evaporator coil. An insulated case houses these components.

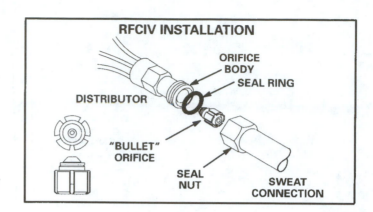

Figure 2–21 Fixed-orifice piston device. *(Courtesy of Lennox Industries)*

CAPILLARY TUBE/FIXED ORIFICE

A capillary tube, or cap tube, flow-control device is a series of small, but long, copper tubes (Figure 2–23). The outside diameter of the cap tube may be one-sixteenth ($\frac{1}{16}$) to one-eighth ($\frac{1}{8}$) inch in size, but the cap tube capacity is sized by its *inside* diameter and length.

Most A/C systems that use cap tubes have two or more of these small, long copper lines. Small-capacity systems, such as window A/C units, may have just one capillary tube.

The length and inside diameter of the cap tube create the pressure drop needed to feed the evaporator with a mixture of cold liquid and gas refrigerant. A kink or blocked cap tube reduces the cooling capacity of the evaporator.

The flow of refrigerant through the cap tube is controlled by the condenser pressure. The warmer it

Figure 2–22 Heat pump style "A" inside coil. *(Courtesy of Bard Manufacturing Company)*

is outside, the higher the condensing pressure will be. This translates into greater refrigerant flow through cap tube into the evaporator.

Capacity tubes are sized by inside diameter and length. Smaller inside diameters and longer tube lengths will create a greater pressure drop. Too much pressure drop will reduce system capacity. Properly sizing replacement capillary tubing is important.

The fixed-orifice metering device may be easily overlooked because of its short, compact construction (Figure 2–21). The body of the fixed orifice device is about 2 inches long and appears to be a brass fitting joining the liquid line to the evaporator. The fixed orifice can be identified by its similarity to a brass fitting with copper-line distributors feeding the evaporator. Some fixed orifices are designed as hollow brass shells with a piston device located inside the shell. The size and type of piston determines the tonnage, or capacity, of the air-conditioning system. Other types or fixed orifices are simply brass fittings with a small predrilled hole that creates the correct pressure drop in order to produce cold, low-pressure refrigerant flow.

Like the cap tube, the condenser pressure controls the refrigerant flow. High condenser pressure will produce greater refrigerant flow. Two common types of fixed-orifice devices are known as the flow rater and accurator.

Figure 2–23 Capillary tube connection to evaporator.

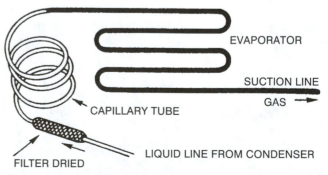

 EVAPORATOR

PURPOSE

The evaporator removes heat from the indoor air or water passing over it and removes moisture from indoor air. Inside the evaporator the cold liquid refrigerant boils and is converted into a refrigerant gas. Heat is absorbed from the air or water when the refrigerant boils inside the evaporator.

LOCATION

The evaporator is located in a metal case in an attic, closet, or basement. In the case of chilled water, the chiller or evaporator is located in a metal shell. In many installations the furnace is attached to the evaporator. In some installations the evaporator and furnace are located in the same case or housing. On window air-conditioning units, the evaporator is located directly behind the indoor air filter.

ALTERNATIVE NAME

It is also called an indoor coil or cooling coil.

PHYSICAL DESCRIPTION

The evaporator coil is made of copper or aluminum tubing surrounded by aluminum fins that aid in heat absorption.

OPERATION

When the air-conditioning system is operating, the evaporator receives a mixture of cold liquid and gaseous low-pressure refrigerant from the metering device. The cold liquid refrigerant removes the greatest amount of heat and moisture from the air; therefore, we ignore the gaseous (vapor) part of this mixture and consider only the cold liquid refrigerant received from the flow control (Figure 2–24).

Liquid substances boil at different temperature points. Water at sea level boils at 212°F; the common residential and light commercial refrigerant R-22 boils at –40°F at the same (atmospheric) pressure. When R-22 is placed in a pressurized, sealed system, such as an air-conditioner evaporator, it will boil at a much higher temperature. When operating normally, the refrigerant will boil and absorb heat at a temperature range of 32°F to 50°F. The boiling or evaporation temperature will depend on

Figure 2–24   Operation of evaporator.

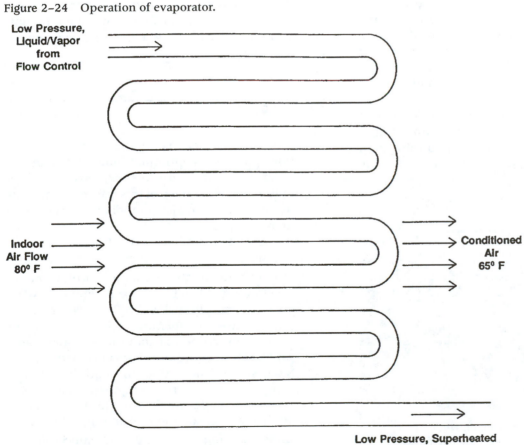

Low Pressure, Liquid/Vapor from Flow Control

Indoor Air Flow 80° F

Conditioned Air 65° F

Low Pressure, Superheated Vapor to Compressor

the evaporator pressure. The higher the pressure, the higher the boiling point.

When a liquid substance boils, it absorbs heat. When liquid refrigerant boils, it absorbs heat from the air or water passing around its surface. The refrigerant is cold enough inside the evaporator to cause the air coming in contact with its surfaces to condense moisture from the air. The cold evaporator surfaces will begin to sweat, or condense moisture on their surface, thereby removing moisture from the air as it passes around the fins and coils. This temperature is known as the dew point of the air. Dew point is the temperature at which moisture will begin to condense from the air.

The evaporator vaporizes the liquid refrigerant inside the evaporator tubing. To prevent liquid slugging of the compressor, the entire liquid refrigerant should be returned to vapor or gas prior to leaving the evaporator.

From the evaporator the cool gas returns to the compressor to be recompressed and recirculated. An important occurence in the evaporator is suction superheat, or superheat. Superheat is heat added to the refrigerant above its boiling point. Understanding superheat and how to measure superheat is important in checking the charge of an air conditioning unit. This idea is covered in the chapter on superheat charging. Figure 2–25 shows an evaporator outside its case. Notice the TXV metering devices on the evaporator coil.

SUMMARY

- This is an elementary chapter on the refrigeration cycle and its important component parts. In order to proceed with this textbook the reader must have a firm grasp of this information. Identifications of the major components are essential to properly checking a charge. Determining the correct metering device will effect how a technician selects the correct superheat temperature.

- One way to enhance skills in learning the refrigeration cycle is to draw a diagram showing the refrigerant flow, labeling general pressures (high or low), temperatures (hot, cold, cool or warm) and refrigerant conditions (superheated vapor, saturated mixture, or subcooled liquid). Next, select functioning air conditioning systems and identify the compressor, condenser, evaporator, metering device type as well as the discharge line, suction line and liquid line.

- Understanding superheat and subcooling is critical in order to fine-tune a charge; it is also

Figure 2–25 Slab type evaporator coil. Notice TXV metering devices. *(Courtesy of Domestic Manufacturing, Houston, TX)*

valuable when troubleshooting refrigerant-side problems. This chapter discussed two types of superheat—suction and discharge superheat. Superheat is heat added to the refrigerant above it boiling or vapor temperature. Superheat occurs in the evaporator after all the liquid refrigerant boils off. Further temperature rise of the refrigerant is known as superheat and is measured in degrees Fahrenheit, not BTU's. Suction superheat is the difference in temperature between the suction line temperature and the evaporating temperature. Discharge superheat is additional heat added to the refrigerant by the process of compression and friction in the compressor. Discharge superheat is calculated by subtracting the condensing temperature from the discharge line temperature.

- Subcooling is heat removed from the refrigerant below its condensing temperature. Subcooling occurs after the condenser develops a steady stream of liquid refrigerant. Subcooling is calculated by subtracting the liquid line temperature from the condensing temperature.

PROBLEMS AND QUESTIONS

1. What is superheat?
2. Define subcooling.
3. Draw the refrigeration cycle. Label all major components and show the direction of refrigerant flow.
4. Explain the operation of the compressor.
5. Describe the *total* operation of an air-cooled condenser. Describe what happens to the pressures, temperatures, and changes in the refrigerant inside the refrigerant tubing. Discuss refrigerant and air-flow conditions.
6. List the refrigerant conditions entering and leaving a metering device. Discuss the pressures, temperatures, and refrigerant changes.
7. Describe the *total* operation of the evaporator inside and outside the refrigerant tubing.
8. The high-side gauge measures what two refrigerant conditions?
9. The compound gauge measures what two refrigerant conditions?
10. What is the purpose of the process tube?
11. How can you determine the difference between the discharge line, suction line, and process tube on an operating compressor?

12. List the two types of air-cooled condensers.
13. List the four types of water-cooled condensers.
14. The discharge side of a compressor measures 235 psi. Approximately what pressure would you expect at the liquid line of the condenser?
15. What is the condensing temperature of a unit that uses R-22 and registers 275 psi on the high-side gauge?
16. R-22 has a condensing temperature of 100°F. What is the corresponding refrigerant pressure?
17. Name the three most common metering devices.
18. What is dewpoint? Where does the dewpoint occur in an air-conditioning system?
19. Does the refrigerant cycle consume refrigerant? Explain your answer.
20. What is the normal operating range of the evaporating temperatures?
21. While you are servicing a residential system, the customer wants you to describe the operation of the system. Describe what you would tell your customer.
22. You and a fellow technician are starting a compressor that has been giving a customer problems. Your partner positions himself in front of the compressor terminals to hook the Amprobe on the common of the compressor to measure starting amperage. He asks you to turn on the disconnect while he watches the amperage reading. What safety recommendation would you make to your coworker regarding his position at the compressor terminals? Why?
23. While servicing a unit on an annual check-up, you record a liquid-line pressure of 260 psi and a discharge-line pressure of 275 psi. Is there a problem with these readings? If so, what is the problem?
24. Your company has the annual maintenance contract on the city hall annex air-conditioning system. Everything seemed normal except the condensing temperature and pressure were higher than the previous year's readings. Last year's high-side reading was recorded at 240 psi with a corresponding condensing temperature of 137°F. The outside temperature was 97°F. This year's readings are 255 psi with an outdoor temperature of 95°F. What is the most likely problem?
25. What part of the refrigeration cycle do you have difficulty in understanding? Explain your answer.

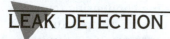

CHAPTER THREE

LEAK-DETECTION PROCEDURES

AFTER STUDYING THIS CHAPTER, THE STUDENT WILL BE ABLE TO:

• List the ways to leak-check an air-conditioning system.
• Describe each leak-detection method.
• Discuss what EPA regulations relate to leak repairs.
• Discuss why it is important to know various leak-detection methods.

• List the indications of a refrigerant leak.
• Describe why leaks are more common in field connections.
• Describe some of the common leak areas.
• State the purposes of gas identifiers and monitors.
• Discuss the sequence of operation in the case of a large refrigerant leak in a machine room.

INTRODUCTION

This chapter discusses various leak-detection methods that are available to the air conditioning technician. Some leak-detection methods are simple, and others are very sophisticated. One leak detector works well for one kind of leak, and another type of leak detector works for other types of leaks. It is beneficial to use and know as many leak-detection methods as possible.

LEAK DETECTION

Refrigerant leaks are said to be the leading cause of undercharged systems. Leaks can occur from poor-quality work; piping vibrations that can wear holes in tubing; and expansion and contraction of tubing or fittings that cause cracks or fractures. The first part of this chapter deals with no-cost or low-cost ways of detecting leaks. Next, other leak-detection devices

are discussed, such as using test instruments involving electronics, flames, dyes, smoke, ultrasonic sound, or ultraviolet lamps. The following leak-detection methods are not ranked by any particular popularity or priority. It is important to know as many different leak-detection methods as possible to be successful at finding leaks.

DETECTION BY THE HUMAN SENSES

Leak detection can be accomplished by using human senses. A technician can hear a leak if it is large enough. A leak can be seen in the form of liquid refrigerant spewing from the equipment. Seeing a liquid leak is not common, because you would need to be at the equipment when the leak occurred. A liquid leak will drain the charge before you arrive on the scene.

Sometimes you can feel a vapor leak. Care should be taken when feeling for a leak. Many of the refrigerants are under high pressure, and the leaking refrigerant could damage your eyes or ears or cut into your skin.

Small leaks can sometimes be seen in the form of oil stains or oil droplets on refrigerant tubing, coils, or valves. Dust collects on the oil around the leak. These leaks are especially detectable around the condenser coil that has outside, unfiltered air circulating around it. The dust collects around the oily, wet spot near the source of the leak. Oil and dust can collect around a nonleaking access port because attaching and removing hoses causes a small amount of oil and refrigerant mixture to spray around the fitting.

Most common refrigerants cannot be detected by odor. However, ammonia is used in absorption air-conditioning and refrigeration systems and is easily detected by it pungent odor. Many times the odor is so strong that the technician knows that there is a leak, but it is difficult to pinpoint without using other leak-detection devices. About the only sense you do not use is your sense of taste. The other four senses can be useful in locating some specific leaks.

SOAP SOLUTION

A soap solution or soap bubbles provide another way to detect leaks. Liquid soap is applied to the area that has a suspect leak. A refrigerant leak is detected when the soap solution bubbles up (Figure 3–1). The bubble(s) may slowly grow or a number of smaller bubbles many appear at the point of a leak. In the case of a large leak, the bubbles may blow off the soap solution so quickly that it is difficult to pinpoint the leak. Soap bubbles can be used to locate the exact point of smaller leaks.

A soap solution may be a homemade concoction or a commercially available soap or liquid designed for the purpose of refrigerant-leak detection. Some technicians have successful soap-solution recipes, such as a three to one (3:1) ratio of water to liquid dishwashing soap. Placed in a spray bottle, the soap solution makes a good leak detector. The cleaning agent Formula 409 is used undiluted by some air-conditioning contractors as their leak detector of choice. Purchase a gallon of the 409 product and put it in smaller spray bottles. It is close to the correct formula and does not need to be diluted with water. 409 can also be used to clean up after the job.

The liquid-soap solutions that are available at air-conditioning supply houses do a good job. The supply-house variety is usually thicker than the homemade variety and is found in various fluorescent colors. If the soap bubbles are applied to refrigerant lines that operate below 32°F, it is important

that they are formulated with a lower freezing point. A frozen solution will not detect leaks. If you are working on a heat pump in the winter, part of the system that is being serviced may be below freezing. The outdoor temperature may also be below freezing, which may cause conventional soap to thicken or solidify. Supply houses have soap-based leak detectors that can work below freezing temperatures. Soap-solution leak detectors are classified as a good and economical way to find leaks. Other leak-detection methods will be needed to find leaks that soap bubbles cannot locate. This chapter discusses using nitrogen or carbon dioxide pressure. These inert gases rely heavily on a soap solution to pinpoint the exact source of the leak. When the job is completed, wash off any soap-based leak detector with water.

ELECTRONIC LEAK DETECTORS

One of most popular leak-detection devices is the electronic leak detector (Figure 3–2). A good-quality electronic leak detector will pull in an air sample. The air sample is pulled through a filter that is located in the wand of the detector. The detector has an air pump in the handle of the wand or in the body of the unit. The filter needs to be changed when it becomes slightly dirty. A dirty filter keeps adequate air samples from getting to the detector, thus preventing the detection of possible refrigerant in the air. Detectors that do not use sampling air pumps should be avoided, because they rely on a chance that refrigerant might come in contact with the sensing element.

Check the power requirements of the leak detector. Some electronic leak detectors require a 120-V power supply. The more versatile units operate on batteries or have a rechargeable battery pack. Portability of a detector is important. Portability is second in importance only to refrigerant sensitivity.

A quality electronic leak detector will cost about twice as much as a lesser unit. Purchasing a low-quality detector is a waste of money, because a low-quality detector is like not having a detector or having a detector that continually gives false signals. A quality unit will detect small quantities of refrigerant and will not give false leak indications. In laboratory tests most quality and nonquality electronic leak detectors have the same leak-detection rate. For example, two identical units might claim to be able to find refrigerant leaks as small as ½ oz per year. Under laboratory conditions this might be true. In field conditions, the quality detector will have a higher success rate at detecting small leaks and will give fewer false indications. In electronic leak detectors you get what you pay for. Save your money until you can afford a quality unit.

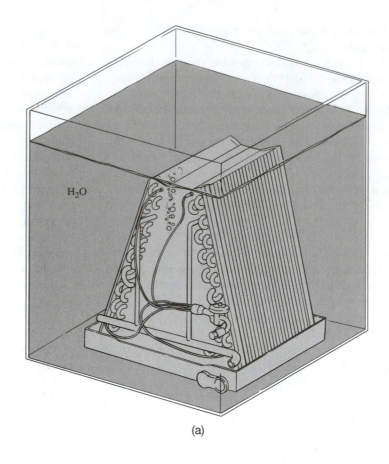

(a)

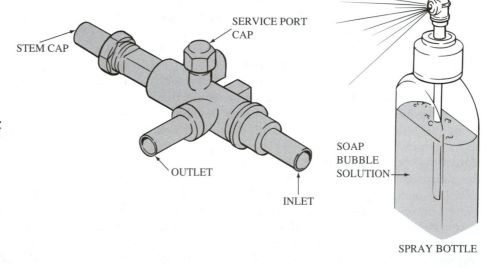

(b)

Figure 3–1 (a) Submersion test. Bubbles form at the point of a leak. The component must be capped off and charged with nitrogen. Do not exceed the test pressure of the component. (b) When checking for and isolating refrigerant leaks a soap bubble solution can be used. Remove service port and stem caps and spray solution. Check field connections such as brazed, soldered, or flared fitting to the inlet and outlet of a service valve *(Courtesy of Lennox Industries)*

LEAK DETECTOR

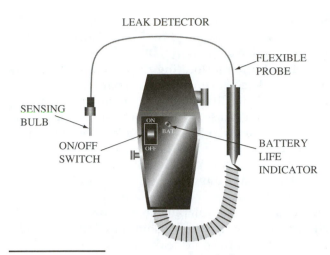

SENSING BULB

ON/OFF SWITCH

FLEXIBLE PROBE

BATTERY LIFE INDICATOR

Figure 3–2   Electronic refrigerant leak detector.

Figure 3–3   Pump-style leak detector using corona discharge technology. *(Courtesy of TIF Instruments, Inc.)*

Finally, purchase an electronic leak detector that has a sensitive setting or one that has an automatic calibration feature. When an area is experiencing a large refrigerant leak, the air is saturated with enough refrigerant to make the cheapest detector signal a leak. Units with sensitivity calibration settings are more successful in finding a leak in refrigerant-laden air. Some leak detectors have a built-in self-calibration feature. As a safety precaution, enclosed areas with large refrigerant leaks need to be ventilated before entering. Enclosed areas with large leaks require continuous ventilation for the safety of technicians and so they can use the electronic leak detector without it continuously sensing refrigerant.

PUMP-STYLE ELECTRONIC LEAK DETECTOR

Figure 3–3 shows a pump-style electronic halogen leak detector. This electronic leak detector draws air over a platinum diode. This unit is capable of detecting leaks as low as 0.4 oz per year. It can be used for HCFC, CFC, and HFC gases. The pump located in the handle draws air directly to the sensing tip. No calibration is required. It is battery operated. It has both visual and audible signals, which increase in frequency as the leak source is approached.

This technology creates a high-voltage corona between the inner tip and the surrounding shell. When refrigerant interrupts the electronic field, the alarm is triggered. The instrument, as shown in Figure 3–4, senses HFC/CFC/HCFC refrigerants by operating a selection switch on the face of the instrument.

Purchase a detector that will sense the refrigerants that are most commonly used in your day-to-day servicing. With the advent of all the new refrigerants, a technician needs to be certain that the purchase of a quality instrument will meet the needs

and daily use patterns of the changing refrigerant market.

Once a refrigerant leak is isolated, use soap bubbles or other measures to pinpoint the leak for repair. The major disadvantage of the electron leak detector is the number of false alarms it registers. Also, this type of leak detector does not work well in areas with high wind velocity. In this instance, the leaking refrigerant is blown away too quickly to be detected. The equipment's blowers and fans may need to be stopped in order to search for leaks.

Figure 3–4   Automatic halogen leak detector uses the corona technology. *(Courtesy of Robinair, SPX Corporation)*

ULTRASONIC LEAK DETECTOR

The ultrasonic leak detector is one of the latest advances in technology. An ultrasonic leak detector (Figure 3–5) will detect sound that is outside the narrow range of human hearing. Pressure leaks and vacuum leaks make a sound that may not be detectable by the human ear. The value of the ultrasonic leak detector is that it can "hear" leaks that the human ear cannot sense.

When purchasing a basic ultrasonic leak-detector, you receive a detector and a set of headphones. The headphones provided in the basic set allow the technician to hear the leak before the detector lights on the unit begin to indicate a leak. The sensitivity of the detector can be set to eliminate background noise. The detector is battery operated, so the device is very portable. Purchasing a detector kit offers more options to the technician.

Included with an electronic leak-detector kit is a ½-in. pipe extension. The pipe extension is 18 in. long and is attached to the detector. The extension allows the technician to get the detector into tight areas not reachable by hand.

The kit also includes a signal generator. Due to the size of the signal generator, it is not useful in small refrigerant tubing. The generator can be placed in lines larger than 4 in. in diameter and communicates a signal that is sensed by the leak detector. The signal generator can be placed in a walk-in cooler or domestic refrigerator to detect air leaks in the gaskets or seams of these appliances. The signal generator can be used to trace pipes.

In addition to locating refrigerant and vacuum leaks this detector can also be used to detect leaks in compressed air lines used in pneumatic systems, motor bearing noise, high-voltage arcing, relay arcing, and steam trap leakage. The kit also includes a handy carrying case that is lined with foam to protect the instrument and accessories.

The disadvantage to this detector is that it picks up too many sounds in a noisy environment. The detector can tolerate some noise, but noisy equipment will need to be shut down during detector use. The ultrasonic leak detector will find the general area of a leak; soap bubbles will be needed to pinpoint the exact location of the leak for repair. The detector is directional and will pick up noise echos. The ultrasonic leak detector is one piece of equipment that a technician needs in the arsenal of leak-detection equipment.

ULTRAVIOLET LAMP

Another new type of leak detector gaining popularity uses a high-intensity ultraviolet lamp or light to detect leaks. A fluorescent additive is introduced into the system and mixes with the refrigerant oil (Figure 3–6). The additive must be circulated in the system for a specific period of running time. The time frame is usually 24 h of normal equipment runtime. When refrigerant leaks, it purges trace amounts of oil with the refrigerant. The fluorescent additive mixes with the oil and can be seen by the ultraviolet lamp. The area of leakage is seen as a green-blue fluorescent glow. A standard ultraviolet black light will not work with this fluorescent additive. The lamp must be a high-intensity variety. The lamp will need several minutes to warm up prior to use. The lamp assembly is about the size of a heat lamp. Smaller versions, about the size of a flashlight, are available.

The ultraviolet light has enough intensity to show up under normal indoor light. Direct sunlight prevents the ultraviolet light from reflecting the additive in the oil. This equipment will work in daylight if the area to be checked is shaded or under cloud cover. The darker it is, the more apparent the fluorescent glow will be.

Another advantage of this leak detector is that the additive will stay in the system as long as the oil is not changed. The ultraviolet detector can be used to check future leaks. Leaks can be checked in a routine preventive maintenance checkup. Using a compressor with an oil sight glass will make the oil level easy to read when the fluorescent additive is in

Figure 3–5   Ultrasonic leak detector. *(Courtesy of Amprobe-Promax Instruments)*

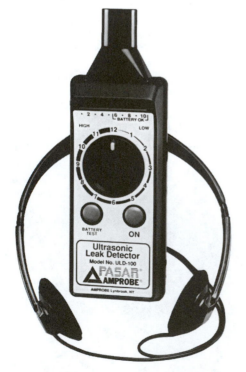

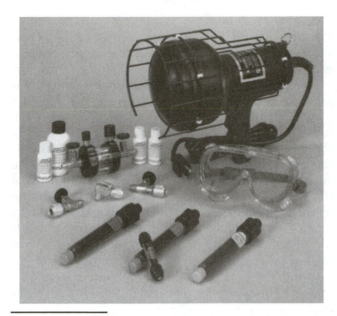

Figure 3–6a An ultraviolet high-intensity leak-detector set. *(Courtesy of Spectronics Corporation)*

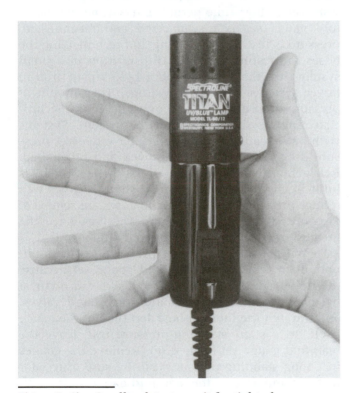

Figure 3–6b Smaller detector unit for tight places.

the system. When installing a replacement compressor inject the additive for future leak detection.

All these leak-detection methods have disadvantages, and the ultraviolet lamp system is no exception. This method requires time for the oil additive to circulate and surface as a leak. The circulation time may be 24 to 48 h. A minor disadvantage is that the detector additive is washed out in bright sun-

light conditions. This can be solved with a minimal amount of shading.

Leaks cannot be detected if the lamp cannot gain access to the leaking area. The size of the lamp assembly can reduce access to areas for leak checking. Finally, every time the gauges are connected and disconnected, a spray of refrigerant and oil coats the area around the service valves. The oil additive around the service fittings will show a fluorescent glow. If the service valve area is leaking, it will be difficult to detect because the area around the service valve has a fluorescent glow.

HALIDE TORCH

A halide torch detects leaks by burning a sample of refrigerant-laden air. The halide torch (Figure 3–7) consists of a propane or acetylene bottle, a copper element located in a burner head, a sampling hose that is used to sniff the air for refrigerant leaks, and a striker to ignite the torch.

The principle of operation is simple. The sampling hose pulls in (primary) air that mixes with the flammable gas and produces a flame. The flame burns across a copper reactor plate and causes the flame to change color as a refrigerant is detected. A normal flame color is blue, indicating no refrigerant in the air. A pale yellow flame indicates a small leak. A yellow-green flame indicates a moderate leak. A purple-blue flame shows that the halide detector has discovered a major leak.

The effective use of a halide torch requires a few important steps. A halide torch should not be used in an environment that is heavily contaminated with refrigerant. First of all, the torch would continually indicate a heavy leak without isolating the area of the refrigerant leak. An open flame in a

Figure 3–7 Halide refrigerant leak detector.

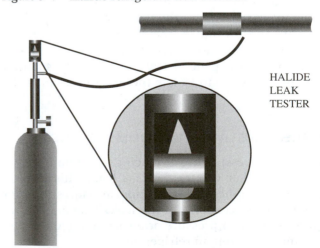

HALIDE LEAK TESTER

highly refrigerant-contaminated atmosphere creates poisonous phosgene gas. Ventilate heavy refrigeration concentrations before entering an enclosed area. A self-contained breathing apparatus (SCBA) may be required if the area cannot be ventilated prior to entering.

A technician must also take precautions not to ignite combustible construction. Using a halide torch in a tight attic or crawl space leads to extra precautions to prevent an accidental fire. Use a halide torch in a well-ventilated area. Large refrigerant leaks produce phosgene gas when burned by the torch. Working in a confined area with a halide torch presents a greater hazard to the technician.

Next, the hose is the "sniffer" that samples the air around areas that have potential refrigerant leaks. A technician must slowly move the hose inlet around the areas that are suspect. There is a time delay between the time the hose picks up the air sample to the time it mixes with the gas and is burned. The user must be patient; if not, a leak will be bypassed. It is best to slowly go over all areas containing refrigerant at least twice, whether using this method or others mentioned in this chapter.

Care must be taken to prevent the sampling hose from being blocked with dirt, grease, insulation, or other debris. A bright yellow torch flame is an indication that the hose is completely blocked. Block the hose with a finger to test the halide detector. The bright yellow flame will appear. A partial blockage may not change the characteristics of the flame. However, partial blockage will impede the performance of the halide detector because it reduces the amount of air sampled. Reduction in air sampling means that a leak may be missed. The hose and the fitting into the torch should be checked for restrictions prior to each use.

Finally, the sampling hose can be extended to reach hard-to-access areas. An extended sampling hose causes a longer delay between the time the refrigerant is sensed and the flame reacts. The technician must slow down the hose movement when using hose extensions. The halide leak detector is used to find the general leak area. A soap solution may be necessary to isolate the exact location of the leak.

Pressure and Vacuum Method

The use of pressurizing or placing a system in a vacuum is another way to detect a leak. This method will require other leak-detection measures in order to find the exact source of the leak. This method requires the use of pressure or a vacuum, and the system must not contain refrigerant.

PRESSURE METHOD

The pressure method uses nitrogen or carbon dioxide to pressurize a system. Never use oxygen or compressed air as a pressurizing gas. This mixture has caused systems to explode. Additionally, using compressed air can introduce moisture into the system.

Add nitrogen or carbon dioxide through the low side and high side simultaneously. After adding the gas, hold the pressure for a period of time. If there is no leak, the pressure will be nearly the same at the beginning of the time period as at the end of the time period. This is called a standing pressure test. The pressure test can be as short as 20 min to as much as several days or more. The longer the standing pressure test, the more accurate the results will be. For example, a small leak may not be detected in a 20-min pressure test. Testing the same system may indicate a pressure drop or leak if left for a 12-h period or longer.

A technician can leave the manifold gauge set hooked up when conducting a standing pressure test over a short time period. Every time the manifold gauge set is disconnected and reconnected, pressure is lost. Disconnecting and reconnecting the hoses can result in a 5- to 10-psig drop in a small system (less than 3 tons) and a 1- to 3-psig drop in larger systems with more volume. The amount of pressure drop depends on how quickly the technician removes the hoses from the system. Systems with back-seating service valves have a lower pressure drop when they are removed and reinstalled. Refrigerant hoses with low-loss fittings such as built-in check valves or manual shutoff valves will reduce pressure losses when the hoses are removed. Pressure charging with Schrader valves have the greatest pressure drops when removing hoses.

When using a standing pressure test it is important to remove the manifold gauge set because the hoses can be a leak source. Leaks can occur at hose connections, gauge connections, hand valves, and through the internal bourdon tube of the gauge itself. In a short standing-pressure test, leave the manifold gauge set hooked up. Remove the gauges in a long standing pressure test. Be sure to test for leaks in and around the manifold gauge set as well as the air-conditioning system itself. Use several different types of leak-detection methods.

When using electronic leak detectors, a technician will need to introduce a small amount of R-22 refrigerant into the system (Figure 3–8). Electronic leak detectors will not detect nitrogen or carbon dioxide that may be leaking from a system. The EPA, through the Clean Air Act, allows the introduction of 10 psig of R-22 refrigerant into a system that has its refrigerant recovered or is an empty system. The mixture may be legally purged to the

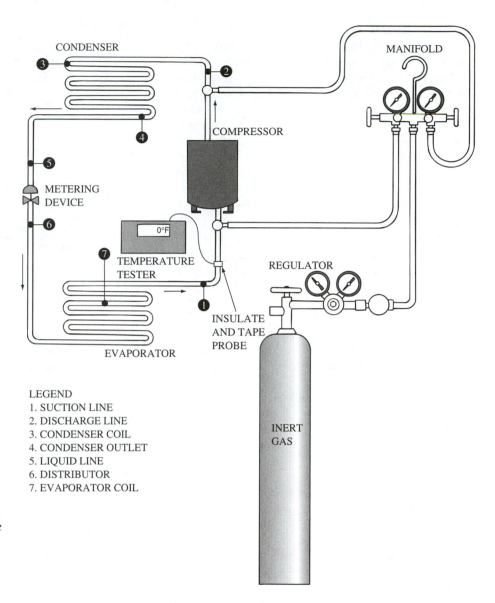

LEGEND
1. SUCTION LINE
2. DISCHARGE LINE
3. CONDENSER COIL
4. CONDENSER OUTLET
5. LIQUID LINE
6. DISTRIBUTOR
7. EVAPORATOR COIL

Figure 3–8a When testing for leaks, use nitrogen or carbon dioxide gas. The EPA will allow a trace gas of R-22 up to 10 PSI. The trace gas can be pressurized by increasing the pressure with inert gas. The mixture can be purged to the atmosphere.

Figure 3–8b Leak-testing using nitrogen to pressurize the piping.

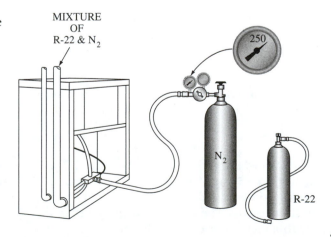

atmosphere after completion of the leak test. The EPA recommends that the technician recover this refrigerant-nitrogen mixture, but it is not required.

The recommended way to do this procedure is to first charge the system with 10 psig of R-22. Next, increase the pressure with nitrogen or carbon dioxide, being careful not to exceed the test pressure of the lowest component. The test pressure of an evaporator that uses R-22 is usually 150 psig, and that of the condenser is 400 psig. In this example, the technician would not exceed a total pressure of 150 psig. Check the evaporator or condenser nameplate to determine maximum test pressures.

A technician can test the condenser pressure to the recommended limit if the condenser is totally isolated from the compressor, metering device, and evaporator. Pressurizing a system to 400 psig may damage the compressor or the thermostatic expansion valve. Shutting off the compressor service valves may not be an adequate safeguard. Some service valves leak in a pressure test and the high pressure will build up inside the compressor, possibly causing damage to it.

Other advantages of using an inert gas are as follows:

- It is low cost.
- It can be purged to the atmosphere.
- It will not fluctuate under a normal range of day and night temperatures.
- It can obtain a higher test pressure when compared to refrigerant.

The major obstacle to using these gases is the initial cost of the regulator and cylinder. Smaller bottles are proportionately more expensive than larger bottles and in many instances cost the same to refill. For the best value, purchase the largest bottle that you can handle.

Vacuum Method

Using the vacuum method is similar to the pressure method except that a technician uses a vacuum to check for a tight system. This is called a standing vacuum test. The vacuum check can be conducted in two ways:

- Using a compound gauge
- Using a thermistor vacuum gauge (micron gauge)

Compound gauge First, the system is pulled into a vacuum until 29.5 in. of mercury is obtained. The vacuum should be pulled from both sides of the system simultaneously. The compound gauge is monitored for this test (Figure 3–9). The high-side gauge

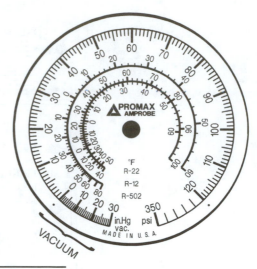

Figure 3–9 Compound gauge for measuring vacuum and pressure. *(Courtesy Amprobe/Promax)*

should drop below 0 psig, but the exact vacuum is not known because there is no high-side-gauge indication below 0 psig. At least the technician can feel comfortable that the vacuum extends to the high side of the system. Pulling a vacuum solely on the low side or solely on the high side does not mean that the total system is in a vacuum. Compressor valves, solenoid valves, and thermostatic expansion valves can be closed, preventing a total vacuum from both low and high sides of the system. For best results, pull the vacuum from the high and low sides simultaneously. Another advantage of this procedure is a quicker vacuum can be obtained when pulling it from both sides of the system.

Once the vacuum is obtained, the manifold gauge set should be closed, isolating the system from the vacuum pump. When the manifold gauge valves are first closed, the vacuum may rise or fall slightly. This occurs because the vacuum is equalizing in the hoses and in the system. The vacuum should hold for 30 min or longer. A vacuum drifting toward 0 psig means there is a leak in the system or in the manifold gauge set. The only way to check this vacuum leak is with an ultrasonic leak detector. Applying soap bubbles to a vacuum leak will pull soap into the system. Introducing a soap solution into the air-conditioning system will become a major source of contamination that will damage the compressor.

This method of leak detection notifies the technician that the system (or manifold) has a leak. The ultrasonic leak detector shows the general area of the leak. The system needs to be pressured with an inert gas in order to find the exact location of the leak, and soap and bubbles are once again recommended for this procedure.

Thermistor Vacuum Gauge A better way to conduct a vacuum leak check is to use a thermistor vacuum gauge in conjunction with the vacuum pump. A thermistor vacuum gauge (Figure 3–10) measures a vacuum in microns. Some technicians refer to this instrument as a micron gauge. A micron is a much more accurate way to measure a vacuum. A compound gauge has gradients between 0 psig and 30 in. of mercury. A thermistor vacuum gauge has a wider and deeper range of vacuum readings. An accurate micron gauge's reading starts when the compound gauge measures near 29.5 in. of mercury. Measured microns start at 2000 microns and decrease towards 0 microns. Most micron gauge instruments with readings higher than 2500 microns are not accurate. Their accuracy begins at the 2000-micron level. The target vacuum is 500 microns. A 1000-micron vacuum level is adequate to determine if the system is leak-tight. A system leak may prevent the micron gauge from reaching the 1000-micron level after the vacuum pump is isolated and stopped. If the micron gauge does not hold a vacuum for at least 15 min, the system or gauge set has a leak. The higher the micron reading, the less a system is in a vacuum. Remember to check the hoses and manifold gauge set for leaks. Hose o-rings may need to be changed. It is a good practice to pressurize a system and leak-check everything, including the manifold gauge set; then follow with the standing vacuum check. Do not apply a soap solution to a system in a vacuum.

The micron or vacuum gauge is hooked up according to the diagram in Figure 3–11. It is important the vacuum gauge be placed in parallel with the low-side hose. This can be accomplished with a ¼-in. flare tee connected to the low side of the manifold gauge set. The vacuum gauge needs to monitor the vacuum of the system after the vacuum pump has been isolated and turned off.

A system that has a leak may not allow the micron gauge to obtain the 1000-micron level. Small leaks will allow the vacuum to pull down to 1000 microns or lower. When a system has reached the target vacuum level, it is important to follow this sequence of steps:

1. Close off the manifold gauge set.
2. Disconnect the hose to the vacuum pump.
3. Turn off the power to the vacuum pump.

This sequence will assure the technician that the vacuum is not compromised by an improper shutdown sequence. A system will lose its vacuum through the vacuum pump, so it is important that the valves on the manifold gauge set are closed prior to shutting off the vacuum pump. Leaving the hose on a vacuum pump after shutdown may siphon the oil out of the vacuum pump. On some vacuum pump models, the pump oil will be pulled into the hoses because the hoses are in a vacuum and the pump is exposed to atmospheric pressure. This causes unwanted oil in the hose set and a loss of oil in the vacuum pump. Oil drawn into hoses can also contaminate the vacuum gauge sensor.

Better-quality vacuum pumps have an isolation hand valve to shut off the pump from the hoses (Figure 3–12). This valve should be closed before the vacuum pump is turned off and after the manifold gauge valves are closed.

Once the target micron level has been obtained and the proper shutdown sequence used, the micron reading may begin to drift higher or lower. This drift, or movement, should stabilize in a couple of minutes. The reason for this instability is because the vacuum is equalizing in the system. The micron gauge may go higher or lower. If, after 2 min the micron reading continues to rise, it is an indication of a leak. Watch a few minutes longer; if the micron reading continues to rise, it is time to investigate the leak(s).

REFRIGERANT DYES

Refrigerant dyes are injected into the refrigeration system and surface as a red or blue spot at the source of the leak. This visual method aids a technician in

Figure 3–10 Digital micron gauge with 9-V battery or AC option. *(Courtesy of TIF Instruments.)*

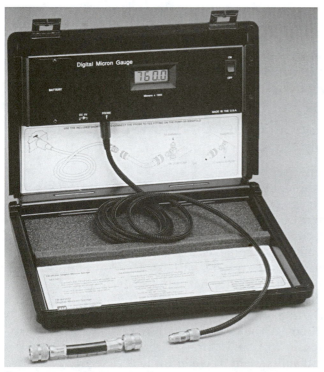

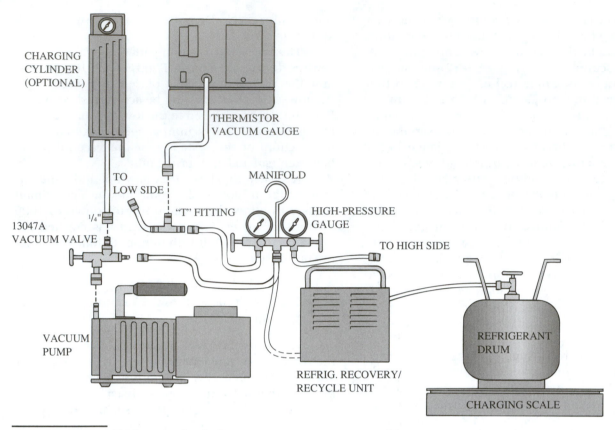

CHARGING
CYLINDER
(OPTIONAL)

THERMISTOR
VACUUM GAUGE

TO
LOW SIDE

MANIFOLD

"T" FITTING

HIGH-PRESSURE
GAUGE

1/4"

13047A
VACUUM VALVE

TO HIGH SIDE

VACUUM
PUMP

REFRIGERANT
DRUM

REFRIG. RECOVERY/
RECYCLE UNIT

CHARGING SCALE

Figure 3–11   Manifold gauges hooked up to a vacuum pump or a refrigerant recovering unit.

finding a leak. These dyes are not to be confused with the florescent additive used with the ultraviolet leak detector. The red or blue dye mixes with the oil and is seen at the point of the leak.

The dye must circulate and mix in the refrigeration system for some time period. The circulation time could be as short as a few hours and as long as 48 h. The dye is pushed out with the leaking refrigerant. Large leaks show more of the dye. Most dyes can be left in the system and can be useful in detecting future leaks.

As with the ultraviolet leak detector, the major disadvantages are the time the leak takes to become visible. Isolated parts of the refrigeration system may be difficult to visually inspect for a leak. The dye can be checked during routine maintenance check-ups.

MISCELLANEOUS LEAK-DETECTION METHODS

SUBMERSION

New-equipment or new-coil manufacturers sometimes use submersion. After the evaporator and condenser coils are manufactured, they are sealed, pressurized with an inert gas, and placed in a vat of water. If a coil has a leak, bubbles will rise and break at the water's surface. The coil is pressurized to a test pressure that is normally higher than that allowed when the coil is in a refrigeration system. The maximum test pressure of an R-22 system is 150 psig. The uninstalled, newly manufactured condenser coil can be pressurized to 450 psig or greater and submerged in water for leak detection.

Figure 3–12   Two-stage vacuum pump, 6-cfm capacity. *(Courtesy of Robinair, SPX Corporation)*

As a last resort a coil may be removed from a system, capped off, pressured, and submerged in water for leak detection. This is a drastic action that is time consuming and very costly. Prior to going to this extent it is best to isolate the system into the high side and low side. Do not use valves to isolate the high side from the low side. Valves may leak between the high and low side, preventing the discovery of the leaking side. Instead, recover the refrigerant; cut and cap the refrigerant lines; and pressure each side with 150 psig of inert gas to see which side leaks down. The leak check should be done as a standing pressure test over a period of hours. Check the capped refrigerant lines for leaks and remove the hoses to prevent loss of pressure through the manifold gauge set.

SULFUR STICKS

Sulfur sticks are used to find leaks in ammonia refrigeration systems. The sulfur stick is lit and burns slowly. A white smoke develops when the ignited sulfur stick comes in contact with a significant ammonia leak. The sulfur stick creates a strong, pungent sulfur odor as it burns.

LITMUS PAPER

A special litmus paper is also used to detect leaks in ammonia systems.

LEAK INDICATIONS

The major indication that a system has a leak is its lack of cooling capacity. A technician may find a leak when seeing an oil spot that is covered with dirt. The area will look wet compared to the rest of the condenser coil. This is not as apparent on the evaporator coil because condensate is continually washing the surface of the coil.

Suspect a leak when finding a low charge on a preventative-maintenance checkup. The unit could be undercharged because the last person working on the equipment did not know the correct charging procedures, but check the system for leaks prior to recharging.

MAJOR SOURCES OF LEAKS

An air-conditioning system can leak anywhere. It is important to take off the blinders and realize that anything in the sealed system can leak and is a potential source of a leak. New copper tubing can leak. Compressor shells can leak. Metering devices can leak. Electrical terminal plates can leak. This section focuses on the most common sources of leaks.

The most common sources of leaks are caused by field installation of tubing, accessories, and replacement parts. Another common source of a leak is

around service valves such as two-way valves, schrader valves, saddle valves, or line-tap valves (Figure 3–13). These leaks should be investigated first.

When one leak is found it is best to continue the investigation for other leaks in the system. Many systems have multiple refrigerant leaks. A professional technician does not stop when the first leak is

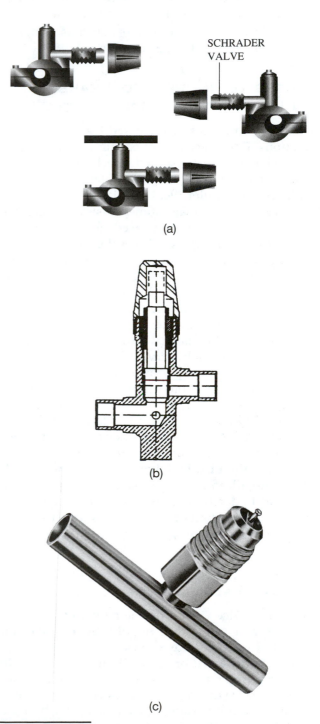

Figure 3–13 (a) The use of piercing valves for access to a hermetic system. (b) Manual shut-off valve. (c) ¼-in. refrigerant access valve. (*Courtesy of J/B Industries Inc.*)

found. A second, undetected leak will cause a recall and an unhappy customer. How many customers use a company again if they do not do the job correctly the first time?

Major sources of leaks in field installed equipment are as follows:

- Brazed, soldered, or flared joints on tubing
- Field installation of accessories such as driers, sight glass, accumulators, or receivers
- Replacement of components such as compressors, metering devices, or coils

CAUSES OF INCREASED LEAKS IN FIELD INSTALLATION

There are two causes of increased leaks in field installations. First, factory connections are brazed, soldered, or flared by personnel, machines, or equipment that do the job day after day with trained employees and under ideal conditions. The connections are tested at maximum pressures. Leaks are repaired before being released for sale to air-conditioning contractors, wholesalers, or the public.

Second, air-conditioning technicians have various degrees of training and experience. Many air-conditioning technicians learn the trade in the field rather than learning the correct procedure and practicing before entering the field. Many air-conditioning technicians who become experienced are removed from field service or installation teams and placed in other supervisory-type positions. The experienced are promoted from fieldwork, thus depleting that pool of talent. Working in the field is not like working in a lab or manufacturing facility. Field-service work finds a technician working in adverse conditions: in hot and cold extremes; in sunshine and rain; upside down, sideways, or on their back; doing blind brazing or soldering; and in tight attics and under floor crawl spaces that can be accessed only by crawling on their stomach. Under these conditions it may be a wonder that the leak rate is not higher.

ACCESS VALVE LEAKS

Access or service valves, by the nature of their construction and operation, should be checked for leaks. Two-way service valves or king valves are opened (front seated or cracked) and closed to gain access to refrigerant pressures. Sometimes the seating surfaces are damaged from system contaminants or from overtightening. Such damage will cause valve leakage. Valves may not be closed (back seated) adequately to prevent a refrigerant leak from the ¼-in. hose port. This is the reason why all service valves should be leak checked after the refrigeration hoses are removed and leak-checked prior to installing hoses. A system may be low on charge because the access valve leaked after the last service call. The technician finds the system low on charge and searches for a leak to no avail. If the technician checks the service valve before installing the gauge set, it would discover that the valve has a leak. The technician should continue to leak-check the system even after finding a leaking service valve.

Schrader valves have a needle valve that can stick open and leak refrigerant. Line-tap valves are temporary valves that should be replaced soon after they are installed. The rubber O-ring in a line tap valve can fail and cause a loss of refrigerant. Unnoticed, the O-ring can fall out in the course of installation, leaving the valve to leak immediately after it is pierced. Line-tap valves are temporary valves used on copper or aluminum tubing. Line-tap valves should not be used on steel tubing. Line taps used on steel tubing may cause an inadequate puncture or just mash the steel tubing without a puncture. Steel tubing is commonly used to make refrigerant lines in domestic refrigerators, freezers, and other small appliances.

It is important to leak-check the service valves before and after servicing. This may be the source of a leak. Install a service valve cap with an O-ring or beveled top to reduce refrigerant loss should a leak occur. These are good service practices recommended by all professional technicians.

LEAK-REPAIR REQUIREMENTS

The Clean Air Act as enforced by the EPA *recommends* that all leaks be repaired. The EPA *requires* repair of certain types of leaks.

Leaks in small appliances do not have a mandatory repair timetable. Professional technicians repair all refrigerant leaks whether the law or a customer requires the repair. The EPA requires leak repair of commercial, industrial process refrigeration, and air-conditioning appliances that contain more than 50 lb of refrigerant.

By law, any size commercial and industrial process refrigeration appliances must be repaired if the annual leak rate exceeds 35%. Leaks must be repaired in large air-conditioning systems with more than 50 pounds of refrigerant, that have an annual leak rate exceeding 15% of the total system capacity. Leaks must be repaired within 30 days of discovery, or a 1-y replacement plan must be available at the site should the EPA make a visit.

GAS IDENTIFIERS AND MONITORS

There are a number of uses for gas identifiers and monitors:

1. To notify the owner of a refrigerant leak.
2. To protect anyone entering a room where refrigerants are stored or used. Refrigerants can displace oxygen and therefore become dangerous. Most refrigerants do not have an odor and are not normally detected by human senses.
3. To identify the CO_2 content in the air and, in some cases, to automatically control the addition of outside air. There is an increasing amount of concern about indoor air quality (IAQ) and—particularly—the need for ventilation to prevent the buildup of CO_2 in enclosed spaces.

The refrigerant identifier shown in Figure 3–14 is a microprocessor-based instrument that extracts a sample of the refrigerant and analyzes it to determine the type of refrigerant. In some cases, the system has been charged with a new refrigerant without properly tagging the installation with the refrigerant number. The service technician *must* be certain of the refrigerant that has been installed in order to perform any services necessary.

REFRIGERANT LEAK-DETECTION SYSTEM

Figure 3–15 shows a refrigerant leak-detection system for monitoring specific refrigerants. The system shown identifies R-11, R-12, R-22, ammonia, HCFC-123, and HCFC-134a. This instrument complies with ASHRAE Standard 15-1992, which requires the use of an instrument of this type where air-conditioning and refrigeration systems are installed in machine rooms. The unit has a visible alarm and LCD readout showing the actual gas concentration. It relays each alarm level. Using a multiport sequencer it can be used to monitor six different locations. The power requirement is 120 V AC.

OXYGEN-DEPLETION MONITOR

The oxygen-depletion monitor shown in Figure 3–16 has a remote oxygen sensor and a controller, which are normally mounted outside the door of the mechanical-equipment room. It is designed to monitor continuously the oxygen level in mechanical-equipment rooms where class A, Group I refrigerants are used. Oxygen is normally 20% of the

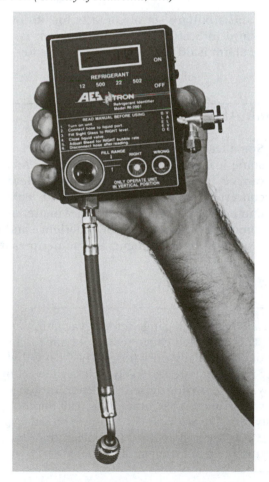

Figure 3–14 Refrigerant identifier for R-12, R-22, R-500, and R-502 *(Courtesy of Neutronics, Inc.)*

Figure 3–15 Refrigerant leak-detection system and monitor for refrigerants R-11, R-12, R-22, ammonia, HCFC-123, and HCFC-134a. *(Courtesy of MSA Instrument Division)*

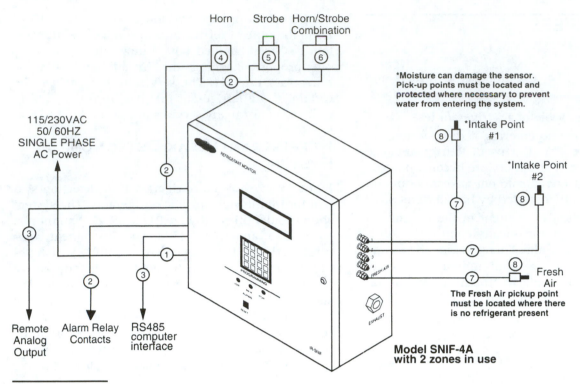

Horn   Strobe   Horn/Strobe
                  Combination

*Moisture can damage the sensor.
Pick-up points must be located and
protected where necessary to prevent
water from entering the system.

115/230VAC
50/ 60HZ
SINGLE PHASE
AC Power

*Intake Point
#1

*Intake Point
#2

Fresh
Air
The Fresh Air pickup point
must be located where there
is no refrigerant present

Remote   Alarm Relay   RS485
Analog   Contacts   computer
Output   interface

Model SNIF-4A
with 2 zones in use

Figure 3–16 Oxygen-depletion monitor and refrigeration sensor is required by ASHRAE 15-1992. *(Courtesy of Sentench Corporation)*

atmosphere, and when levels drop to 19.5% an amber warning light goes on, starting a mechanical exhaust. If the oxygen level continues downward to 18.5%, a red warning light goes on and an alarm sounds. Relays provide an option for sending the warning to remote locations. A "purge" switch permits the fan to run continuously or only when signaled by the monitor. The alarm turns off automatically when normal levels of oxygen return or may be turned off manually—at which time there is automatic reset.

The first stage, or warning state, activates an amber LED and a set of relay contacts to operate a fan or other mechanical equipment. The second stage, or alarm stage, activates a red LED, an audible

alarm, and auxiliary set of contacts for optional remote-alarm indication. After the problem is solved and the alarm is off, the system automatically resets.

SUMMARY

- Some leaks are easy to detect, and it seems that some leaks are impossible to locate. Any leak can be found. As far as refrigerant leaks are concerned, one technician can find leaks that another cannot. Sometimes it is a matter of luck; sometimes it is a matter of experience and having the right equipment; and sometimes it is a

ON THE JOB

It is a good idea to have a dedicated manifold gauge set for evacuation and pressure-checking procedures. This dedicated gauge set should have tight connections and use a 5-ft set of black, heavy-duty hoses. The heavy-duty hoses have a high permeability rating and are less likely to lose the vacuum or pressure through the hose wall. The suction-hose connection should have a ¼-in. flared tee for use with a micron vacuum gauge. The manifold gauge set should be periodically tested to withstand a 350-psig pressure and also hold a 50-micron vacuum. This test will assure a technician that the leak is in the system under scrutiny and not the manifold gauge set.

matter of perseverance. It is up to owners of the equipment to determine how much they are willing to spend to find the leak. The EPA places leak-repair requirements on owners of larger pieces of air-conditioning and refrigeration systems. These owners are responsible for having leaks repaired within 30 days or having a plan drafted for its replacement within 1 y. The technician should include this warning on the customer's invoice. Such a warning releases the technician from any liability regarding having leaks repaired in a timely manner.

- The air-conditioning technician must have a number of leak-detection options to be successful in finding leaks. No one leak detector will find all leaks. Do not give up. All leaks can be found. If you cannot find the leak, the next technician will become the hero and find what you were not able to locate—the leak.

PROBLEMS AND QUESTIONS

1. According to EPA regulations, what is the leak rate that requires repair in a window air-conditioning unit?

2. What is the leak rate that requires repair in a chilled-water air-conditioning system that uses 100 lb of R-22?

3. What is the leak rate that requires repair in an industrial-process cooling system that contains 25 lb of R-134a?

4. Why are there more leaks associated with field connections than with factory-made connections?

5. What are the human senses that can be used for detecting refrigerant leaks?

6. What is the major indication of a refrigerant leak?

7. What does the following statement mean?

   Most available electronic leak detectors are rated to locate a refrigerant leak that is less an ounce per year. When purchasing an electronic leak detector, it is important to select one that has the quality to get the job done.

8. How does an ultrasonic leak detector work?

9. What valuable options are available for an ultrasonic leak detector?

10. How does an ultraviolet leak detector work? Describe the setup procedures.

11. What are the advantages and disadvantages of an ultraviolet leak detector?

12. How does a halide torch leak detector work?

13. What precautions must be exercised when using a halide leak detector?

14. How are refrigerant dyes used to detect leaks?

15. Describe how to find leaks using the pressure method of leak detection.

16. Describe how to find leaks using the vacuum method of leak detection.

17. The manifold gauge set can be a source of leaks when used with the pressure or vacuum leak-detection methods. What leaks can be experienced in a manifold gauge set?

18. Describe the advantage and disadvantage of the submersion leak-detection method.

19. Access valves are a source of leaks. Why should access valves be checked before and after installing refrigeration hoses?

20. Line-tap valves can be used to gain temporary access to a refrigeration system. What precautions must be exercised when using these valves?

21. Discuss the sequence of operations in the case where a large quantity of refrigerant is released in a mechanical room.

22. What is the purpose of a refrigerant identifier?

23. Why are soap-based leak detectors used as leak detectors?

24. What detectors are used to locate leaks in ammonia systems?

25. What is the EPA regulation regarding use of R-22 as a leak indicator?

CHAPTER FOUR

HOW TO CHARGE AIR-CONDITIONING SYSTEMS

AFTER STUDYING THIS CHAPTER, THE STUDENT WILL BE ABLE TO:

•Describe how to access a refrigeration system.
•List the types of access fitting and valves.
•Discuss how to charge with liquid and vapor.
•List the different methods of charging.
•Read and interpret the manifold gauges.
•Maintain and check the operation of the manifold gauge set.

•List the equipment and materials necessary for charging.
•Discuss equipment preparations for the charge.
•Describe how to remove and add a compressor oil charge.

INTRODUCTION

This chapter is the lead chapter in the series of how to charge air-conditioning systems. It is an elementary chapter on how to gain access to the refrigeration system, how to operate and read the manifold gauge, and how to add refrigerant should the system be undercharged.

This chapter summarizes the conditions necessary to check a charge. The equipment should be cleaned and lubricated. The airflow should be at the correct volume. A thorough explanation of the cleaning procedures is given in the chapter on preventive maintenance. This chapter does not discuss the various ways to check the charge. That information is provided in future chapters.

The subsequent chapters discuss the following charging methods:

1. Pressure-temperature charging method
2. Suction and discharge superheat charging methods

3. Subcooling, amperage draw, and manufacturer charging methods
4. Temperature drop, sweat-back, and sight-glass charging methods
5. Weigh-in, frost-back, and approach charging methods

In all, twelve charging methods are discussed in the following chapters. Some of the charging methods are excellent and some are poor or even dangerous to the equipment.

It is important to know as many charging methods as possible. One reason it is important is that it will arm a technician with many different ways to charge a system. Many of these charging methods can be used as refrigerant-side troubleshooting tools.

The other benefit to knowing these charging methods is that you will be exposed to most of the charging procedures that are used in the field. *Not* knowing the various charging methods makes a technician vulnerable to influence from technicians who may be presenting a charging method in an incorrect manner or presenting one that is dangerous to the technician or the equipment. The air-conditioning trade is full of practices that are *not* considered professional, even if they seem reasonable. Being introduced to all the charging methods, both good and bad, provides information that a technician can use to analyze and validate a "new" method before accepting it as valuable. New ideas are easily adopted if a person does not have the informational base on which to process the ideas. Having a broad base of knowledge of charging will make you better able to judge new charging methods as to their value for daily work activities.

EQUIPMENT AND INSTRUMENT REQUIREMENTS

In order to complete the charging procedures discussed in this book, a technician needs the following equipment or instruments:

- Refrigeration manifold gauge set with hoses
- Digital temperature tester with multiprobe capabilities
- Sling psychrometer
- Accurate digital scale
- Charging cylinder
- Amprobe clamp-on ammeter

An optional piece of equipment that is useful is an airflow hood or other airflow-measuring device, such as a velometer or anemometer. These devices are used to measure airflow in cubic feet per minute (cfm) or feet per minute (fpm). If this equipment is not available, an airflow chart is provided in this chapter to gain the necessary information to calculate proper airflow. It is important that the air-conditioning system generate the correct airflow through the duct system, or charging the unit will be difficult and inaccurate.

SUPPLIES AND INFORMATION

The following supplies and information will be needed to complete the charging methods listed in this textbook:

- Duct tape
- Pipe insulation
- SEER rating of the system
- Rated load amperage of the compressor
- Manufacturers charging recommendations

- Superheat charging chart for fixed metering devices
- Sling psychrometer
- Accurate thermometer

PREPARING TO CHECK THE CHARGE

Before a technician can conduct an accurate charging procedure, several things are required. The equipment must be clean and lubricated. The condenser should automatically be washed. The evaporator should be inspected and cleaned if dirty. This is a good time to determine the type of metering device. Identifying the type of metering device is necessary when checking suction superheat.

Check the condition of the air filter and remove or replace it if necessary. Next, check the airflow. The required airflow is 400 cfm per ton of air conditioning. One of the best instruments for checking the airflow is the airflow hood (Figure 4–1). The hood measures the quantity of air directly in cfm. The total air quantity can be measured in one of two ways. The technician can measure and combine the cfm output of each supply air register or measure the cfm of the return air. The air supplied by the duct system should equal the air returned through the filter. If the difference between the combined supply air cfm is not within 5% of the return airflow, then the duct system has an air leak. The leak or leaks could be in the return air or the supply air duct

Figure 4–1   Electronic flow hood. *(Courtesy of Shortridge Instruments)*

system. There could be leaks in both supply and return ducts.

Another way to measure airflow is to use the airflow chart, Chart 4–1. In order to use this chart, a technician needs to know the indoor wet-bulb and dry-bulb air temperatures and the supply-air temperature. These can be measured with a sling or a digital psychrometer. It is best to measure this temperature at the return-air grille or near the air entering the evaporator coil. The return air wet-bulb and dry-bulb temperatures will be measured.

The indoor entering air (return-air) wet bulb is found across the top of the chart. The indoor entering-air dry bulb is found on the left side of the chart. The temperature at which the wet bulb and dry bulb intersect is the supply-air target temperature. If the leaving-air temperature is lower than the target by more that 3°F, then the airflow through the evaporator is too low. Low air flow through the evaporator allows the air to stay in contact with the air longer, thereby cooling it more than normal.

WORKING EXAMPLE

The indoor wet-bulb temperature is 67°F, and the dry-bulb temperature is 76°F. The supply-air temperature is 59°F. The indoor temperatures intersect at 60°F. This reading is within ±3°F of the normal airflow reading; therefore, the air flow is within the 400-cfm per ton rating.

SOLUTION: See Chart 4–1.   ∎

Caution When using Chart 4–1, the charge on the unit should be close to normal. The unit should be operating at least 15 min. Systems very low on charge will show a higher-than-normal supply-air reading. Systems that are undercharged enough to create short-term very cold or frosty air may fool the technician into believing that the airflow is low. In some low-charge conditions the air leaving the evaporator will produce air temperatures in the 30° range. Eventually the coil will freeze over and block the flow of the air.

Using this chart is valuable provided that the charge is close to being correct. Using the airflow hood is the best alternative. The airflow hood can be used in heating systems and in balancing the airflow in residential and commercial buildings.

Finally, a technician should make sure that the equipment evaporator and condenser match. The refrigerant lines should be sized correctly. Mismatched equipment or improper refrigerant line sizes will affect the performance and the charge on the system.

In summary, a technician needs to go though the following checklist before finalizing any charging procedure:

- Equipment must be cleaned, especially the condenser, evaporator, and filters.
- All motors should be lubricated.
- The type of metering device should be determined.

Chart 4–1

PROPER AIRFLOW RANGE (Cooling)

	Indoor Entering Air Wet Bulb °F																		
	57	58	59	60	61	62	63	64	65	66	67	68	69	70	71	72	73	74	75
	⇩Proper Evaporator Coil Leaving Air Dry Bulb °F⇩																		
70	51	51	52	52	53	53	54	55	55	56	57	58	59	60	—	—	—	—	—
72	52	52	53	53	54	55	55	56	57	57	58	59	60	61	62	63	—	—	—
74	53	53	53	54	55	55	56	57	58	58	59	60	61	62	63	64	65	—	—
76	54	54	54	55	55	56	57	57	58	59	60	61	62	63	64	65	66	67	—
78	55	55	55	56	56	57	57	58	59	60	61	62	63	64	65	66	67	68	69
80	56	56	56	56	57	58	58	59	60	61	62	63	64	65	66	67	68	69	69
82	57	57	57	57	58	59	60	60	61	62	63	64	65	66	67	68	69	70	71
84	—	58	59	59	60	60	61	61	62	63	63	64	65	66	67	68	69	70	71

Indoor Entering Air Dry Bulb °F (left axis label)

Note: This table is based on 400 CFM per Ton of cooling capacity

- The condenser and evaporator must be determined to match.
- Proper airflow requirements must be checked.
- The system SEER rating must be established.

GAINING ACCESS TO THE SEALED SYSTEM

MANIFOLD GAUGE SETS AND ACCESS VALVES

This section covers the types and uses of the manifold gauge set. In the air-conditioning trade they are referred to as the *gauges*. This basic test instrument tells an air-conditioning technician what is happening inside a refrigeration system. The gauges give operation pressures and show the evaporating and condensing temperatures. The manifold gauge set is universally recognized as the instrument used to test air-conditioning equipment. Various types of access valves are covered following the manifold gauge set explanation.

TYPES OF MANIFOLD GAUGE SETS

Three types of manifold gauge sets are illustrated in Figure 4–2. Figure 4–2(a) shows the digital type, which can be calibrated. The combination low-side gauge reads from 30 in. of mercury (Hg) to 99.9 psig. The high-side gauge reads from 0 to 500 psig. The four-way block is equipped with three refrigerant hoses and one evacuation hose.

Figure 4–2(b) shows a different manifold gauge set. These liquid-filled analog gauges are filled with glycerine to dampen pulsations, lengthen service life, and improve accuracy. The low-side gauge reads pressures from 30 in. of Hg to 120 psi. The high-pressure gauge reads from 0 to 500 psi. The four-way block is equipped with three refrigerant hoses and one evacuation hose.

Figure 4–2(c) shows a gauge manifold set with dry gauges mounted on a two-way refrigerant block.

All digital, glycerin-tilled and standard dry gauges are also available with Metric Bar, kPa and kg/cm² scales.

The gauge manifold is one of the handiest tools in the technician's kit, used for checking operating pressures, adding or removing refrigerant, adding oil, and performing other necessary operations.

The manifold has five connections. The two top connections hold the compound and the high-pressure gauges. The compound gauge, placed on the left side, reads pressures on the low pressure side of the system from 30 in. Hg vacuum to usually 250 psig pressure. The high-pressure gauge reads pressures on the high side of the system from 0 to 500 psig. Gauges used to measure R410A will have higher pressure gauges up to 800 psig. Older gauges may have lower maximum pressures.

The bottom of the manifold has three connections. To these openings are attached high-vacuum hoses, usually capable of being leak-tight down to 50 microns or less. The left hose is connected to the low side of the refrigeration system being serviced.

Figure 4–2 Three types of gauge manifolds. *(Courtesy of TIF Instruments)*

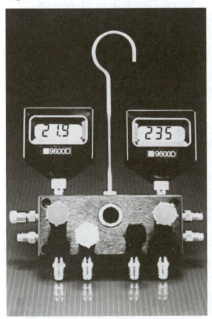

(a)   (b)   (c)

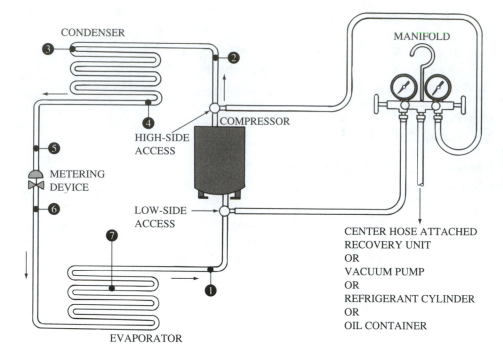

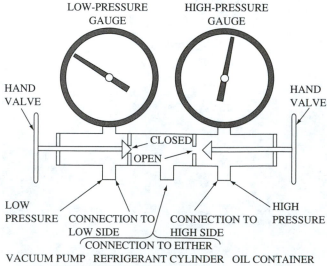

Figure 4–3 Gauges attached to a typical system. The high-side hose can also be connected to the liquid line if an access valve is available. It is rare to find an access valve on the distribution (6) tube feeding the evaporator.

LEGEND
1. SUCTION LINE
2. DISCHARGE LINE
3. CONDENSER COIL
4. CONDENSER OUTLET
5. LIQUID LINE
6. DISTRIBUTOR
7. EVAPORATOR COIL

The right hose is connected to the high side of the system. The center hose has a number of uses, all associated with servicing the system.

Refrigerant hoses are available with anti-blow-back or check valves to prevent venting of refrigerant.

A gauge manifold attached to a typical system is shown in Figure 4–3.

The exteriors of gauge manifolds are often color-coded. The compound gauge and low-side hoses are blue. The high-side gauge and high-side hose are red. The center utility hose is usually white or yellow. The center hose is useful for connecting to the charging cylinder, vacuum pump, vacuum indicator, recovery/recycling unit or container. The container may be a cylinder for collecting the refrigerant during the recovery process or an oil container used for charging the system with lubricating oil.

Two positions of the service valves for adjusting the gauge manifold are shown in Figure 4–4.

The valve on the left in Figure 4–4 is front-seated. In this position the compound gauge reads the pressure in the low side of the system. The opening to the center port is shut off.

The valve on the right side is back-seated or in the open position. In this position the high side of

Figure 4–4 Manifold valve set with low-side valve closed and high-side valve open. Turning valve handle clockwise will close the valve. This is also known as a front-seated valve. A back-seated valve will be in the open position. The valve will be backed out as far as possible.

the system is open to the center port for whatever use is desired.

There is also a third possible position of the valve (not shown) where the connection to the center port is only partially open—called the cracked position. In this position the flow through the center port can be controlled by minor adjustments of the valve.

When both service valves are front-seated (Figure 4–5), the following service operations can be performed:

1. Connecting or disconnecting the gauge manifold from the system.
2. After the gauge manifold is in place, measuring both the low-side and the high-side system pressures without disturbing the system (no flow).

Figure 4–6 shows the connections at the bottom of the gauge manifold when set for a charging operation.

The left valve in Figure 4–6 is back-seated; the right valve is front-seated. The hose from the center connection goes to a refrigerant cylinder. When the valve on the cylinder is opened, the refrigerant can flow from the cylinder into the low side of the system. This is a typical arrangement for charging refrigerant vapor with or without the compressor running.

The low pressure gauge reads the pressure at the outlet of the cylinder. The high-pressure gauge reads the pressure on the high-side of the system.

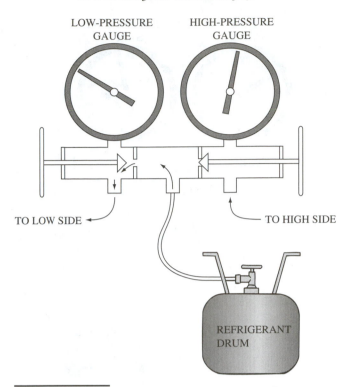

Figure 4–6   Set up manifold gauge set for vapor charging by connecting the refrigerant cylinder to the center hose. Purge the hose at the manifold. Open the cylinder and back-seat the low-side valve. The refrigerant will leave the cylinder provided it is at a higher pressure than the low-side pressure.

Figure 4–7 shows the service valve settings for removing the refrigerant from the high side of the system.

The left valve in Figure 4–7 is front-seated and the right valve is back-seated. These are the settings used when a vacuum pump or a recovery unit is evacuating the high side of a system.

The student/technician should be cautioned not to operate a hermetic compressor while evacuating the system because the loss of refrigerant would reduce the normal cooling of the seated-in motor.

With the compressor off and the refrigerant cylinder connected to the center opening, the refrigerant would migrate to the cylinder, provided it was at a lower temperature. When both service valves are back-seated and either a vacuum pump or a recovery unit is attached to the center connection (Figure 4–8), both gauges register the pressure in the center hose.

This setting allows the evacuation system to remove refrigerant from both the low side and the high side at the same time. Based on there being a restriction at the metering device, such as a closed expansion valve, this arrangement is effective in performing complete evacuation. Use high- and low-side pressure hookup to speed evacuation.

Figure 4–5   Both valves front-seated, or closed. The manifold gauge set is used to connect gauges to the system, allow temperature and pressure readings, and charge, recover and evacuate.

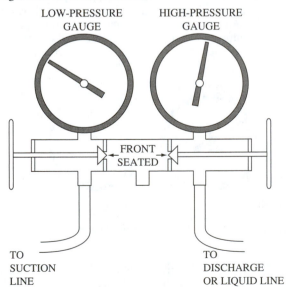

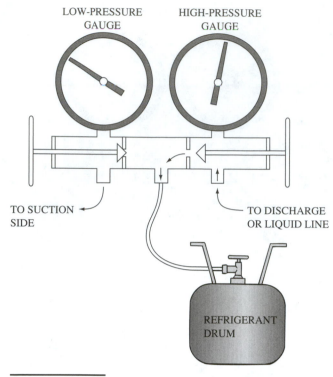

Figure 4-7   Removing refrigerant from the high side. Liquid or vapor can be recovered. Liquid will be recovered from the liquid line and vapor from the discharge line. The low-side valve must be front-seated and the high-side valve must be back-seated. This process can also be used to pull a vacuum from the high side.

When both of the service valves are back-seated with the center connection capped (Figure 4-9), the difference in pressure between the two

Figure 4-8   Manifold gauge hookup with both valves back-seated. The recovery unit must be capable of handling liquid for this hookup. The system should not have any pressure on it prior to hooking up a vacuum pump.

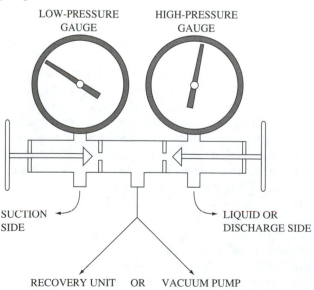

sides causes a flow through the valve until the pressures equalize. This can occur accidentally. If it does, both gauges will read the same pressure.

Figure 4-10 shows the position of the valves for liquid charging into the receiver on the high side of the system, with the compressor shut off. The left gauge is front-seated and the right valve is back-seated.

It is also feasible to charge liquid to the high side of the system, entering on the inlet side of the expansion valve, with the compressor running. For this arrangement, the valve positions are the same as shown in Figure 4-10. Again, the pressure in the liquid line must be lower than the pressure in the refrigerant bottle. This can be accomplished when the system is low on charge or is being pumped down. Do not charge liquid into the discharged line.

A modified gauge manifold that has been designed especially for evacuating and dehydrating a system (Figure 4-11) has larger hose connections and uses larger hose. It has multiple openings on the center connection to accommodate various devices using this connection. These innovations serve to speed up the evacuation process.

INSTALLATION OF A MANIFOLD GAUGE TEST SET

A manifold gauge set is a valuable service tool for a refrigeration servicer. It affords a service technician a quick method of installing pressure gauges for checking system conditions, charging, and adding oil to the compressor.

Just as a doctor uses blood tests to diagnose many illnesses, a qualified technician may use a

Figure 4-9   The manifold is in the bypass position. The high-pressure side will equalize with the low-pressure side. This can be used to equalize pressures quickly. Do not allow liquid refrigerant to transfer from the liquid line into the low side. Liquid refrigerant will dilute compressor lubrication.

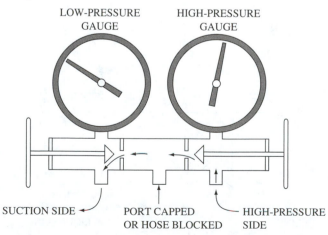

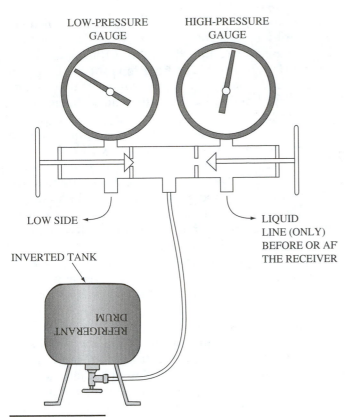

Figure 4–10   Manifold gauge hookup for liquid charging into the liquid line or liquid receiver. The pressure in the cylinder must be higher than the pressure in the system or refrigerant will be removed from the system back into the tank.

gauge to diagnose trouble in a refrigeration system. A test set allows an operator to watch both gauges simultaneously during purging or charging operations and saves time on almost any work that must be done on the system. A cutaway view of a gauge manifold is shown in Figure 4–12.

Figure 4–11   Manifold gauges hooked up to a vacuum pump or a refrigerant recovery unit. A multiple-port utility connection can be fabricated in the field that would facilitate use of the standard gauge manifold. Some manufacturers offer these utility connections as an accessory for the gauge manifold.

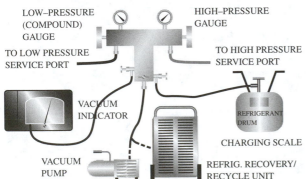

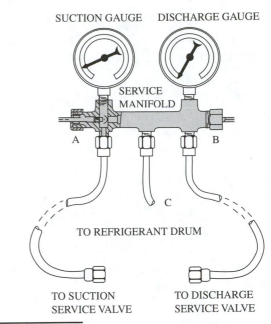

Figure 4–12   Cut-away view of a gauge manifold.

To determine system operation, add charge, purge, equalize, or evacuate, service gauges should have the best degree accuracy that is available.

This is a test instrument and should be treated as such:

1. Never drop or abuse the gauge manifold.
2. Keep ports or charging lines capped when not in use.
3. Never use any fluid other than clean oil and refrigerant.
4. Zero the gauge before each use.

Figure 4–13 shows the pressure gauges on the manifold.

The high-pressure gauge (Figure 4–14) has a single continuous scale, usually calibrated (marked off) to read 0 to 500 psi. The scale may be marked in either 2-lb or 5-lb increments and is usually connected to the high side of the refrigeration system. The outer scale is the pressure scale and the inner scales indicate the temperature of the various refrigerants at respective pressures. For example, if the gauge pointer indicated 200 psi pressure for R-12, the condensing temperature of the refrigerant would be approximately 139°F. This can also be calculated from a P-T chart.

The compound gauge (Figure 4–15) measures both pressure and vacuum. It is usually calibrated from 0 to 30 in. of mercury and from 0 to 350 psi. Like the high-pressure gauge, the compound gauge also has scale calibrated to read temperatures of various refrigerants such as R-12, R-22, and R-502. With these scales it is not necessary to refer to

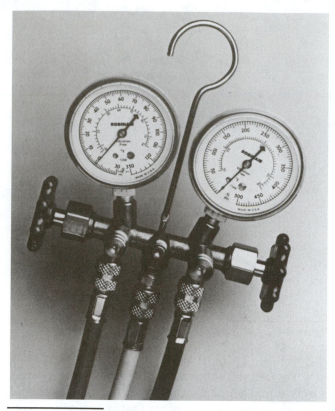

Figure 4–13   Testing manifold with gauges. *(Courtesy of Robinair, SPX Corporation)*

pressure-temperature tables or curves to calculate pressure-temperature relationships.

Gauges use a Bourdon tube as the operating element (Figure 4–16). The Bourdon tube is a flattened metal tube (usually a copper alloy) sealed at one end, curved and soldered to the gauge fitting at the other end.

A pressure rise in a Bourdon tube tends to make it straighten. This movement will pull on the link, which will turn the gear sector counterclockwise. The pointer shaft will move clockwise to move the needle. On a decrease in pressure, the Bourdon tube moves towards its original (clockwise) position, and the pointer moves counterclockwise to indicate a decrease in pressure.

Many systems have a high- and low-pressure service tap point for checking pressures and charging. They may be located on the compressor shut off valve, liquid valves, or as independent points. If the refrigeration system does not have "installed" service valves, the technician must tap the suction or discharge lines. Service access ports (Figure 4–17) are sometimes used for this. Note that Schrader service valve (Figure 4–19) ports require an adaptor fitting or a core remover between the service valve and the hose. These ports should always be leakproof and capped or plugged when not in use. Leak check access valves when removing hoses.

Figure 4–14   High-pressure gauge reads increments of 5 psig. The outer ring is the pressure ring and the inner three rings are condensing temperature rings. *(Courtesy of Robinair, SPX Corporation)*

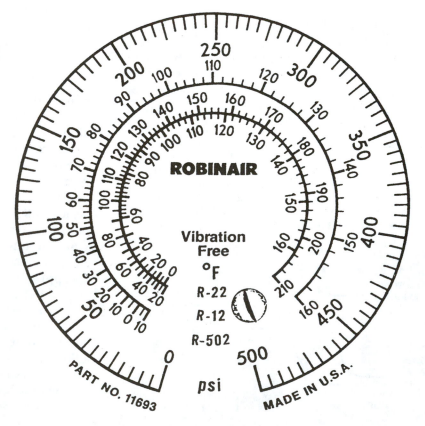

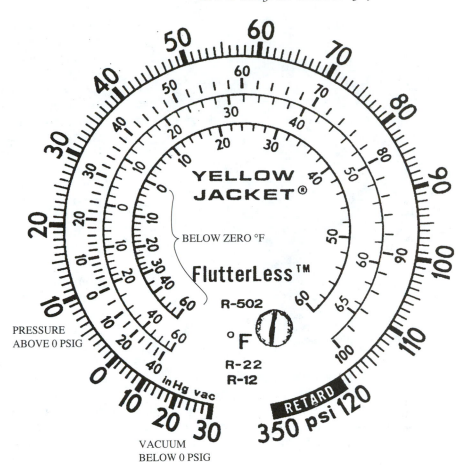

Figure 4–15 When the compressor is running, the compound gauge measures both pressure and vacuum on the outer ring. The outer ring is the pressure ring, and the inner three rings measure the evaporating temperature. *(Courtesy Ritchie Manufacturing)*

PRACTICING PROFESSIONAL SERVICES

Some gauge readings will fluctuate wildly making it difficult to determine the correct pressure or temperature reading. This fluctuation can be reduced by closing or back seating the service valve followed by slightly cracking it open. This will reduce the pulsations and extend the life of your gauges. Purge the hose to make sure the gauge is reading system pressure.

If there is ever any oil in your gauges after removing them, be sure to drain the oil into an approved container to avoid reintroducing the wrong oil into a new system with different lubricants.

Whenever you are working with Schrader valves, use a quick connect line fitting on the discharge line. This will eliminate a large loss of refrigerant on a critically charged unit and avoid having refrigerant burn your hands or face.

PREVENTIVE MAINTENANCE

Gauge manifolds should be routinely calibrated for accuracy. A simple test is to install your gauges on a new refrigerant cylinder. Convert the pressure to temperature using your PT chart and then measure the tank temperature. The tank should be at a stable temperature for 24 h. The two temperatures should match if not, remove the cover from your gauges and calibrate the dial accordingly. Zero the gauges before each use.

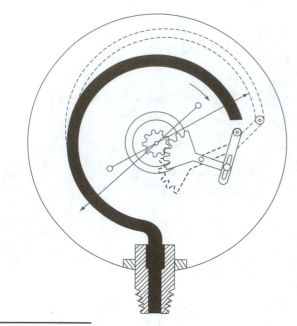

Figure 4–16 Bourdon tube.

The low side of gauge manifold test set is attached to the suction side of the system and air is purged out of the line with refrigerant. You can also evacuate the lines up to the service valve. Purge hoses each time they are installed.

The high side of the gauge-manifold test set is attached to the discharge side of the system, and air is purged out of the line again as before. Purge hoses each time they are hooked to the system.

Pressure readings are taken with the system operating. (The compressor is running.) Normal pressure readings of an air-conditioning R-22 unit may be 65–80 psi pressure on the low side (compound gauge), and 175–300 psi on the high side (Figure 4–18).

The temperature of the refrigerant may be calculated by observing the inner scale for that particular refrigerant, in this case R-22, and the corresponding pressure (Figure 4–15). The pointer will indicate its corresponding pressure and temperature. Let's look at the compound gauge. Assuming the gauge indicates 70 psig, the evaporating temperature of R-22 scale or inner scale is 41°F. The high-pressure gauge indicates 250 psig; the corresponding temperature for R-22 at this pressure is 117°F (Figure 4–14).

Now to remove the gauge-manifold test set from the system: If the refrigeration system has two-way service valves installed or the technician has used a core removal tool, or has installed a valve to the flexible hoses connected to the suction and discharge service valves, shut off the service ports (suction and discharge), and purge the refrigerant remaining in the gauge lines, out through the center port of the gauge manifold. Remove the hoses and valves on the system and return to the storage position.

The evaporating and condensing temperatures can be obtained from a pressure-temperature (P-T) chart. Find the evaporating and condensing pressure and find the corresponding temperature on a P-T chart. Many manufacturers are deleting the temperature rings because of the large proliferation of refrigerants now on the market (Figure 4–20).

Figure 4–17 Piercing valve *(Courtesy of Robinair, SPX Corporation)*

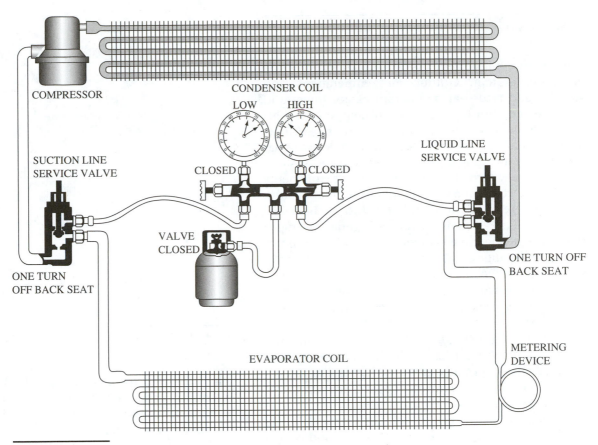

COMPRESSOR

CONDENSER COIL

LOW   HIGH

SUCTION LINE
SERVICE VALVE

CLOSED   CLOSED

LIQUID LINE
SERVICE VALVE

VALVE
CLOSED

ONE TURN OFF
BACK SEAT

ONE TURN
OFF BACK SEAT

METERING
DEVICE

EVAPORATOR COIL

Figure 4–18 Typical gauge readings for an air-conditioner R-22 unit.

Figure 4–20 Pressure gauges that can be used for R-410A refrigerant. *(Courtesy of Robinair, SPX Corporation)*

Figure 4–19 ¼″ refrigerant access valve. *(J/B Industries Inc.)*

PRESSURE MEASUREMENTS

Temperature measurements are taken outside the operating system. Outdoor-air temperatures and supply- and return-air temperatures are important. Suction-line, liquid-line, and discharge-line temperatures help in determining a good charge. It is also necessary for a service technician to know what the refrigerant temperatures are inside the system. This can be learned from pressure measurements that have a corresponding temperature relationship. This pressure-temperature relationship can be determined from a P-T chart or from refrigeration gauges that have refrigerant temperatures on their inner rings. The gauges can tell a technician the corresponding condensing and evaporating temperatures.

Two pressure gauges are necessary to determine the performance of the system (Figures 4–14 and 4–15).

On the right (Figure 4–13) is the high-pressure gauge, which measures high-side or condensing pressures. It is normally graduated from 0 to 500 psi in 5-lb increments. Pictured is a 0 to 800 psi gauge for R-410A refrigerant (Figure 4–20).

The compound gauge (left, in Figure 4–20) is used on the low side (suction pressures) and is normally graduated from a 30-in. vacuum to 350 psi; thus, it can measure pressure above and below atmospheric pressure (Figure 4–15). This compound gauge is calibrated in 2-lb increments. Other pressure ranges are available for both gauges, but these two are the most common. Readings above 120 psi are not accurate on this gauge.

Figure 4–20 shows the left low-pressure gauge with vacuum readings of 0–30 inches of mercury and pressures from 0 to 500 psig.

Note that on some of the dial faces there is an *intra-scale,* which gives the corresponding saturated refrigerant temperatures at a particular pressure. In Figures 14–14 and 14–15 the dials are marked for R-12 and R-22 and R502 refrigerants. Gauges for other refrigerants and for SI measurements are also available.

By opening and closing the refrigerant valves on gauge manifold A and B (Figure 4–22), different refrigerant flow patterns can be obtained.

The valving is so arranged that when the valves are closed (front seated), the center port on the manifold is closed to the gauges (Figure 4–22(a)). When the valves are in the closed position, gauge ports 1 and 2 are still open to the gauges, permitting the gauges to register system pressures. In most instances this will be the hose and valve arrangement.

With the low-side valve (1) open and the high-side valve (2) closed (Figure 4–22(b)), the refriger-

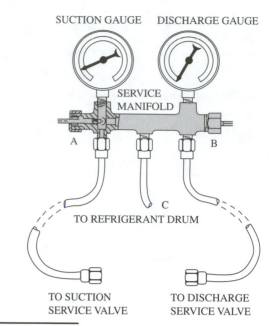

Figure 4–21 Testing manifold.

ant is allowed to pass through the low side of the manifold and the center port connection. This arrangement might be used when refrigerant or oil is added to the system.

Figure 4–22(c) illustrates the procedure for bypassing refrigerant from the high side to the low side. Both valves are open and the center port is capped. Refrigerant will always flow from the high-pressure area to a lower-pressure area. Do not operate like this with the compressor running.

Figure 4–22 Manifold valve operation.

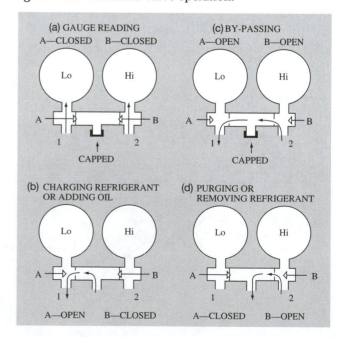

Figure 4–22(d) shows the valving arrangement for purging or removing refrigerant. The low-side valve is closed. The center port is connected to an empty recovery refrigerant drum. The high-side valve is opened, permitting a flow of high pressure refrigerant out of the center port.

The method of connecting the gauge manifold to a refrigerant system depends on the state of the system—that is, whether the system is operating or just being installed. For example, let's assume that the system is operating and equipped with back seating in-line service valves (Figure 4–23).

The first step is to purge the gauge manifold of contaminants before connecting it to the system. Purging and connecting is done with the following procedure:

1. Check to be sure that both service valves are back-seated, then remove the valve stem caps from the equipment service valves.
2. Remove hoses from hose rack.
3. Connect the center hose from the gauge manifold to a refrigerant cylinder, using the same type of refrigerant that is in the system, and open both valves on the gauge manifold.
4. Attach hoses loosely on access service valves. Open the valve on the refrigerant cylinder for about two seconds, and then close it. This will purge any air from the gauge manifold and hoses.

5. Quickly tighten the gauge manifold hoses to the gauge ports—the low-pressure compound gauge to the suction service valve and the high-pressure gauge to the liquid-line service valve, as illustrated in Figure 4–23.
6. Front-seat or close both valves on the gauge manifold. Crack (turn clockwise) both equipment service valves one turn off the back seat. The system is now allowed to register on each gauge. With the gauge manifold and hoses purged and connected to the system, we are free to perform whatever service function is necessary within the refrigeration cycle.

To remove the gauge manifold from the system, follow this procedure:

1. Back-seat (counterclockwise) both the liquid and suction service valves on the compressor.
2. Remove hoses from gauge ports and seal ends of hoses with ¼-in. flare plugs to prevent hoses from being contaminated. (Some manifold assemblies have built-in hose racks that hold the hoses when not in use).
3. Leak-check access valves.
4. Replace all gauge-port and valve-stem caps. Make sure that all caps contain the O-rings provided with them and are tight.

The manifold and gauges are necessary tools to perform many system operations. Once the system

---

Figure 4–23 Purging gauge manifold.

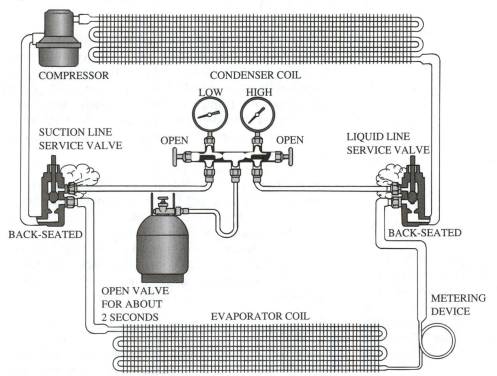

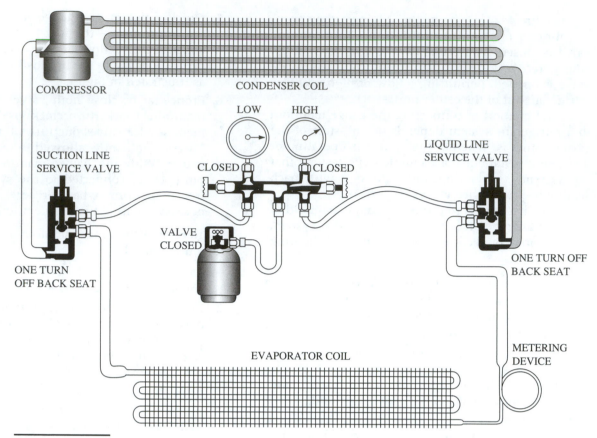

Figure 4–24 Connecting manifold.

has been completed and cleaned of most of the air by purging, it must be tested for leaks. Also, whenever a component has been repaired or replaced, it is imperative that the entire system be checked for leaks. Leak testing was discussed in a previous chapter.

PURGING

Purging the entire system with pure CFC or HCFC refrigerants, following servicing the system, is not allowed under EPA regulations. If refrigerant is removed from the system, it must be recovered. If the system has been opened for any reason, and air and moisture permitted to enter, the system should be evacuated before recharging. The drier should be changed.

Whenever a defective component such as an expansion valve is to be removed, the system should be pumped down and that part of the system should be isolated by means of the service valves. Then when the new component is installed the lines should be evacuated from both sides. Pumping down means to store the refrigerant in the receiver or condenser.

ENTERING THE SEALED SYSTEM

Refrigeration problems can be categorized as electrical failures or mechanical failures or a combination of both. Approximately 80% of the problems are electrical in nature, and of the remaining 20%, only a small portion require extensive sealed-system service. One of the first things to make quite clear is to never enter the sealed system unless it is absolutely necessary.

If one or more of the electrical load devices is not functioning, the problem is electrical. If all components are functioning but the unit does not refrigerate, the problem is mechanical.

Mechanical problems include: poor-capacity compressor (does not pump properly, bad valves), low-side leak, low-side restriction, high-side leak, high-side restriction, dirty condenser, and dirty evaporator.

FIELD-INSTALLED SERVICE VALVES

One of the easiest devices available for sealed-system access is the saddle, piercing valve, or line tap.

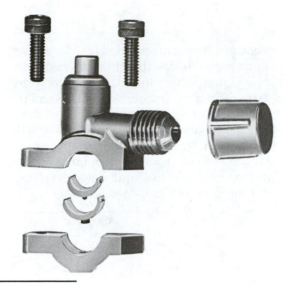

Figure 4–25 Piercing valve. *(Courtesy of Robinair, SPX Corporation)*

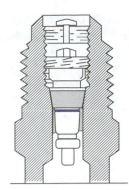

Figure 4–27 Schrader valve core.

using a core-removal tool to unscrew it. Some tools allow this to be done while the system is under pressure (see Figure 4–28).

USING THE PROCESS TUBE FOR SERVICE

Another accepted method of access to the refrigeration sealed system is the process tube adaptor. The process tube is located on the hermetically sealed compressor. The equipment manufacturer uses this tube to evacuate and charge the system. This generally gains access to the low side of the system. A process tube adapter kit is shown in Figure 4–29, and a process tube adaptor in Figure 4–30.

Cut the process tube at the brazed pinched end. Use the proper size adaptor (these are marked according to outside diameters of the tubing to fit directly on the process tube). Tighten the handles of the tool, which will tighten the gasket ring to the tubing, and provide a compression seal.

The charging hose is attached to the process tube adaptor, and the system can be repaired, evacuated, and charged. When the system has been completely serviced, a pinch-off tool (see Figure 4–31) can be used to seal the hermetic system from the

Other terms for this are used by some manufacturers. These piercing valves are clamped to the tubing, sealed by a bushing gasket, and pierce the tube with a tapered needle. Most contain some sort of shut-off control. The technician should keep in mind that these valves should be used to gain temporary access to a hermetically sealed system for checking system-operating pressures or for pressurizing for leak testing. The piercing valve (Figure 4–25) allows quick access to system pressures to immediately start diagnosing the refrigeration problem. Piercing valves are used only on copper and aluminum tubing, never on steel tubing.

Access piercing valves should be removed once the source of the sealed system malfunction has been located. Saddle valves, with Schrader valve cores that are brazed on the tubing, are often used (Figure 4–26).

The Schrader valve core is a spring-loaded device, for position seating (see Figure 4–27).

The valve is like those used on automobile tires. The stem must be depressed to force the valve's seat open against spring pressure. The manifold hose should have a valve depressor in the brass connector. If a valve core leaks, it can be replaced by

Figure 4–28 Valve core remover/installer. *(Courtesy of Robinair, SPX Corporation)*

Figure 4–26 Saddle valve.

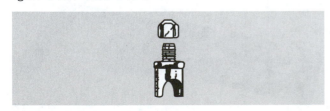

ON THE JOB

Remember to always leak test your manifold gauges when leak testing any refrigeration system. They are more prone to leak than the system due to frequent use. If leaks occur, they will probably be found around the o-rings at the end of the hoses. Carry extra o-rings in your service vehicle and repair as needed. Over tightening of these o-rings is often the cause of the problem. Snug but not tight is a better policy to prevent gauges leaking. If the valve or stem on the gauge is found to be leaking, then order a repair kit. It is often much cheaper to repair your own gauges than buy a new pair. Proper care and maintenance will extend the life of your gauge over many years. Repair kits are available for the manifold gauge set.

adaptor. The tool is left in place while the end of the process tube is brazed shut again.

Be sure to use the temporary access valves to determine operating pressure prior to using the process tube adaptor, as these are important to accurate diagnosis. Recover the refrigerant before cutting off the pinched end of the process tube to attach the adaptor; otherwise, system refrigerant would be lost. This would make it impossible to determine operating pressures at that line. When attaching gauge hoses, be sure that they are purged with refrigerant to remove air in the lines.

FACTORY-INSTALLED SERVICE VALVES

With the implementation of the Clean Air Act, all manufacturers, with the exception of Type I equipment manufacturers, are required to install factory-installed service valves. Many commercial refrigeration and air-conditioning systems have had factory-installed service valves for years.

Figure 4–29   Process tube adapter kit. *(Courtesy of Robinair, SPX Corporation)*

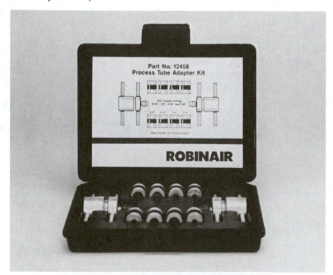

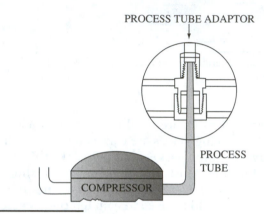

Figure 4–30   Process tube adapter installed.

Factory-installed service valves may be either a manually operated stem shutoff valve (see Figure 4–32), or a Schrader type valve (see Figure 4–33). Installed valves are usually located at the compressor as suction and discharge valves, and at the outlet of the receiver (the "King" valve). These service valves are equipped with a gauge service port. Operating refrigerant pressures may be observed on the service gauge manifold when hoses are connected to these ports.

The valve will be in the back-seated (the stem turned all the way out, counterclockwise) position

Figure 4–31   Pinch-off tool. *(Courtesy of Robinair, SPX Corporation)*

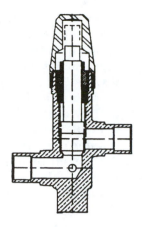

Figure 4–32 Manual shut-off valve. *(Courtesy of Henry Valve Company)*

when you attach your gauges. This closes the gauge port, and the valve is open to the line connection. The valve is front-seated (the stem is turned all the way in, clockwise) to isolate the compressor from the system. The gauge port is open to the compressor but the line connection is closed. In order to read system pressures, a technician first checks to ensure that the service valve is back-seated and then turns the stem in one or two turns (the service position, cracked), in order to slightly open the connection to both line and gauge port. Service valves are returned to the back-seated position after service and the gauges are removed. All caps are reinstalled on the system. Note that the compressor is always open to either the line connection or the gauge port, or both, if the valve is in the cracked position.

> 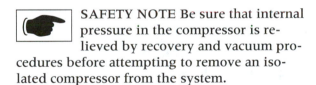 SAFETY NOTE Be sure that internal pressure in the compressor is relieved by recovery and vacuum procedures before attempting to remove an isolated compressor from the system.

Schrader valves provide a convenient method of checking system pressures or servicing the system, where it is not economical or convenient to use the compressor service valves with gauge ports. Some manufacturers provide Allen wrench stop

Figure 4–33 Schrader valve. *(Courtesy of Robinair, SPX Corporation)*

valves used in conjunction with the Schrader valve. The Schrader valve is similar to the air valves used on bicycle and automobile tires; however, they are not the same. The rubber used in tire valves is not compatible with refrigerants and would dissolve. The Schrader valve used in refrigeration and air-conditioning systems must also have a cap for the fitting to ensure a leakproof operation.

This type of service valve enables a technician to quickly check system operation without disrupting the unit's operation. Technicians should use "quick-connect" low-loss hose adapters to attach to the Schrader valves, greatly reducing the refrigerant loss, when connecting or disconnecting the service hoses.

▼ REFRIGERANT CHARGING

Whether the system is a new one or an existing one that has been repaired, the final step in putting the system in operation is to charge it with refrigerant. In any event, the process is the same.

HOW MUCH REFRIGERANT?

Some systems are more critical than others about the amount of the refrigerant charge. Systems that have a receiver—usually the larger systems—are not as critical since extra refrigerant can be stored in the receiver, and the expansion valve feeds the refrigerant into the evaporator as required to match the load.

In the smaller systems that do not have a receiver, any excess refrigerant will be stored in some part of the system where it reduces the effectiveness of that part and reduces the capacity of the system. If the system is short of refrigerant, the metering device is not supplied with a solid stream of refrigerant on full load and the evaporator will be starved for refrigerant. This will also reduce the capacity of the system. So, in the smaller systems it is critical that the proper charge be determined. Units that have capilliary tubes or fixed-orifice devices are critical charged; units that have thermostatic expansion valves are not.

One way to determine the proper charge is to read it on the nameplate as shown in Figure 4–34. This is the charge specified by the manufacturer.

On some systems, the length of the refrigerant lines is determined by the conditions of the installation. For example, on a split system the amount of charge required is affected by the length of the refrigerant lines. Most manufacturers give specific information for determining the charge on this basis. For example, the manufacturer may state, "charge adequate for matched system including up to 25 ft.

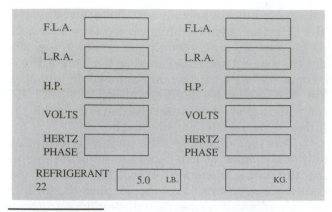

F.L.A.		F.L.A.	
L.R.A.		L.R.A.	
H.P.		H.P.	
VOLTS		VOLTS	
HERTZ PHASE		HERTZ PHASE	
REFRIGERANT 22	5.0 LB.		KG.

Figure 4–34 Unit nameplate.

of refrigerant line." Be sure to refer to the manufacturers' information. This is discussed in more detail in the charging chapters.

A LIQUID OR VAPOR CHARGE

The unit can be charged with either liquid or vapor refrigerant. Whichever method is used, it is important to charge the system with the right amount of refrigerant and to protect the compressor from any damage that might be caused by liquid slugging the compressor.

Since the refrigerant cylinder is not filled over 80%, in an upright position the vapor is at the top and the liquid at the bottom. To charge with vapor,

the refrigerant cylinder must be in the upright position. To charge with liquid, the cylinder is turned upside down, as shown in Figure 4–35.

It is usually considered good practice to charge with vapor (Figure 4–36) rather than liquid to prevent any danger of slugging the compressor with liquid. However, when using refrigerant blends, sometimes the tank will be constructed so that when the cylinder is right-side-up, it will discharge liquid. Refrigerant blends should not be charged as a vapor because the refrigerant will fractionate. Fractionated refrigerant will not cool satisfactorily.

VAPOR CHARGING

Prior to charging, the system must be leak-tested and evacuated. When the charging is started, the system is under vacuum so that when the refrigerant enters the system it will be drawn into the unit due to the difference in pressures.

To charge a system in vacuum with vapor refrigerant, the system is connected to the gauge manifold in the usual way, as shown in Figure 4–37. The hoses are purged up to the manifold. Both valves on the gauge manifold are backseated. When the refrigerant is released from the cylinder, the vapor flows into both sides of the system.

The refrigerant will stop flowing when the pressures equalize and no more refrigerant will enter the system.

Figure 4–35 Liquid charging.

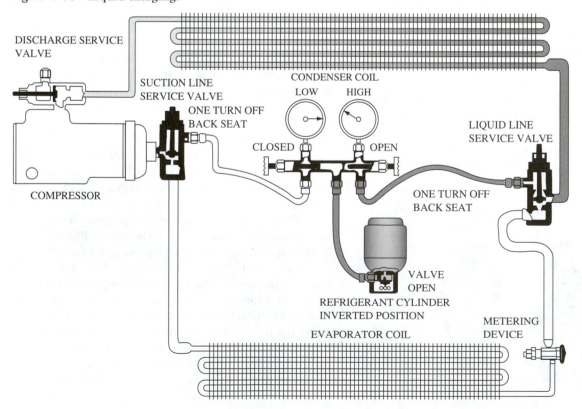

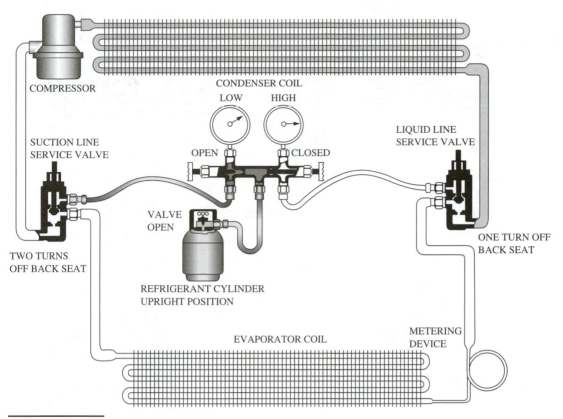

Figure 4-36 Vapor charging.

Figure 4-37 Both valves back-seated.

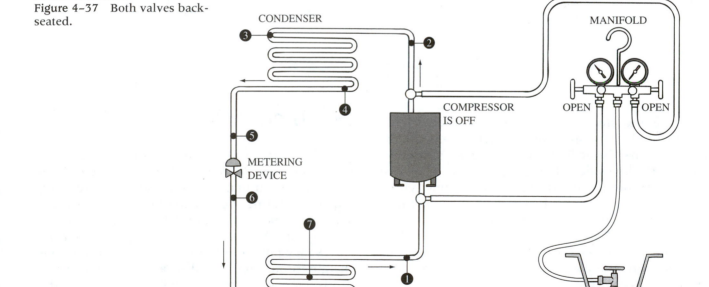

LEGEND
1. SUCTION LINE
2. DISCHARGE LINE
3. CONDENSER COIL
4. CONDENSER OUTLET
5. LIQUID LINE
6. DISTRIBUTOR
7. EVAPORATOR COIL

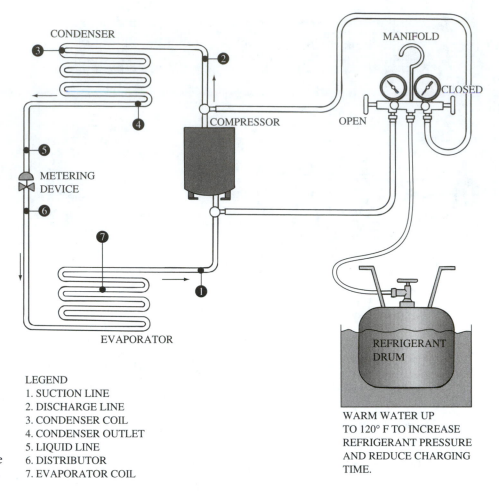

CONDENSER

MANIFOLD

CLOSED

COMPRESSOR

OPEN

METERING
DEVICE

EVAPORATOR

REFRIGERANT
DRUM

LEGEND
1. SUCTION LINE
2. DISCHARGE LINE
3. CONDENSER COIL
4. CONDENSER OUTLET
5. LIQUID LINE
6. DISTRIBUTOR
7. EVAPORATOR COIL

WARM WATER UP
TO 120° F TO INCREASE
REFRIGERANT PRESSURE
AND REDUCE CHARGING
TIME.

Figure 4–38   High-side valve
closed. Low-side valve open.

The next step is to close (front-seat) the high-side manifold valve as shown in Figure 4–38 and to continue the balancing of the vapor charging through the low side of the system.

At this point check the scales or charging cylinder to see how much of the charge has entered the system and how much more is needed. Normally less than 50% of the required charge is completed.

The balance of the vapor refrigerant will be charged with the compressor running. Even with the compressor running the process slows down. One way to speed up the process is to add heat to the cylinder. This is done by placing the cylinder in a water bath at a temperature not to exceed 120°F.

The refrigerant measuring device should be watched closely. When the full charge has been added, remove the low-side gauge. Turn off the compressor and disconnect the high-side gauge. This procedure will minimize refrigerant loss when removing hoses. Service valves should be closed or backseated before hoses are removed.

LIQUID CHARGING

Liquid charging is always much faster than charging with vapor. Liquid can be charged into the liquid line of a system that is in a vacuum (Figure 4–39). Liquid should not be charged into the discharge or suction side because liquid refrigerant can dilute the compressor oil or cause liquid slugging. Discharge valves may not totally seal. Under these conditions, charging liquid into the discharge side of a system that is in a vacuum may allow liquid refrigerant into the compressor cylinder. When the compressor starts it will try to compress liquid refrigerant and possibly do mechanical damage to the compressor.

Systems that employ a liquid-line service valve or King valve can accept a liquid charge with the system running. The service valve needs to be front-seated after the system is operating. This creates a low pressure in the liquid line and allows a technician to make a liquid charge. The gauge set should be hooked up and purged prior to charging. *Never*

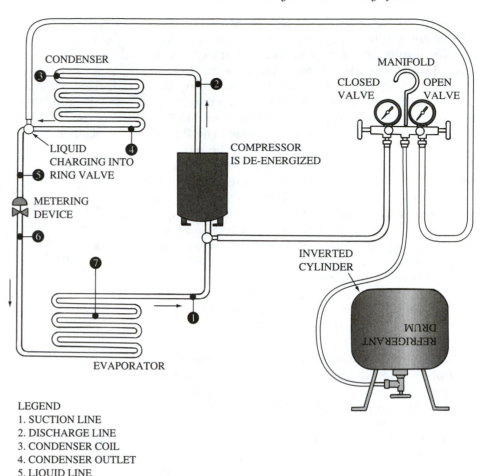

LEGEND
1. SUCTION LINE
2. DISCHARGE LINE
3. CONDENSER COIL
4. CONDENSER OUTLET
5. LIQUID LINE
6. DISTRIBUTOR
7. EVAPORATOR COIL

Figure 4-39   Liquid charging into the liquid line. The high-pressure manifold valve is open. The low-pressure valve is closed.

charge into the liquid line if the pressure is higher in the system than it is in the refrigerant cylinder. In this example, the refrigerant would leave the system and collect in the refrigerant cylinder. The refrigerant cylinder is at a lower pressure compared to the operating high side. In most instances, unless the system is in a vacuum, the operating high-side pressure will be higher than the refrigerant cylinder pressure.

Prior to closing the liquid line valve, a technician should jumper or bypass the low-pressure switch. The low-side pressure may get low enough to cause this device to open up the control circuit and stop the operation of the compressor.

Liquid can also be changed into the liquid line of a system that is in a vacuum. Charge liquid into the liquid line until the refrigerant pressure in the system equalizes with the bottle pressure. Charging this way may cause an overcharge that will need to be recovered. It is a quick way of introducing a charge into the system.

Liquid charging is faster than vapor charging. A technician must be careful not to overcharge using the liquid-charging method. An overcharge is less likely when vapor charging. The following precautions must be taken when liquid charging:

- Charge only into the liquid line.
- The liquid-line pressure must be less than the refrigerant-cylinder pressure.
- Liquid can be charged into the liquid line of a system that is in a standing vacuum.
- Liquid can be charged into an operating system if the proper service valves are installed in the liquid line. Front-seat the liquid-line service valve (King valve) to reduce line pressure; then begin the liquid-line charge when the line pressure drops below the refrigerant cylinder pressure.
- Liquid charging can create an overcharged condition that will need recovery.

▶ CHARGE FASTER AND SAFER

There are devices to safely charge liquid refrigerant into the suction line. These devices, such as the ChargeFaster™, create a restriction that allows the liquid to flash off to a vapor, (see Figure 4–40). The results are a quick charge without the risk of damage that liquid refrigerant can create. When charging with liquid, it is easy to overcharge the system. A technician will need to spend time recovering the overcharged refrigerant. On a slight overcharge it is best to remove some liquid refrigerant from the liquid line while the system is operating. The liquid refrigerant can be pushed directly into the recovery cylinder. This is a passive recovery procedure. No recovery unit is needed for this procedure. Just as it is easy to overcharge with liquid, it is easy to remove too much liquid refrigerant, resulting in an undercharge. Remember to purge all hoses of air prior to recovering or charging with refrigerant. Hose-purging should become an automatic part of your behavior every time the gauges are hooked to a system.

Finally, some technicians prefer to charge with liquid through the manifold gauge set. They hook up the hoses to the air-conditioning system, invert the refrigerant bottle, and begin the liquid charge into the suction side of the system. Some technicians control the liquid flow by throttling the low-side service valve to cause the refrigerant to flash before it enters the compressor. It is difficult to determine if the liquid is flashed to a vapor. An in-line sight glass installed on the suction-line hose may be an indication of the presence of liquid refrigerant. Liquid refrigerant entering the compressor will dilute the oil's lubrication quality and possibly cause slugging of the compressor.

A technician can charge liquid into a suction-line accumulator. The liquid can be charged between the outlet of the evaporator and the inlet to the accumulator. Again, a technician must be careful not to overcharge the system or all the time saved in a liquid charge will be lost in recovering the refrigerant. When a system is charged with liquid, it takes 15 to 20 min to obtain a stable pressure and superheat reading. Liquid charging is best accomplished when the system is in a vacuum and the correct

weight of the charge is known to the technician. A digital scale will be required to accurately weigh in the charge.

Charging liquid into the low-pressure side can damage the compressor by diluting the oil or causing liquid slugging. There are special techniques that can be used to liquid-charge into the low side. A technician can

- Control liquid-refrigerant flow by using a special metering device that creates a pressure drop and flashes the refrigerant to a vapor.
- Control liquid refrigerant flow by throttling the manifold gauge valve. A sight glass should be installed in the low-side hose and used to monitor the refrigerant condition.
- Charge liquid into the inlet of a suction accumulator.

The next section concludes the discussion on basis charging information. Oil charging is not used as much as refrigerant charging. Oil charging can be more difficult, especially with sealed, welded hermetic compressors.

CHARGING THE SYSTEM WITH OIL

The proper amount of oil can be measured into the system in several ways:

1. In a new system, it can be measured or weighed in. Unit installation instructions include the compressor oil requirements in either weight or liquid measurements. This method is also applicable following a compressor overhaul, when all of the oil has been removed from the compressor; however, it should be used only when the system has no oil in it.

2. The dipstick method is used primarily with small, vertical-shafted, hermetic compressors. Some larger, open types of compressors may also have openings designed for the use of a dip stick. The manufacturer's recommendations of the correct level should always be followed.

3. The compressor crankcase sight glass (Figure 4–41) is used after the system has operated for a period of time under normal conditions to determine the proper oil level. This procedure will assure proper oil return to the crankcase. It will also allow the oil lines and reservoirs to fill and give the refrigerant an opportunity to absorb its normal operating oil content, if applicable. The correct oil level will vary among compressor models.

Figure 4–40 ChargeFaster™ liquid charging device. *(Courtesy A-1 Components, Inc.)*

SIGHT GLASS
(OIL LEVEL)

Figure 4–41　Charging to the oil sight glass. Some require ⅛, ½, or ⅞ of a sight glass. Check with the manufacturer.

When a compressor is replaced, the new unit should be charged with the same amount of oil as the old unit.

Oil is normally introduced into a refrigerant system by one of three methods (Figure 4–42). It may be poured in as shown on the left, providing the compressor crankcase is at atmospheric pressure. This method is normally used prior to dehydration since it will expose the compressor crankcase interior to air and the moisture the air contains.

On the right in Figure 4–42 the method normally used with an operating unit is shown. In this case the crankcase is pumped down below atmospheric pressure and the oil is drawn in. When this method is used, the tube in the container subjected to air pressure should never be allowed to get close enough to the surface of the oil to draw air. As shown, the tube is well below the level of the oil in the container.

The third method is to use a hand oil pump designed to add or remove oil from the crankcase. The pump can be reversed to add or remove an oil charge.

Oil charging of replacement welded hermetic compressors should be done according to the manufacturer's recommendations, and depends on whether the replacement compressor has been shipped with or without an oil charge.

Figure 4–42　Two methods of charging oil into the compressor.

There are four precautions to take in charging or removing oil:

1. Use clean, dry oil. Hermetically sealed oil containers are available and should be used.
2. Pressure must be controlled when the crankcase is opened to the atmosphere. Too much pressure can force oil out through the opening rapidly and create quite a mess.
3. System overcharging should be avoided. Not only will this create the possibility of oil slugs damaging the compressor, but it also can hinder the performance of the refrigerant in the evaporator. Oil overcharging will also cause liquid refrigerant to return to the compressor from the evaporator.
4. On a semi-hermetic compressor, replace the oil plug with a service valve. This valve will allow easy charging and removal options.

SUMMARY

- You have learned basic information that is needed before you can progress into the chapters on the various charging methods. Understand how to use the manifold gauge set. Practice using the gauges in various applications. The refrigeration gauges will become the most-used tool to create success as a professional. Several sets of gauges are recommended. One heavy-duty set can be designated for vacuum checks. Other sets are necessary to allow a technician to service multiple units at the same time. Purchase a repair kit in order to keep your equipment in the best working shape. These are the tools of the trade; they should be treated with the respect they deserve.

CHAPTER QUESTIONS

1. The inner rings of the compound and pressure gauge tell the technician the _____ temperature and the _____ temperature of the system.
2. What is the pressure range of a standard compound gauge?
3. What is the pressure range of a pressure gauge?
4. What does front-seated mean?
5. What does back-seated mean?
6. What is a Schrader valve?
7. Describe three ways a technician can safely charge liquid refrigerant into a system.

8. What damage(s) can a direct charge of liquid refrigerant do to a compressor?

9. Describe three ways to charge oil into a compressor.

10. What equipment and materials are necessary for charging an air-conditioning system?

11. What three ways can a technician use to determine proper airflow through the system?

12. What are the necessary preparations prior to checking the charge on a system?

13. Why is it important to have more than one manifold gauge set?

14. Using a manifold gauge, what is the corresponding temperature of R-22 at 200 psig?

15. Using a manifold gauge, what is the corresponding temperature of R-22 at 10 psig?

16. Using a manifold gauge, what is the corresponding temperature of R-22 at 70 psig?

17. What type of manifold gauge set would you purchase? Why?

18. Using the airflow chart provided in this chapter, and provided the charge is correct, what is the supply air temperature if the return-air dry bulb is 78°F and the wet bulb is 70°F?

19. What does a technician need to check prior to beginning the charging procedures?

20. What is the airflow requirement per ton of air conditioning? How many CFM are required on a 3½-ton system?

CHAPTER FIVE

CHARGING PROCEDURES: PRESSURE-TEMPERATURE METHOD

AFTER STUDYING THIS CHAPTER, THE STUDENT WILL BE ABLE TO:

• Describe the benefits of proper air-conditioning charging procedures.

• Discuss the effects of an improperly charged system.

• Describe safety practices used on the job.

• List the information needed to charge a system.

• List other charging methods.

• Describe the three pressure-temperature methods of charging.

• Discuss the cleaning activities that must be done prior to charging.

• Be able to locate the condensing temperature and pressure on a gauge.

• Be able to locate the evaporating temperature and pressure on a gauge.

• Use a P-T chart to find the condensing and evaporating temperatures.

• Use the P-T method to determine if the system is over- or undercharged or has some mechanical problem.

WHY LEARN PROPER CHARGING TECHNIQUES?

Why bother to learn to charge an air conditioner properly? Many (some say most) conditioning system are not charged properly, yet they cool adequately. If the goal is to cool air enough to keep a place cool, then all that is necessary is a cylinder of the correct refrigerant and a hose connection from the refrigerant to the air-conditioning equipment. In essence, this is what happens when a technician relies only on a manifold gauge set to charge an air-conditioning system. If this is the only test instrument a technician uses, then they might as well save their money and purchase just a single hose. That single refrigerant hose can be hooked to the system and it can be charged until the air *feels* cold. This may seem like a joke, but using only the manifold gauge set to charge a system is equivalent to using a hose and charging until the air feels cold. A truly *professional* air-conditioning technician uses various methods to check the charge and verify the correct operating conditions of an air-conditioning system. A correctly charged air-conditioning system will

1. Reduce the systems operating cost.
2. Extend the life of the compressor.
3. Provide the designed comfort level.

SYSTEM OPERATING COST

An air-conditioning system's operating costs are influenced by the

- Efficient design of the equipment
- Correct installation procedures
- Correct refrigerant charge

The higher the SEER (seasonal energy efficiency rating), the lower are the operating costs. A quality installation is the most important aspect of the equipment's longevity and operating cost. A high SEER rating does not guarantee lower operating cost if the installer uses improper sizing and poor installation techniques. These include checking the charge on the system.

COMPRESSOR LIFE

Compressor life is the function of many factors. One major contributor to long compressor life is a correct charge. A system that is undercharged may cool, but the compressor motor windings will run hot as compared to a system that is charged properly. A motor winding will eventually burn up. The cool returning suction gas is used to cool the compressor motor windings. Compressor manufacturers design a compressor to last beyond the warranty period. Compressors are designed to handle tough conditions. An undercharged system that is cooling will not normally burn out during the warranty period. An undercharged compressor will not last its 15-y life expectancy. The average compressor lasts 9 to 10 y because of poor installation techniques and improper charging procedures. An undercharged system will need to operate longer to achieve proper cooling, thus reducing its mechanical life.

An overcharged system will not last as long either. A severely overcharged system will cause compressor damage in the form of liquid flooding. Liquid flooding or slugging is found in the form of liquid refrigerant, oil, or a combination of liquid refrigerant and oil. The compressor that tries to compress a liquid may damage its valves, crankshaft, piston, or connecting rods. At best a flooded compressor may stall and cause the compressor to cycle on its overload protection. In most cases overcharging does not cause immediate compressor damage. Instead, overcharging dilutes the oil's lubricating properties. Low superheat is a danger to the compressor. A superheat of 4°F may be sending liquid refrigerant to the compressor. In an overcharged condition, with superheat between 1° and 5°, liquid travels in the high-velocity stream in the middle of the suction line. The liquid refrigerant mixes with the oil and becomes part of the compressor's lubrication. In an overcharged condition some of the liquid refrigerant boils to a vapor in the warm oil and some of the refrigerant remains in the liquid state and mixes with the oil. Refrigerant is a good cleaning agent and is not a lubricant. This mixture of refrigerant and oil causes mechanical parts to wear out prematurely. Damage occurs over a long period of time.

It is slow and progressive, but the compressor will not normally cease until the warranty period expires. Compressor warranty periods are generous, but poor installation, over- and undercharging procedures, and lack of preventative maintenance need to be addressed to extend the life of a compressor. Again, as with an uncharged system, an overcharged system will reduce system capacity. Reduced system capacity translates into longer run cycles, thus reducing the compressor's mechanical life.

DESIGNED COMFORT

A system that is undercharged may cool the air, but a lack of refrigerant in the evaporator will reduce the moisture-removal capabilities of the system. An undercharge will cause the system to operate longer because there is less liquid refrigerant to absorb heat in the evaporator. An undercharged system will result in longer compressor run times and higher operating costs. Longer run times translate into shorter equipment life for the compressor, fans, and all electrical parts in general.

An overcharged system will also result in higher operating cost and shorter compressor life. In some systems, a slight overcharge will increase the capacity of a system. Increased capacity on an already oversized system will create short cycling. Short cycling will result in shorter cooling operation cycles, thus reducing the system's capability to remove moisture from the air. Short cycling creates a cool building with a high humidity level. The system cools the air quickly without the opportunity to dehumidify the air. The occupants feel uncomfortable with the high moisture level. They turn the thermostat down, which causes the system to operate even longer, further driving up the operating cost.

In most instances, when the system is overcharged, it causes a reduction in the system's capacity. Reduced capacity translates into longer operating cycles, higher utility cost, and shorter equipment life.

Before continuing, review the following safety information. Your life and your career depend on your safety habits.

SAFETY FIRST

 — Wear safety glasses and gloves.
— Beware of electrical shock.
— Wear dry clothes and shoes.
— Do not stand on wet ground.

— If you think the electricity is disconnected, measure the voltage and double-check by grounding the terminal with a wire or screwdriver before touching.
— Lift with your back straight using your legs as leverage.
— Review the chapter on safety.
— Remove watches and jewelry, even plastic watches.
— Close the cover on the condensing unit prior to starting compressor.

PRESSURE-TEMPERATURE METHOD OF CHECKING A CHARGE

This chapter explains how to use the pressure-temperature (P-T) method of checking the charge of an air-conditioning system. The other charging methods are discussed in subsequent chapters.

ASSUMPTIONS

The following assumptions are necessary to proceed with the pressure-temperature charging method. Operating a system outside of these conditions will result in an inaccurate charge result.

• A building near normal comfort zone temperatures (72°F–82°F).
• Clean filters, condenser, and evaporator coils.
• Airflow at least 400 CFM per ton of cooling
• Properly sized refrigerant lines
• Correct capacity match between evaporator and condenser
• No line kinks or restrictions
• Familiarity with the correct installation of refrigeration gauge set

Remember that prior to using this charging method it is important to know the efficiency rating of the system. The air filter should be clean. Check the evaporator to make sure it is clean. Clean the condenser unless you know that it has been cleaned in the last month. These preparatory checks are mandatory for any charging methods used in this textbook.

The equipment needed for this charging method are a manifold gauge set and two digital temperature testers. Knowing the SEER rating of the system is important in determining the correct factors used to calculate the correct refrigerant charge.

A dirty filter, evaporator coil, or condenser coil will not allow a correct charge of the system using any of the methods discussed in this textbook. In most instances, it will take longer to check and clean the filter and coils than it will to check or adjust the charge of system. When you check the charge, consider it a preventative maintenance checkup and you will be less likely to cut corners to get the job done. Doing these proper procedures will take in excess of 1 h and in some cases 2 h. You can make a dirty system cool; however, you and your company's reputation is not worth rushing through the charging sequence. Be a professional! Do it right!

The pressure-temperature method is the most commonly used method of charging an A/C system. Its accuracy is fair at best. This method will obtain cooling for your building but not at optimal efficiency. In most cases, using only the P-T method will translate into a cool building with higher-than-normal operating costs and shorter compressor life.

Other methods of checking the charge should be used in conjunction with the pressure-temperature (P-T) method of charging. Consider the P-T method as a rough method of checking and adjusting the charge. Other charging methods should be used to verify the correct charge. These will also determine whether the unit has any refrigerant-side problems.

If you elect to use only the P-T method, the unit's efficiency could be off as much as 25% and the BTU heat removal capabilities could be reduced as much as 15%. By using additional charging methods you will be able to fine-tune the charge and verify that the unit is properly charged as well as operating at peak performance.

The P-T charging procedures are applied to three categories:

1. Systems with SEER ratings less than 10.5
2. Systems with SEER ratings between 10.5 and 11.9
3. Systems with SEER ratings greater that 12.0

Knowing the correct SEER rating is important for the use of the P-T method. The system must also be clean.

SEER ratings are sometimes a challenge to locate. The SEER rating is not normally found on the condenser nameplate. The SEER rating is found on the yellow energy tag on new condensing units (Figure 5–1). On split systems with high SEER ratings, the evaporator must be changed to match the condenser. SEER ratings can also be obtained from installation instructions, the equipment manufacturer, or the air-conditioning supply house.

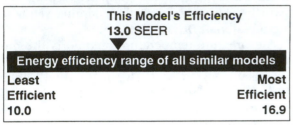

Compare the Energy Efficiency of this Air
Conditioner with Others Before You Buy.

This Model's Efficiency		
13.0 SEER		
▼		
Energy efficiency range of all similar models		
Least		**Most**
Efficient		**Efficient**
10.0		**16.9**

The **SEER**, Seasonal Energy Efficiency Ratio, is the measure of energy
efficiency for central air conditioners.

**Central air conditioners with higher SEERs are
more energy efficient.**

■ This energy rating is based on U.S. Government standard tests of this
condenser model combined with the most common coil. The rating may
vary slightly with different coils.

■ Federal law requires the seller or installer of this appliance to make
available a fact sheet or directory giving further information about the
efficiency and operating cost of this equipment. Ask for this information.

Important: Removal of this label before consumer purchase is a violation of Federal law (42 U.S.C. 6302).

Figure 5–1 SEER rating label required on all new
residential and light commercial condensing units.

PRESSURE-TEMPERATURE METHOD FOR UNITS WITH LESS THAN A 10.5 SEER

DETERMINING THE CONDENSING TEMPERATURE OR PRESSURE

1. Hook up the refrigerant gauge set. Purge hoses. Start unit. Allow unit to run continually for 15 minutes. Setting thermostat temperature to 65°F will assure that the unit does not cut off during your checkout.

2. Measure the temperature of the air entering the outdoor condenser coil. Measure the airflow close to the condenser coil. Do not touch the condenser coil with the temperature probe because it will cause a higher-than-normal temperature reading (Figure 5–2).

3. Measure the temperature of the air entering the return-air grille. Do not guess at any of these temperatures. Use your electronic temperature tester.

The thermometer on the thermostat may be inaccurate (Figure 5–2).

4. Take the temperature of the air entering the condenser and add 30°F to this temperature. This value, 30°F, is a constant added to the outdoor temperature for systems with a SEER rating less than 10.5. Look at the high-pressure gauge and find the temperature that corresponds to the refrigerant type on the gauge. This is called the *condensing temperature*. The gauge indicator (needle) should be within 10% of this reading (Figure 5–3). If the gauge indicator is below the condensing temperature desired (calculated temperature of the outdoor air temperature plus 30°F), then refrigerant may be added to the system provided the low-side pressure is not exceeded. This is stated as the following formula:

$$\begin{array}{ll} \text{Target condensing} & \text{outdoor-air} \\ \text{temperature} & = \text{temperature} \\ & + 30°F \end{array}$$

Being familiar with the condensing and evaporating temperatures will be useful when discussing other charging methods. Many technicians use pressures as an indicator of the correct charge. When learning to check a charge, it is best to use the condensing and evaporating temperature because they are necessary when checking superheat and subcooling.

If the refrigerant gauge indicator (needle) is 10% above the target condensing temperature, then the unit may be overcharged or have a dirty condenser, bad condenser fan motor, or some type of high-side pressure restriction. If the unit is overcharged, recover some refrigerant. Venting refrigerant to the atmosphere is now illegal. Use an approved refrigerant-recovery method to capture the excess charge.

The refrigerant gauge temperature corresponds to a gauge pressure. The outer ring of the gauge indicates the gauge pressure (Figure 5–4). Some technicians feel more comfortable reading the pressure measurement as opposed to reading the inner condensing temperature ring. Having determined the correct condensing temperature, follow the gauge indicator outward until you reach the outer pressure ring. This will be the pressure you desire for the high-pressure side of the air-conditioning unit. This is also called the condensing pressure, the pressure at which the refrigerant is condensing in the middle of the condenser.

WORKING EXAMPLE

Air entering the outdoor condenser is measured at 80°F. What is the target condensing temperature or condensing pressure?

First, clean the filter, condenser, evaporator, and blower wheel. The system is using the typical

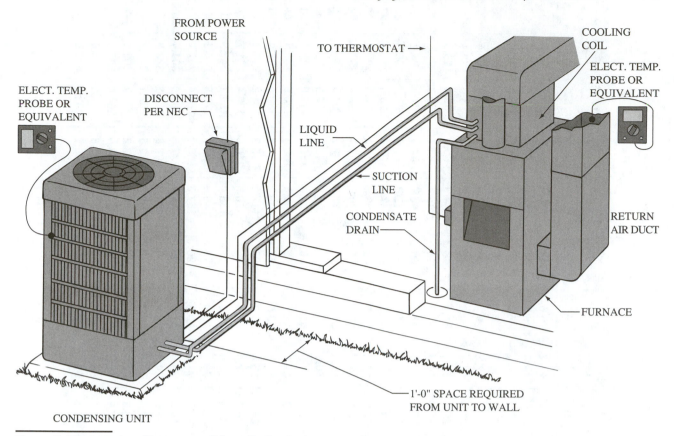

FROM POWER
SOURCE

TO THERMOSTAT →

COOLING
COIL

ELECT. TEMP.
PROBE OR
EQUIVALENT

ELECT. TEMP.
PROBE OR
EQUIVALENT

DISCONNECT
PER NEC

LIQUID
LINE

← SUCTION
LINE

CONDENSATE
DRAIN

RETURN
AIR DUCT

FURNACE

1'-0" SPACE REQUIRED
FROM UNIT TO WALL

CONDENSING UNIT

Figure 5–2   Generic split system with probe hooked up to measure air temperatures.

R-22 refrigerant and has a SEER rating of 10. The air entering the condenser is 80°F.

SOLUTION:

Target condensing temp. = outdoor-air temp.
+ 30°F constant
110°F = 80°F + 30°F

110°F is the targeted condensing temperature.

Figure 5–3   High-side, high-pressure gauge with outer pressure ring and inner temperature ring for R-22, R-12, and R-502. *(Courtesy of Amprobe Promax)*

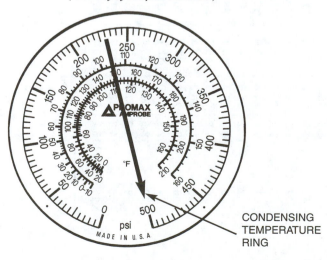

CONDENSING
TEMPERATURE
RING

Look at the high-side gauge (Figure 5–4) and locate the R-22 temperature ring. Next, locate 110°F on the temperature ring. The outer pressure ring indicates 226 psi. Figure 5–4 shows a 111°F condensing temperature, which corresponds to a 226-pSIG condensing pressure. The condensing temperature may be difficult to read on the high-side gauge.

Figure 5–4   High-side gauge. *(Courtesy of Amprobe Promax)*

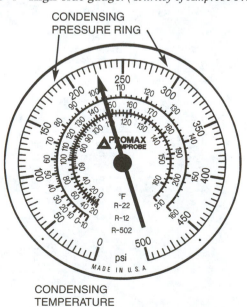

CONDENSING
PRESSURE RING

CONDENSING
TEMPERATURE

Table — Pressure-temperature chart

Temperature °F.	12-F	22-V	500-D	502-R	717-A
−60	19.0	12.0	17.0	7.2	18.6
−55	17.3	9.2	15.0	3.8	16.6
−50	15.4	6.2	12.8	0.2	14.3
−45	13.3	2.7	10.4	1.9	11.7
−40	11.0	0.5	7.6	4.1	8.7
−35	8.4	2.6	4.6	6.5	5.4
−30	5.5	4.9	1.2	9.2	1.6
−25	2.3	7.4	1.2	12.1	1.3
−20	0.6	10.1	3.2	15.3	3.6
−18	1.3	11.3	4.1	16.7	4.6
−16	2.0	12.5	5.0	18.1	5.6
−14	2.8	13.8	5.9	19.5	6.7
−12	3.6	15.1	6.8	21.0	7.9
−10	4.5	16.5	7.8	22.6	9.0
−8	5.4	17.9	8.8	24.2	10.3
−6	6.3	19.3	9.9	25.8	11.6
−4	7.2	20.8	11.0	27.5	12.9
−2	8.2	22.4	12.1	29.3	14.3
0	9.2	24.0	13.3	31.1	15.7
1	9.7	24.8	13.9	32.0	16.5
2	10.2	25.6	14.5	32.9	17.2
3	10.7	26.4	15.1	33.9	18.0
4	11.2	27.3	15.7	34.9	18.8
5	11.8	28.2	16.4	35.8	19.6
6	12.3	29.1	17.0	36.8	20.4
7	12.9	30.0	17.7	37.9	21.2
8	13.5	30.9	18.4	38.9	22.1
9	14.0	31.8	19.0	39.9	22.9
10	14.6	32.8	19.7	41.0	23.8
11	15.2	33.7	20.4	42.1	24.7

Temperature °F.	12-F	22-V	500-D	502-R	717-A
12	15.8	34.7	21.2	43.2	25.6
13	16.4	35.7	21.9	44.3	26.5
14	17.1	36.7	22.6	45.4	27.5
15	17.7	37.7	23.4	46.5	28.4
16	18.4	38.7	24.1	47.7	29.4
17	19.0	39.8	24.9	48.8	30.4
18	19.7	40.8	25.7	50.0	31.4
19	20.4	41.9	26.5	51.2	32.5
20	21.0	43.0	27.3	52.4	33.5
21	21.7	44.1	28.1	53.7	34.6
22	22.4	45.3	28.9	54.9	35.7
23	23.2	46.4	29.8	56.2	36.8
24	23.9	47.6	30.6	57.5	37.9
25	24.6	48.8	31.5	58.8	39.0
26	25.4	49.9	32.4	60.1	40.2
27	26.1	51.2	33.2	61.5	41.4
28	26.9	52.4	34.2	62.8	42.6
29	27.7	53.6	35.1	64.2	43.8
30	28.4	54.9	36.0	65.6	45.0
31	29.2	56.2	36.9	67.0	46.3
32	30.1	57.5	37.9	68.4	47.6
33	30.9	58.8	38.9	69.9	48.9
34	31.7	60.1	39.9	71.3	50.2
35	32.6	61.5	40.9	72.8	51.6
36	33.4	62.8	41.9	74.3	52.9
37	34.3	64.2	42.9	75.8	54.3
38	35.2	65.6	43.9	77.4	55.7
39	36.1	67.1	45.0	79.0	57.2
40	37.0	68.5	46.1	80.5	58.6
41	37.9	70.0	47.1	82.1	60.1

Temperature °F.	12-F	22-V	500-D	502-R	717-A
42	38.8	71.4	48.2	83.8	61.6
43	39.8	73.0	49.4	85.4	63.1
44	40.7	74.5	50.5	87.0	64.7
45	41.7	76.0	51.6	88.7	66.3
46	42.6	77.6	52.8	90.4	67.9
47	43.6	79.2	54.0	92.1	69.5
48	44.6	80.8	55.1	93.9	71.1
49	45.7	82.4	56.3	95.6	72.8
50	46.7	84.0	57.6	97.4	74.5
55	52.0	92.6	63.9	106.6	83.4
60	57.7	101.6	70.6	116.4	92.9
65	63.8	111.2	77.8	126.7	103.1
70	70.2	121.4	85.4	137.6	114.1
75	77.0	132.2	93.5	149.1	125.8
80	84.2	143.6	102.0	161.2	138.3
85	91.8	155.7	111.0	174.0	151.7
90	99.8	168.4	120.6	187.4	165.9
95	108.2	181.8	130.6	201.4	181.1
100	117.2	195.9	141.2	216.2	197.2
105	126.6	210.8	152.4	231.7	214.2
110	136.4	226.4	164.1	247.9	232.3
115	146.8	242.7	176.5	264.9	251.5
120	157.6	259.9	189.4	282.7	271.7
125	169.1	277.9	203.0	301.4	293.1
130	181.0	296.8	217.2	320.8	—
135	193.5	316.6	232.1	341.2	—
140	206.6	337.2	247.7	362.6	—
145	220.3	358.9	264.0	385.0	—
150	234.6	381.5	281.1	408.4	—
155	249.5	405.1	298.9	432.9	—

Figure 5-5   Pressure-temperature chart.

Some gauge manufacturers no longer include the condensing and evaporating temperatures on the inner rings of their gauges. Use a pressure-temperature chart, as shown in Figure 5–5, to determine the condensing pressure. In the working example, the target condensing temperature is 110°F, which corresponds to a condensing temperature of 226 psig.

When the unit has been operating for 15 min, the gauge indicator should be within ±10% of 110°F or 226 psi.

$$10\% = 0.10$$
$$0.10 \times 110°F = 11°F$$

The condensing temperature should be plus or minus 11° (+11° or −11°) from the target condensing temperature.

110°F target condensing temperature
−11°F (−10%)
  99°F minimum condensing temperature

110°F target condensing temperature
+11°F (+10%)
121°F maximum condensing temperature

The range of acceptable condensing temperatures is 99°F through 121°F. The reading on Figure 5–4 is 111°F; therefore, the condensing temperature is in the range of the target temperature and the charge seems good according to this method of charging. ■

After the unit has run 15 min, the condensing temperature/pressure should be within ±10% of the desired reading. If it is less than 10% less, add refrigerant. Recover refrigerant if it is more than 10% above the target. Remember that the unit's coils, filters, and blower wheel must be clean before adjusting the charge. A visual inspection of the evaporator and blower wheel is adequate. In most instances, the condenser coil will need to be flushed with coil cleaner and water unless it has been cleaned in the past month. Condenser coils may look clean, but a coil cleaning may reveal embedded dirt and debris.

After the condenser coil is cleaned, it should be allowed to dry for 20 to 30 min by operating the condensing unit. The movement of the air through the coil and the warm refrigerant will dry the coil. A wet condenser coil will operate at a lower condensing temperature and pressure, thereby giving the impression of an undercharged condition. Monitor the gauges during this operation. Figure 5–6 is a flowchart that shows the sequence for using this charging procedure.

DETERMINING THE CORRECT EVAPORATING TEMPERATURE

1. To determine the correct evaporating temperature and pressure, take the temperature of the air en-

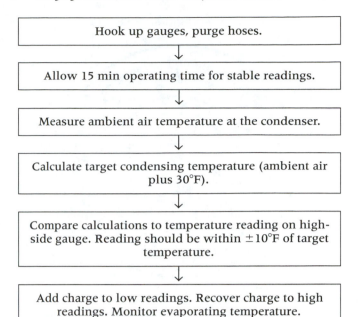

Figure 5–6 Flowchart for determining condensing temperature on less than 10.5 SEER

tering the return air filter. Use an accurate temperature tester. Subtract 40°F from this temperature reading. This formula is used to calculate the evaporating temperature:

Evaporating temp. = return-air temp.
  − 40°F constant

2. Look at the low-pressure (suction-side) gauge and locate the R-22 temperature ring. Find the temperature that corresponds to the return-air temperature minus 40°F. The evaporating temperature should be within ±10% of this reading but not below 32°F. An evaporator operating below 32°F will cause the evaporator coil to freeze up and block the airflow after a long period of operating. The frost accumulation on the evaporator will also reduce the heat-removal capabilities of the evaporator. The frost coating acts as an insulator between the air and refrigerant inside the evaporator.

WORKING EXAMPLE

The air entering the filter grille is 78°F. What is the target evaporating temperature?

Clean filter, evaporator, blower wheel, and condenser.

- R-22 refrigerant
- 10 SEER rating

SOLUTION:

Evaporating temp. = return-air temp.
  − 40°F constant
38°F = 78°F − 40°F

EVAPORATING PRESSURE

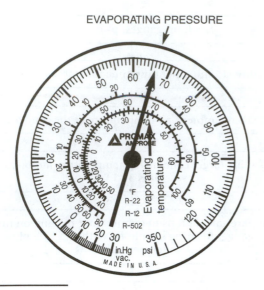

Figure 5–7 Low-side, compound gauge with outer pressure ring and inner evaporating temperature ring for R-22, R-12, and R-502. *(Courtesy of Amprobe Promax)*

A temperature of 38°F is the targeted, or desired, evaporating temperature. Look at the low-side pressure gauge and locate the R-22 temperature ring. Next, locate 38°F on the temperature ring. The outer pressure ring will indicate 66 psi (Figure 5–7).

When the unit has been operating for 15 min the gauge indicator should be within ±10% of 38°F.

$$10\% = 0.10$$
$$0.10 \times 38°F = 3.8°F, \text{ or } 4°F$$

$$
\begin{array}{l}
38°F \text{ target evaporating temperature} \\
\underline{-4°F \ (-10\%)} \\
34°F \text{ minimum evaporating} \\
\quad \text{temperature (never below 32°F)}
\end{array}
$$

$$
\begin{array}{l}
38°F \text{ target evaporating temperature} \\
\underline{+4°F \ (+10\%)} \\
42°F \text{ maximum evaporating} \\
\quad \text{temperature}
\end{array}
$$

The range of acceptable evaporating temperatures is 34°F to 42°F. If a technician feels more comfortable using pressure readings, these temperatures can be converted to pressures using the pressure-temperature chart in Figure 5–5. In the example shown in Figure 5–7, the evaporating temperature is in range, or within the targeted temperature required for a correct charge. Recheck the outdoor and indoor temperatures and recalculate the information if the readings have changed. Other charging methods should be used in conjunction with the P-T method.

Figure 5–8 is a flowchart that shows the sequence for using this charging procedure.

This example is for a system with a SEER rating of 10.4 or less. Checking the charge on systems with higher SEER ratings is discussed in the next section.

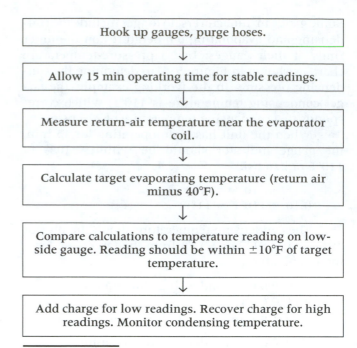

Figure 5–8 Flowchart for determining evaporating temperature on systems with a SEER rating of 10.4 or less.

CHECKLIST FOR DETERMINING REASONS FOR LOW EVAPORATING TEMPERATURES
Assumptions

- Condensing temperature is in range.
- Adding refrigerant will cause no rise in evaporating temperature.
- Adding refrigerant will cause the condensing temperature to become excessive.

  Check the design airflow:

- Is the evaporator coil clean?
- Is the evaporator coil iced?
- Is the air filter clean?
- Is the duct system restricted?

  Check refrigerant restriction:

- Is the metering device restricted?
- Is the liquid line kinked?
- Is there a measurable Δ-T across the drier?

Troubleshooting and diagnosing low-side problems are covered in detail in the chapters on mechanical troubleshooting.

PRESSURE-TEMPERATURE METHOD FOR UNITS BETWEEN 10.5 AND 11.9 SEER

1. Hook up the refrigerant gauge set. Purge hoses. Start the unit. Allow the unit to run continually for

15 min. Setting the thermostat to 65°F will assure that the unit does not cut off during your checkout.

2. Measure the temperature of the air entering the outdoor condenser coil. Measure the temperature of the air entering the return-air grille.

3. Take the temperature of the air entering the condenser and add 25°F to this reading. Look at the high-pressure gauge and find the temperature that corresponds to the refrigerant type on the gauge. This is the condensing temperature. The gauge indicator (needle) should be within ±10% of this reading. If the gauge indicator is below the calculated temperature of the outdoor air temperature plus 25°F, then refrigerant may be added to the system, provided the low-side evaporating temperature or pressure is not exceeded. If the refrigerant gauge indicator is above 10% of the calculated outdoor air temperature plus 25°F, then the unit may be overcharged, have a dirty condenser, have a bad condenser fan motor, or have some type of high-side pressure restriction or condenser airflow problem.

If the unit is overcharged, recover some refrigerant. Venting refrigerant is illegal. Use an approved recovery method to remove the excess charge. Use other charging methods to verify that the system is overcharged. High condensing temperatures or pressures may not be an overcharged condition but some other refrigerant-side problem.

The gauge temperature corresponds to a gauge pressure. The outer ring of the gauge indicates the gauge pressure. You may feel more comfortable reading the pressure measurement than the inner temperature ring. Determine the correct condensing temperature; then follow the gauge indicator outward until you reach the outer ring. This will be the desired pressure for the high-pressure side of the air-conditioning system (Figure 5–9).

WORKING EXAMPLE

Air entering the outdoor condenser is 80°F. What is the target condensing temperature or condensing pressure? The system uses R-22 refrigerant.

SOLUTION: The formula is

$$\text{Target condensing temperature} = \text{ambient air} + 25°F$$

$$105°F = 80°F + 25°F$$

105°F is the targeted condensing temperature.

Charge the unit 10% above or below the desired condensing temperature of 105°F. In this case the range would be 94°F to 116°F.

Look at the high-side gauge and locate the R-22 temperature ring. Next locate 105°F on the temperature ring. The outer pressure ring will indicate 211 psi. This is the desired condensing temperature (Figure 5–10).

When the unit has been operating for 15 min, the gauge indicator should be within ±10% of 105°F, which corresponds to 211 psi.

$$10\% = 0.10$$
$$0.10 \times 105°F = 10.5°F, \text{ or } 11°F$$

$$\begin{array}{r} 105°F \text{ target condensing temperature} \\ \underline{-11°F\ (-10\%)} \\ 94°F \text{ minimum condensing} \\ \text{temperature} \end{array}$$

$$\begin{array}{r} 105°F \text{ target condensing temperature} \\ \underline{+11°F\ (+10\%)} \\ 116°F \text{ maximum condensing} \\ \text{temperature} \end{array}$$

The range of acceptable condensing temperatures is 94°F through 116°F. ∎

Figure 5–9   High-pressure gauge. *(Courtesy of Amprobe Promax)*

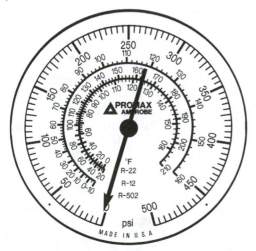

Figure 5–10   High-side gauge measuring 211 psi, which corresponds to 105°F when using R-22 refrigerant. *(Courtesy of Amprobe Promax)*

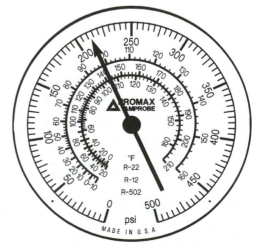

4. Determine the evaporating temperature or pressure by taking the temperature of the air entering the return air. Subtract 35°F from this temperature reading.

5. Look at the low-pressure (suction-side) gauge and locate the R-22 temperature ring. Find the temperature that corresponds to the return-air temperature minus 35°F. The evaporating temperature should be within ±10% of this reading but not below 32°F. If an evaporator operates below 32°F, the evaporator coil will freeze up and block the airflow after a long period of operating. The frost accumulation on the evaporator will also reduce the heat-removal capabilities of the evaporator. The frost coating acts as an insulator between the returning air and refrigerant inside the evaporator.

WORKING EXAMPLE

The air entering the filter grille is 78°F. What is the target evaporating temperature when the equipment uses R-22?

SOLUTION

     78°F return-air temperature
  −35°F constant
     43°F evaporating temperature

43°F is the targeted evaporating temperature. Look at the low-side pressure gauge and locate the R-22 temperature ring. Next, locate 43°F on the temperature ring. The outer pressure ring will indicate 73 psi (Figure 5–11).

When the unit has been operating for 15 min, the gauge indicator should be within ±10% of 43°F. Double-check temperature readings and recalculate when necessary.

    10% = 0.10
 0.10 × 43°F = 4.3°F, or 4°F

    43°F target evaporating temperature
   −4°F (−10%)
    39°F minimum evaporating
       temperature (never below 32°F)

    43°F target evaporating temperature
   +4°F (+10%)
    47°F maximum evaporating
       temperature

The range of acceptable evaporating temperature is 39°F through 47°F.

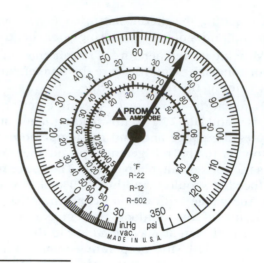

Figure 5–11   Low-side gauge measuring 73 psi, which corresponds to 43°F when using R-22 refrigerant. *(Courtesy of Amprobe Promax)*

High-SEER systems will operate at higher evaporator temperatures and lower condensing temperatures compared to less efficient systems. Guessing at the SEER rating will probably cause an improper charge. The unusual temperature-pressure readings that are found in high-SEER-rated units can cause a misdiagnosis of system problems. This is especially true when it comes to the compressor. Knowing the SEER rating is important!

PRESSURE-TEMPERATURE METHODS FOR UNITS WITH A SEER OF 12 OR HIGHER

Systems with a SEER rating of 12 or higher should use a manufacturer's charging chart to check the charge on the unit. If the information is available, a charging chart should be used when checking the charge on *all* units. Once a unit becomes this efficient, general rules of thumb and pressure-temperature relationships are difficult to establish for all the various types of units on the market. Check inside the condensing unit for a charging chart or charging instructions for that unit.

This information may also be obtained from the installation instructions, the manufacturer of the unit, or from a local supply house that sells that brand of equipment.

 Units with SEER 12 or greater need special charging instructions.

CHARGING EXAMPLES USING THE PRESSURE-TEMPERATURE METHOD

CONDENSING AND EVAPORATING TEMPERATURES LOW

The system probably needs refrigerant if the condensing and evaporating temperatures are both below the minimum targeted charging temperatures.

Hook up the refrigeration gauge set. Purge air from hoses. Add refrigerant in the vapor state through the low-pressure side only. Charge until the condensing and evaporating temperature rises into the correct charging range, as discussed in the previous section. Use other charging methods to check and fine-tune the charge.

CONDENSING TEMPERATURE ACCEPTABLE, EVAPORATING TEMPERATURE LOW

If the condensing temperature is in the targeted range, but the evaporating temperature is below the targeted temperature range, add refrigerant until the upper limit of the condensing temperature is reached. This upper condensing temperature is 10% higher than the air temperature entering the condenser plus 30°F (or plus 25°F for high-SEER units with a rating 10.5 to 11.9). If the evaporating temperature is still lower than the minimum required temperature, it is time to diagnose the problem. This condition may not be a low charge.

Some causes of low evaporating temperature are low airflow, dirty air filter, or dirty evaporator coil; restriction in the low-pressure side; restriction in the metering device; or a restriction in the liquid line feeding the metering device. Check the filter, evaporator coil, and airflow prior to adding refrigerant. An evaporating temperature below 32°F, which freezes the coil, can also cause a low evaporating temperature. Prior to checking the charge, the ice will need to be melted off the evaporator. You may notice this icy condition when inspecting the evaporator prior to beginning the charge-check procedure.

HIGH CONDENSING AND EVAPORATING TEMPERATURES

If the condensing and evaporating temperatures are both in excess of the 10% upper targeted temperature, recover a small amount of refrigerant to reduce these temperatures. Dropping the condensing temperature 1° *will not* drop the evaporating temperature 1°. Eventually, removing a charge will drop both temperatures but not on a one-to-one ratio. Operate the unit for 15 min to assure stable operation after adjusting the charge.

NORMAL CONDENSING TEMPERATURE AND HIGH EVAPORATING TEMPERATURE

When the condensing temperature is normal and the evaporating temperature is higher than the upper limit of the targeted temperature, try recovering refrigerant to bring down the evaporating temperature.

The condensing temperature may drop. The condensing temperature should not drop below the lower limit of the targeted condensing temperature.

Super-high-efficiency (SEER 10.5 or more) air-conditioning systems will have higher evaporating temperatures and lower condensing temperatures. Specifically, for systems with a SEER of 12 or greater, you will need a manufacturer's charging chart to accurately determine the condensing and evaporating temperatures, subcooling, or superheat.

A building that has a very high return air temperature (above 82°F) and high relative humidity (above 60% RH) will have a high evaporating temperature. Allow the air conditioning system to bring the temperature down below 80°F and relative humidity level below 60% prior to recovering the charge when the evaporating temperature is above normal.

LOW CONDENSING TEMPERATURE AND HIGH EVAPORATING TEMPERATURE

Finally, a defective compressor is suspected in cases where the condensing temperature is 40°F lower than normal targeted temperature and the evaporating temperature is 40° above the upper targeted evaporating temperature. Remember the 40-40 rule. These are approximate temperatures that indicate a mechanical malfunction inside the compressor. Basically, the high pressure is equalizing with the low pressure inside the compressor, causing high evaporating temperatures or pressures and low condensing temperatures or pressures. Defective valves or gaskets that separate the high- and low-pressure compressor compartments will cause these readings. An open internal pressure-relief valve will also cause the same symptoms. Replace the compressor if it is a hermetic compressor. Check the valve plate if it is a semi-hermetic compressor.

Also, make sure your refrigerant-gauge hand valves are both closed when making pressure checks. Leaving these valves open will give an unusual reading of low condensing temperatures and high evaporating temperature. In this case, the high-pressure side is equalizing with the low-pressure side through the refrigeration gauge set.

In summary, the pressure-temperature method of checking the charge will provide an air-conditioning system with adequate refrigerant to provide comfort cooling. Other charging methods must be used to fine-tune the charge to assure low operating costs and long compressor life. These charging methods are discussed in the chapters that follow.

SUMMARY

- The pressure-temperature method of checking the charge is the first step in assuring that an air-conditioning system is charged properly. It is necessary to obtain the correct range of condensing and evaporating temperatures before proceeding with any other charging method. The P-T method is a rough charging method that gets the system cooling and allows the technician to fine tune the charge by using superheat, subcooling, etc.

- The technician must be able to use and interpret the reading on the gauges. The outer ring on most gauges is the pressure in the system. The inner rings are the corresponding temperature rings for the refrigerant being checked. When the system is operating the inner ring on the high pressure gauge is the condensing temperature. The inner ring on the low pressure gauge is the evaporating temperature.

- In order to use the P-T method of charging, the technician must know the SEER rating of the system. The SEER rating will determine the range of temperatures or pressures used to check the charge. SEER rating 12 and higher will need a manufacturer's charging chart. These rules-of-thumb do not apply to these high-efficiency systems.

- The chapter reviewed ways to use the P-T chart to determine if the system had charging or mechanical problems. Some technicians use the P-T method as the sole method of checking and adjusting the charge. In these cases the unit may cool but it will not operate efficiently and the compressor motor life will be shortened.

PROBLEMS AND QUESTIONS

1. What is the condensing temperature of an R-22 air-conditioning system if the discharge pressure is 275 psig?

2. What is the evaporating temperature of an R-22 system if the pressure is 75 psig?

3. What is the condensing pressure of an R-22 system if the condensing temperature is 100°F?

4. An air-conditioning technician measures a distributor-tube temperature of 35°F. The distributor tube is used to feed refrigerant from the metering device to the evaporator inlet. What is the evaporating pressure of the evaporator if the system uses R-22 refrigerant?

5. List all the parts of the air-conditioning system that must be clean prior to checking or adjusting the charge on an air-conditioning system.

6. Why must these components be clean?

7. What information is needed to check the charge on an air-conditioning system?

8. What safety practice(s) do you need to improve on when working on an air-conditioning system?

9. What are the benefits of a properly charged air-conditioning system?

10. Name all the effects of an incorrectly charged system.

11. What are the target condensing and evaporating temperatures of an R-22 system that has a return air temperature of 75°F and an outdoor temperature of 100°F?

12. What are the target pressures of the Problem 11?

13. What are the target evaporating and condensing temperatures of an R-22 system with a return-air temperature of 80°F and an outdoor-air temperature of 80°F?

14. On the first service call of the day, a technician cleans a system and records the following information:

    R-22 refrigerant at 11 SEER

    250 psig condensing temperature

    75 psig evaporating temperature

    78°F outdoor temperature

    75°F return-air temperature

    Using only the P-T charging method, is this system charged correctly? If not, what is the problem?

15. On the second service call of the day, a technician finds that everything was cleaned 2 wk earlier and records the following information:

    R-22 refrigerant at 10 SEER

    145°F condensing temperature

    40°F evaporating temperature

    85°F outdoor temperature

    74°F indoor temperature

Using only the P-T charging method, is the system charged correctly? If not, what is the problem?

16. Describe how to check the charge of a system that has a SEER rating of 14.

17. The P-T charging method is considered a rough charging method. Name five other charging methods that could be used to fine-tune the charge on an air-conditioning system.

18. High-SEER-rated equipment has (<u>a higher, a lower, or the same</u>) condensing temperature compared to a system with a standard efficiency.

19. High-SEER-rated equipment has (<u>a higher, a lower, or the same</u>) evaporating pressure compared to a system with a standard efficiency.

20. Where can a technician find the SEER rating of a system? List three sources.

21. What are the target condensing and evaporating temperatures and pressures of an R-22 system with a 9.5 SEER rating, an outdoor temperature of 100°F, and a return-air temperature of 72°F.

ON THE JOB

The SEER rating of an air-conditioning system may be a challenge to locate. With the great variety of high-efficiency equipment, it is not advisable to guess the SEER rating. Do not be fooled by terms equipment manufacturers use to advertise their equipment. Terms such as *high efficiency or super high efficiency* do not divulge the SEER rating of the system. Equipment built in the early 1990s was espoused as being super-efficient systems even though the SEER rating was 10 or 11. SEER 11 was considered quite high at that time.

CHAPTER SIX

CHARGING PROCEDURES: SUPERHEAT METHODS

AFTER STUDYING THIS CHAPTER, THE STUDENT WILL BE ABLE TO:

- Define superheat.
- Describe how to calculate suction superheat.
- Describe how to calculate discharge superheat.
- Discuss what happens if suction superheat is high.
- Discuss what happens if suction superheat is low.
- Determine if a system needs charging or refrigerant recovery based on suction and discharge superheat readings.

- Describe the damage caused by high discharge temperatures.
- List the causes of high and low suction superheat.
- List the causes of high and low discharge superheat.

CHAPTER OVERVIEW

This chapter will discuss the suction and discharge methods of checking the charge on an air-conditioning system. The superheat methods are used to fine-tune the air-conditioning system's charge. The superheat methods can also be used to troubleshoot refrigerant-side problems. The pressure-temperature charging method is used prior to checking superheat. The condensing and evaporating temperatures must be within the ±10% of the target temperatures. The superheat methods, especially the suction superheat method, will fine-tune the charge to allow the equipment to operate at peak cooling and dehumidification performance, with lower operating cost and maximum compressor life.

DEFINITION OF SUPERHEAT

Superheat is heat added to a substance above its boiling point. In an air-conditioning system, superheat is heat added to the refrigerant above its boiling point. In an air-conditioning system, the boiling (or evaporating) point occurs in the evaporator. The evaporator receives a mixture of cold liquid and vapor refrigerant from the metering device. This combination of liquid and vapor is known as a saturated mixture. The temperature of the refrigerant in the evaporator coil is known as the evaporating temperature, or the coil temperature. This is the temperature at which the refrigerant is boiling off and absorbing heat from the air or water passing over it. The evaporating temperature is constant in the evaporator until every last drop of liquid refrigerant evaporates or boils away inside the evaporator coil. The liquid refrigerant boils off to a vapor before leaving the evaporator. The final drops of liquid refrigerant evaporate at a point approximately 80% into the evaporator. This is known as a dry evaporator because no liquid, only vapor, leaves the coil. Another common name is a direct expansion coil, abbreviated as DX. After all the liquid is converted to a vapor the remaining vapor travels through the evaporator, picking up a minimal amount of heat. This additional heat picked up by the vapor is called superheat.

EVAPORATING TEMPERATURE

The evaporating temperature can easily be obtained from the compound gauge on the refrigeration manifold set. The inner temperature ring shows the evaporating temperature when the system has been operating. The system should be operating for 15 min to obtain a stable temperature reading. Figure 6–1 shows an evaporating temperature of 40°F with R-22 refrigerant. Forty degrees (40°F) corresponds to an evaporating pressure of 68 psig. A pressure-temperature chart can be used to determine the evaporating temperature from a compound-gauge pressure reading if the temperatures are not available on the face of the gauge.

Once the last drop of refrigerant evaporates, the superheating of the vapor begins. Most of the heat absorbed by the evaporator is absorbed when the liquid changes states from a liquid to a vapor. To help you understand this concept, review how water changes temperature and state with the addition or removal of heat. One Btu will heat 1 lb of water 1°F. Adding 1 Btu raises the temperature of 210°F water to 211°F. Changing 1 lb of 212°F water to 212°F steam takes 970 Btu of heat energy. It takes a tremendous amount of energy to break the liquid bond and cause it to "explode" into a vapor. Increasing the steam's temperature above the 212°F vapor temperature takes only ½ Btu per degree rise in temperature. In this example, the superheat is the heat added to the steam once all the water evaporates. If the 212°F steam temperature were to rise to 217°F, then the steam would have a superheat of 5°F. Superheat is heat added to the refrigerant above its boiling or evaporating point.

Once all the liquid refrigerant boils off in the evaporator, superheating of the refrigerant begins. The cool vapor will rise in temperature, or superheat, as it makes its way through the final passages of the evaporator and through the suction line to the compressor. Again, suction superheat is heat added to the refrigerant vapor above it boiling point. This is an important concept to learn.

Figure 6–2 shows a saturated mixture entering the evaporator. Liquid refrigerant boils off, absorbing heat. Superheating of the refrigerant begins at the point where the last drop of liquid boils off. Heat is added to vapor as it continues through the evaporator and suction line to the compressor.

In the compressor, more superheat is added to the refrigerant vapor. This superheat is derived from the compressor motor heat and the heat of compression and friction. The refrigerant leaving the compressor is a hot, high-pressure superheated vapor. This is called discharge superheat because it is a heat-laden vapor on the discharge side of the system. The condensing liquid is used as a reference temperature to calculate discharge superheat. Discharge superheat is the difference in temperature between the discharge-vapor temperature and the condensing temperature.

WHAT YOU WILL NEED TO START THE CHARGING PROCEDURE

Before starting this charging procedure you will need the following information, instruments, and materials.

INFORMATIONAL NEEDS

- Type of metering device on the equipment
- Superheat charging chart for fixed orifice/capillary tube devices
- Manufacture's superheat recommendations

Figure 6–1 Low-side compound gauge with evaporating temperature rings labeled. *(Courtesy of Amprobe Promax)*

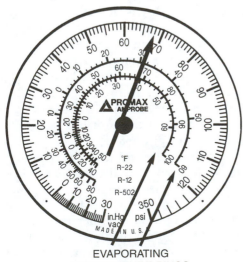

EVAPORATING
TEMPERATURE RINGS

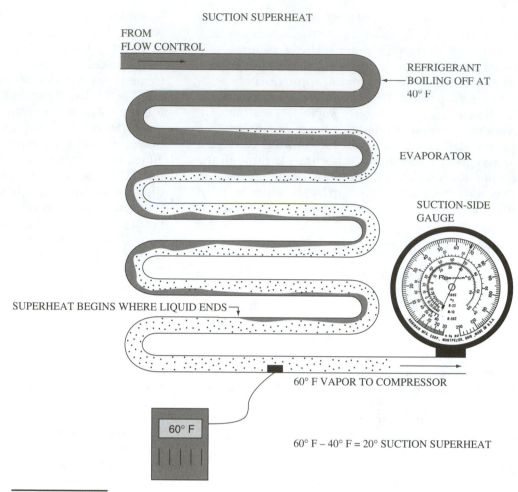

SUCTION SUPERHEAT

FROM
FLOW CONTROL

REFRIGERANT
BOILING OFF AT
40° F

EVAPORATOR

SUCTION-SIDE
GAUGE

SUPERHEAT BEGINS WHERE LIQUID ENDS →

60° F VAPOR TO COMPRESSOR

60° F

60° F – 40° F = 20° SUCTION SUPERHEAT

Figure 6–2   Measuring suction superheat. Liquid refrigerant decreases as it boils off in the tubing. The remaining cool vapor is superheated.

EQUIPMENT NEEDS

- Refrigeration gauge set
- Two electronic temperature testers
- Sling psychrometer (Figures 6–5 or 6–6)
- Recovery equipment if system is overcharged

MATERIAL NEEDS

- Duct tape
- Pipe insulation
- Proper refrigerant (Refrigerant will be used only if the unit is undercharged.)

SUCTION-SUPERHEAT METHOD OF CHECKING A CHARGE

Suction superheat is the amount of heat added to the refrigerant above its boiling point. Superheat is the heat added to and absorbed by the cool gas in the evaporator. The heat comes from the air or water passing over the evaporator.

First, the cold liquid in the evaporator must be boiled off to a cool refrigerant gas or vapor in order for superheating of the gas to occur. The temperature at which the refrigerant boils or evaporates is determined by looking at the temperature registered on the suction side of the compound gauge.

For this example let us record a 40°F evaporating temperature using R-22 refrigerant. This 40°F reading is taken from the suction-side gauge (low-pressure) reading after the unit has been operating for at least 15 min (Figure 6–1).

While the unit is running, tape an accurate electronic thermometer to the suction-side piping near the compressor. Insulate the sensing part of the thermometer to keep it isolated from the effects of the surrounding outdoor temperature. Allow the suction-line thermometer reading to stabilize. Pocket thermometers are not accurate for measuring pipe surface temperatures.

WORKING EXAMPLE

In Figure 6–2, the technician measures 60°F on the suction line near the compressor. The readings from the compound gauge indicate the R-22 refrigerant is boiling at 40°F. What amount of heat must be added?

SOLUTION:

60°F	suction-line temperature
−40°F	refrigerant boiling temperature
20°F	suction superheat

This 20°F represents the amount of heat added to the refrigerant after it is boiled off in the evaporator. The target superheat will be determined by the type of metering device used in the system. ■

The correct superheat reading is important because

- It tells us how much liquid refrigerant is in the evaporator.
- It shows that the refrigerant charge is correct.
- It can be used as a refrigerant-side troubleshooting tool.
- It indicates whether liquid refrigerant is leaving the evaporator and flooding the compressor. This flooding condition could cause compressor damage. A flooding condition exists if the superheat reading is zero degrees, i.e., the suction-line temperature equals the evaporator temperature. Some liquid refrigerant may be returning to the compressor with a superheat reading of 5°F. At 5°F, liquid refrigerant may be travelling down the middle of the suction line. The middle of the refrigerant tubing carries the highest velocity of refrigerant. A temperature probe attached to the suction line tubing will not detect this returning liquid refrigerant.

Figure 6–3   Flowchart for Determining Suction Superheat

Hook up gauges and purge hoses. Attach the temperature probe to the suction line.
↓
Allow 15 min of operating time for stable readings.
↓
Measure suction-line temperature near the compressor and record evaporating temperature.
↓
Subtract suction-line temperature from evaporating temperature.

Flooding can occur at a temperature higher than the 0° superheat.

Too much superheat is an indication of a lack of refrigerant in the evaporator, which translates into poor cooling (reduced capacity) and not enough cool gas entering the compressor to keep it from overheating. High superheat can also be caused by other factors, which are covered in the chapter on refrigerant-side troubleshooting.

High-suction superheat can overheat the internal compressor motor. Electric-motor windings need to be cooled. Most reciprocating compressor motors are located inside the compressor shell in an atmosphere of refrigerant gas. The returning cool gas from the evaporator keeps the motor winding from overheating. An overheated motor will temporarily cut the compressor off. Extreme or persistent overheating will burn out the motor, thereby requiring a compressor replacement. Measuring suction superheat is one of the best ways to verify that an adequate flow of cool gas is returning to the motor windings. Superheat is used to establish a correct charge, monitor the performance of the evaporator and keep the motor windings from burning up.

Knowing the suction superheat will help fine-tune the charge on an air-conditioning system. Knowing the suction superheat is one of the most important charging methods.

USING SUCTION SUPERHEAT TO CHECK A CHARGE

The suction superheat method of checking a charge is used to fine-tune the charge of an air-conditioning system. You know how to measure suction superheat; now we must determine if the superheat is correct.

Remember, conduct several of these charging methods to check the charge of a unit. Using only one charging method may give inaccurate information about a unit even if the system is cooling. *A safety reminder:* Refrigerant can burn your skin and damage your eyes. Wear safety glasses and gloves, and watch for electrical shock!

Before checking the superheat, determine the type of metering or flow-control device installed at the evaporator. Look for a thermostatic expansion valve, capillary tube, or a fixed-orifice device. Review the section on metering devices if help is needed in deciding which flow-control device is installed on the unit that is being checked.

Determine the correct type of flow-control device on the unit. Write this information on the evaporator and inside the condensing unit. This information will then be available the next time the charge is checked on the unit.

COOLING MODE ONLY
CHARGING BY SUPERHEAT METHOD
INDOOR COIL—FIXED ORIFICE METERING

INDOOR CONDITIONS			OUTDOOR <°F> DRY BULB TEMPERATURE									
WB	DB	RH%	65	70	75	80	85	90	95	100	105	110
61	65	80	16	12	8	6	CHARGING AT THESE CONDITIONS CAN RESULT IN DAMAGE TO THE COMPRESSOR. CHARGE TO 5° SUPERHEAT AND CHECK SUPERHEAT WHEN CONDITIONS ARE MORE FAVORABLE.					
	70	60	18	14	10	8						
	75	45	20	16	12	8	6					
	80	33	21	17	14	10	6					
	85	23	23	19	16	12	7					
63	70	68	21	17	13	10	8	6				
	75	52	23	19	16	12	9	6				
	80	39	24	20	17	14	10	7				
	85	29	25	21	18	15	12	8	6			
	90	20	26	22	19	16	13	9	7			
65	70	77	24	20	17	13	10	8	6			
	75	59	26	22	19	15	12	10	7			
	80	45	27	24	21	18	14	11	8	6		
	85	33	28	26	22	19	15	12	9	6		
	90	25	29	26	23	20	16	13	10	7		
67	70	86	27	24	21	17	14	11	8	6		
	75	66	28	26	22	18	15	13	10	8	6	
	80	50	30	27	24	21	18	15	12	10	7	
	85	39	31	28	25	22	19	16	13	11	8	
	90	30	32	29	26	23	20	17	14	12	9	6
69	70	95	30	27	25	21	18	15	12	9	7	6
	75	75	31	28	25	22	19	16	13	10	8	6
	80	58	32	29	27	24	21	19	16	14	11	8
	85	45	34	31	28	26	23	20	17	15	12	9
	90	35	35	32	30	27	24	21	19	16	13	11
71	75	82	34	31	29	26	23	21	18	15	12	9
	80	65	35	32	30	28	24	22	20	18	15	13
	85	51	37	34	31	29	26	24	21	19	16	14
	90	39	38	35	32	30	28	25	22	20	17	16
	95	30	39	35	33	30	29	26	23	21	18	17
73	75	92	37	34	32	29	27	24	22	20	17	15
	80	72	37	35	33	30	28	26	24	22	19	17
	85	58	38	36	34	31	29	27	25	23	20	18
	90	45	38	36	34	31	30	28	25	23	21	19
	95	35	38	36	35	32	31	28	26	24	21	19

NOTES: SUPERHEAT MEASUREMENTS SHOULD BE TAKEN AT CONDENSING UNIT SERVICE VALVES.
CHARGE SYSTEM WITHIN 2° F OF SUPERHEAT INDICATED. RECOMMENDED
MINIMUM SUPERHEAT IS 5° F
WHITE AREA IN THE CHART IS THE OPTIMUM WINDOW FOR CHARGING.

DB = DRY BULB TEMPERATURE DEGREE ° F.   WB = WET BULB TEMPERATURE DEGREE ° F.
RH = APPROX. % OF INDOOR RELATIVE HUMIDITY.

Figure 6–4   Superheat chart. *(Courtesy of Rheem/Rudd Corporation)*

A superheat-charging chart will be required if the unit has a capillary tube or fixed-orifice-metering device. See Figure 6–4. Superheat is the difference in temperature between the boiling point of the refrigerant in the evaporator (temperature is found on the low-side gauge) and the temperature of the suction line near the compressor.

The formula for calculating suction superheat is

$$\frac{\text{Suction}}{\text{superheat}} = \frac{\text{suction-line temp} -}{\text{evaporating temp}}$$

MEASURING SUCTION SUPERHEAT

Prior to checking suction superheat, the system should be clean and operating in the correct range of condensing and evaporating temperatures. Operate the system for 15 min while hooking up gauges and temperature probes. The superheat reading is obtained by first looking at the low-side gauge. Follow the gauge indicator (needle) to the temperature of the refrigerant (Figure 6–1). In most air-conditioning units the refrigerant will be R-22. With the advent of new refrigerants, it is best to be positive which refrigerant is used in the system. This temperature reading, as indicated by the low-side gauge indicator, is the temperature at which the refrigerant evaporates and boils off inside the evaporator. Once you have obtained the evaporating-temperature reading from the gauge, you will need to obtain the suction-line temperature within a few feet of the compressor.

At this point it is worth noting that there are several types of superheat. This section discusses superheat found on the low side. The next section discusses discharge superheat, which is found on the high side. There are two types of low-side superheat. One type of superheat is measured as near as possible to the compressor inlet. This is the most common type, known as *system superheat*. System superheat is used in all the examples in this book. The other type of low-side superheat is known as evaporator superheat, measured very near the outlet of the evaporator. The purpose of this measurement is to check the performance of the metering device and the evaporator coil. It is common for manufacturers of thermostatic expansion valves to relate to superheat at the outlet of the evaporator. TXV manufacturers design their product with specific outlet-superheat characteristics. They cannot control the suction line that connects the outlet of the evaporator to the compressor. An improperly designed or installed suction line can add excess superheat by the time the refrigerant arrives at the compressor.

The suction-line temperature is obtained by tapping the thermometer-sensing bulb to the top or side of the suction-line pipe. Insulate the thermometer-sensing element to keep the surrounding tempera-

ture from interfering with the thermometer for this reading. Do not attach the temperature probe to the bottom of a horizontal suction line. Oil travels on the bottom of the suction line and acts as an insulator between the refrigerant and test probe. Attach the probe to the top or side of the suction line.

Subtract the evaporating temperature as found on the compound gauge from the suction-line temperature. This temperature difference is the suction superheat.

Finally, *double-check* your readings. This *is* important, because they may change during the process. A 2° change is not important.

$$\frac{\text{Suction}}{\text{superheat}} = \frac{\text{suction-line temp} -}{\text{evaporating temp}}$$

WORKING EXAMPLE

60°F	actual suction line temperature
−40°F	evaporating temperature on gauge (Figure 6–2)
20°F	suction superheat ■

SUCTION SUPERHEAT

If your unit has a thermostatic expansion valve (TXV) flow control, charge the unit to a range of 15°F to 20°F suction superheat unless the manufacturer recommends a different superheat reading. Some equipment manufacturers recommend superheat ranges as low as 8°F to 12°F and a few in excess of 20°F. The 15° to 20° range is safe if the manufacturer's information is not available. Always use manufacturers' charging recommendation when available.

If your unit has a capillary tube or fixed-orifice device for flow control, charge the unit to the superheat temperature recommended from the charging table in Figure 6–4. A cap-tube or fixed-orifice device has no set suction superheat. The amount of superheat depends on the outside ambient air temperature as well as the return wet-bulb and dry-bulb air temperatures. A sling psychrometer can be used to make these indoor measurements (Figure 6–5 and 6–6).

Superheat can be determined as follows:

1. Read evaporating temperature from the manifold gauges.
2. Read suction-line temperature from an accurate digital thermometer.
3. Superheat = suction-line temperature − suction line temperature.

Next, use the superheat chart in Figure 6–4 to determine the correct superheat. To use this superheat

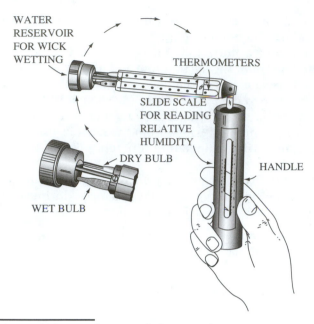

WATER
RESERVOIR
FOR WICK
WETTING

THERMOMETERS

SLIDE SCALE
FOR READING
RELATIVE
HUMIDITY

DRY BULB

HANDLE

WET BULB

Figure 6–5 Hand-operated sling psychrometer used to check wet-bulb and dry-bulb temperatures.

chart, you will need to take three temperature measurements:

- Dry-bulb air temperature entering the condensing coil
- Dry-bulb air temperature of the return air
- Wet-bulb air temperature of the return air

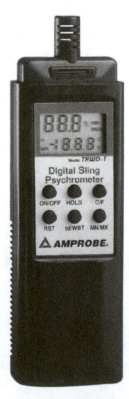

Figure 6–6 Digital psychrometer calculates wet-bulb, dry-bulb, and relative humidity. *(Courtesy Amprobe/Promax Instruments)*

The return-air readings are taken with a sling psychrometer (Figure 6–6). A sling psychrometer measures both dry-bulb and wet-bulb temperatures. The sling psychrometer can be the standard glass thermometer type or a more accurate digital thermometer device. The digital device is the best option.

USING THE SUCTION SUPERHEAT CHART

The superheat chart is used to charge a system with a fixed-orifice device, such as a capillary tube or flow-control piston. The piston device (Figure 6–7) has a trade name such as an accurator, flow rater, or flow device.

To use this chart you will need the following accurate readings:

- Outdoor dry-bulb temperature entering the condensing unit
- Indoor dry-bulb temperature at the return-air grille
- Indoor wet-bulb temperature at the return-air grille

The outdoor dry-bulb temperature is found across the top, horizontal portion of the chart. The outdoor temperature range is 65°F to 110°F. The indoor wet-bulb temperature is found on the far-left column under the heading of "Indoor Conditions, WB."

The indoor dry-bulb temperature is found at the second column from the left, next to the wet-bulb column and is identified as "DB." The "RH%" column (relative humidity) is not used with the superheat charging method, but it is important to note the relative humidity reading inside the building. The human comfort zone is 40% to 60% relative humidity with a temperature range of 70°F to 80°F. High relative humidity readings cause higher suction

Figure 6–7 Fixed-orifice metering device.

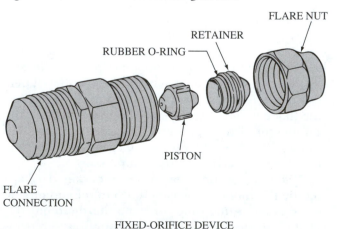

FLARE NUT

RETAINER

RUBBER O-RING

PISTON

FLARE
CONNECTION

FIXED-ORIFICE DEVICE

ON THE JOB

Air-conditioning systems with fixed-bore metering devices should not be used in temperatures below 65°F. Below 65°F the condenser pressure is too low to drive an adequate amount of refrigerant though the metering device. TXVs work better in delivering adequate refrigerant flow in low ambient temperatures. Head-pressure controls may be used to keep the head pressure high enough to push refrigerant through the metering device.

pressures, lower system superheat, and longer equipment run times.

The numbers under the outdoor dry-bulb temperature are the target suction-superheat reading. A ±2°F superheat temperature spread is acceptable. Numbers that are in the center, with the white background, are the best target superheat temperatures. The numbers that are in the shaded areas are marginal superheat temperatures and can be used. Estimated readings off the chart are not recommended.

WORKING EXAMPLE

The following readings are recorded on an air-conditioning system that has been operating for 20 min. The condensing and evaporating temperatures are in range, according to the pressure-temperature method. The system has been cleaned. Outdoor temperature at the air inlet to the condenser is 75°F. The indoor temperature at the filter grille is 63°F wet-bulb and 75°F dry-bulb. What is the target superheat?

SOLUTION The answer is 16°F, ±2°F. An acceptable range of temperatures is 14°F to 18°F superheat. See Figure 6–8. ■

WORKING EXAMPLE

The following readings are obtained in the afternoon on the same unit. The outdoor temperature is 100°F. The indoor temperature is 63°F wet bulb and 75°F dry bulb. What is the target superheat?

SOLUTION The answer is 5°F. These readings fall in the area without numbers. As it states in the upper right-hand corner of the chart, charging at these conditions can result in damage to the compressor. Charge to 5° superheat and check superheat when conditions are more favorable. The damage to the compressor they are talking about is liquid dilution of the compressor lubrication or possible liquid slugging. ■

Compare the two working examples. The indoor conditions are the same, but the outdoor temperature in the second example is much higher. Air-conditioning systems with capillary tubes or fixed-bore devices have lower superheat readings as the outdoor temperature increases. The increase in outdoor temperature causes the condensing pressure to increase, driving more liquid refrigerant through the metering device. This is the time of day that the building is also absorbing more heat and needs additional cooling. Fixed metering devices are designed with this in mind. As the day cools, the condenser pressure drops, with a corresponding drop in refrigerant being pushed through the metering device. On a cool day, higher superheat readings occur. On a hot day, low superheat readings occur.

Other factors that affect the flow of refrigerant through the capillary tube are the length and inside diameter of the tubing. Longer tubing with a small inside diameter reduces the capacity of the refrigerant flow to the evaporator. A fixed piston that has a small opening reduces the amount of refrigerant flow to the evaporator. The piston must be sized to the capacity of the system.

Finally, most outdoor- and indoor-temperature readings fall between the temperatures that are shown on the chart. For example, the outdoor temperature may be 83°F. Do not use the superheat reading for 80° or 85°, but use the estimated superheat temperature between 80° and 85°. Accurately calculate the target superheat temperature or the superheat charging method will do more harm than good.

DETERMINING A GOOD CHARGE

The suction superheat is acceptable under the following conditions:

- The superheat reading is 15° to 20° for a TXV.
- The superheat is in range per manufacturer's recommendations.
- The reading is as calculated on the superheat chart for fixed-bore metering devices.

If the superheat is lower than the recommended level, the system could be overcharged. If the superheat is higher than the recommended level, the system could be undercharged. This seems pretty simple, but other problems can cause low and

COOLING MODE ONLY
CHARGING BY SUPERHEAT METHOD
INDOOR COIL—FIXED ORIFICE METERING

INDOOR CONDITIONS			OUTDOOR °F DRY BULB TEMPERATURE									
WB	DB	RH%	65	70	75	80	85	90	95	100	105	110
61	65	80	16	12	8	6	CHARGING AT THESE CONDITIONS CAN RESULT IN DAMAGE TO THE COMPRESSOR. CHARGE TO 5° SUPERHEAT AND CHECK SUPERHEAT WHEN CONDITIONS ARE MORE FAVORABLE.					
	70	60	18	14	10	8						
	75	45	20	16	12	8	6					
	80	33	21	17	14	10	6					
	85	23	23	19	16	12	7					
63	70	68	21	17	13	10	8	6				
	75	52	23	19	16	12	9	6				
	80	39	24	20	17	14	10	7				
	85	29	25	21	18	15	12	8	6			
	90	20	26	22	19	16	13	9	7			
65	70	77	24	20	17	13	10	8	6			
	75	59	26	22	19	15	12	10	7			
	80	45	27	24	21	18	14	11	8	6		
	85	33	28	26	22	19	15	12	9	6		
	90	25	29	26	23	20	16	13	10	7		
67	70	86	27	24	21	17	14	11	8	6		
	75	66	28	26	22	18	15	13	10	8	6	
	80	50	30	27	24	21	18	15	12	10	7	
	85	39	31	28	25	22	19	16	13	11	8	
	90	30	32	29	26	23	20	17	14	12	9	6
69	70	95	30	27	25	21	18	15	12	9	7	6
	75	75	31	28	25	22	19	16	13	10	8	6
	80	58	32	29	27	24	21	19	16	14	11	8
	85	45	34	31	28	26	23	20	17	15	12	9
	90	35	35	32	30	27	24	21	19	16	13	11
71	75	82	34	31	29	26	23	21	18	15	12	9
	80	65	35	32	30	28	24	22	20	18	15	13
	85	51	37	34	31	29	26	24	21	19	16	14
	90	39	38	35	32	30	28	25	22	20	17	16
	95	30	39	35	33	30	29	26	23	21	18	17
73	75	92	37	34	32	29	27	24	22	20	17	15
	80	72	37	35	33	30	28	26	24	22	19	17
	85	58	38	36	34	31	29	27	25	23	20	18
	90	45	38	36	34	31	30	28	25	23	21	19
	95	35	38	36	35	32	31	28	26	24	21	19

NOTES: SUPERHEAT MEASUREMENTS SHOULD BE TAKEN AT CONDENSING UNIT SERVICE VALVES.
CHARGE SYSTEM WITHIN 2° F OF SUPERHEAT INDICATED. RECOMMENDED
MINIMUM SUPERHEAT IS 5° F
WHITE AREA IN THE CHART IS THE OPTIMUM WINDOW FOR CHARGING.

DB = DRY BULB TEMPERATURE DEGREE ° F.   WB = WET BULB TEMPERATURE DEGREE ° F.
RH = APPROX. % OF INDOOR RELATIVE HUMIDITY.

Figure 6–8   Superheat chart. (*Courtesy of Rheem/Rudd Corporation*)

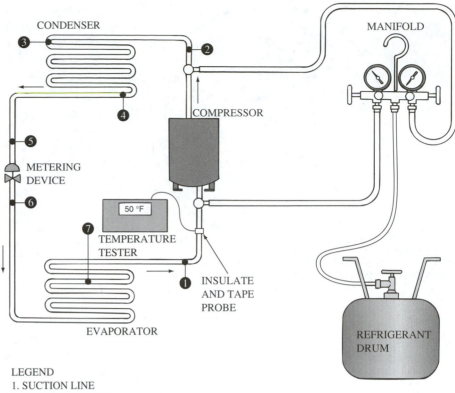

Figure 6-9 Checking suction superheat. 1% change in charge equals 3°F change in superheat. Temperature probe should be taped to the suction line and insulated with pipe insulation. Tape insulation for a good seal.

LEGEND
1. SUCTION LINE
2. DISCHARGE LINE
3. CONDENSER COIL
4. CONDENSER OUTLET
5. LIQUID LINE
6. DISTRIBUTOR
7. EVAPORATOR COIL

high superheat readings. These problems are covered in detail in the chapter on refrigerant-side troubleshooting.

In summary, the suction superheat on fixed-metering-device systems varies greatly under partial load conditions, so it is important to follow the manufacturer's instructions, using the charts provided for the system being tested.

The superheat method is a very accurate means of checking the refrigerant charge (Figure 6-9). A change of about 1% in refrigerant charge will change the superheat 3°F or more.

In making these tests it is important to use a fast-reading resistance thermometer or an electronic temperature probe. Tape the sensing element in good contact with the suction line. Tape and insulate the probe to the line.

The suction-superheat charging method is an excellent tool for checking the charge of a system. Systems with thermostatic expansion valves should have a superheat in a range of 15° to 20°. Systems with fixed metering devices must use a superheat chart that includes the outdoor temperature and indoor wet-bulb and dry-bulb temperatures. Some charts do not include the wet-bulb temperature. The

wet-bulb temperature is more important than the dry-bulb temperature; therefore, do not use a chart unless it includes the wet-bulb temperature for determining superheat on fixed-bore metering devices.

DISCHARGE-SUPERHEAT METHOD OF CHECKING A CHARGE

The discharge-superheat method of checking the charge on an air-conditioning system should be considered a good charging procedure that must be used with other charging methods. It is considered a good charging method even though it has a wide range of acceptable temperatures. Some manufacturers rate the discharge-superheat method very highly. This method is used to verify that the equipment is charged properly. This procedural check, like all other checks previously discussed, gives an indication of the proper operating characteristics as well as a verification that the unit is charged correctly. Again, use as many of these charging methods as possible to verify the accuracy of the charge. The

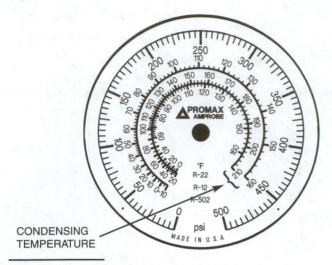

CONDENSING
TEMPERATURE

Figure 6–10 High-side gauge used to determine the condensing temperature. *(Courtesy of Amprobe/Promax)*

true professional technician takes the time and uses various charging methods as troubleshooting tools in determining a problem in an air-conditioning system. It does take time.

DISCHARGE SUPERHEAT DEFINED

Discharge superheat is defined as the difference between the discharge-line temperature and the condensing temperature. The discharge superheat should range from 50°F to 120°F.

To determine the discharge superheat, simply measure the discharge-line temperature and subtract the condensing temperature that is indicated on the high-side gauge (Figure 6–10). Remember to tape the temperature probe to the discharge line and insulate the probe. Be careful; the discharge line will be hot. You need an accurate thermometer that can read up to 250°F. A pocket thermometer will *not* do. Pocket thermometers have their place in air conditioning. They do a good job in measuring air and fluid temperatures, but they cannot make good line contact for measuring refrigerant piping temperatures. The correct gauge and temperature probe hookup is shown in Figure 6–11.

Figure 6–11 Measuring discharge superheat. Subtract the condensing temperature from the discharge line temperature.

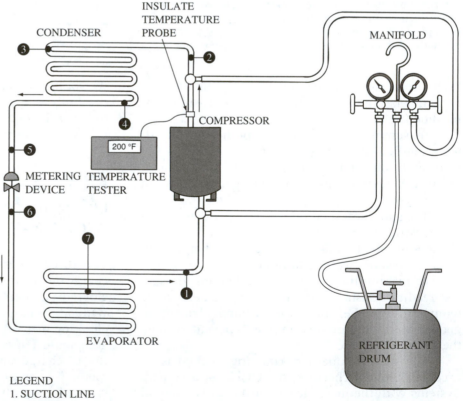

LEGEND
1. SUCTION LINE
2. DISCHARGE LINE
3. CONDENSER COIL
4. CONDENSER OUTLET
5. LIQUID LINE
6. DISTRIBUTOR
7. EVAPORATOR COIL

WORKING EXAMPLE

The discharge line temperature is 200°F and the condensing temperature is 110°F (Figure 6–13). The formula for this is

$$\text{Discharge superheat} = \text{discharge-line temperature} - \text{condensing temperature}$$

SOLUTION

200°F	discharge-line temperature
−110°F	condensing temperature
90°F	discharge superheat

A reading of 90°F discharge superheat falls within the acceptable range of 50°F to 120°F. ■

If the discharge superheat were less than 50°F, the technician would suspect cooler-than-normal refrigerant returning to the compressor, thereby cooling the compression process. Actually, droplets of liquid refrigerant returning to the compressor could be causing this condition of low discharge superheat. So whatever is causing low suction superheat will be causing low discharge superheat.

Low discharge superheat could be caused by

- An overcharge
- A dirty filter or evaporator
- Low airflow across the evaporator
- An overfeeding thermostatic expansion valve

Again, it is important to check the condition of the filter, evaporator, and condenser prior to checking the charge. High discharge superheat, in excess of 120°, could be caused by

- An undercharged condition
- High superheat

Figure 6–12   Flowchart for determining discharge superheat.

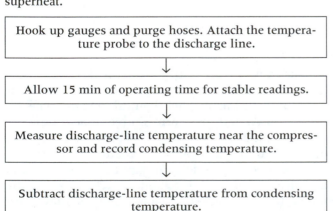

- High condensing temperature
- Dirty condenser
- Underfeeding thermostatic expansion valve
- Lack of lubrication in the compressor
- High compression ratios

WORKING EXAMPLE

The following information is provided on an unidentified 7-ton rooftop package unit:

- The system was cleaned prior to checking the charge, and airflow is 2800 CFM.
- The condensing and evaporating temperatures are in range.
- The system uses a thermostatic expansion valve with a measured suction superheat of 8°F.
- The condensing temperature is 120°F and the discharge-line temperature is 160°F.

According to this information, is the system charged correctly?

SOLUTION The system is overcharged.

Analysis The condensing and evaporating temperatures are in range, which is necessary prior to checking superheat. The 8°F suction superheat is low; it should be in a range of 15°F to 20°F. Low suction superheat is an indication of an overcharge. Low superheat can be caused by conditions other than overcharge. The unit was cleaned and the proper airflow of 400 CFM per ton was established. The discharge superheat is 40°F, which tells the technician that more than enough refrigerant is returning to cool the compressor. As expected, the discharge superheat below 50°F tracks the low suction superheat reading. An overcharged system is the answer. Other charging methods should be used to verify this conclusion. ■

COMPRESSORS DESIGNED FOR LOW DISCHARGE SUPERHEAT

A few semihermetic compressor manufacturers have designed some of their compressor models to keep a cool discharge temperature and, therefore, a low discharge superheat. These few models will have low discharge superheat under normal operating conditioning. Some models have low discharge superheat even if the suction superheat is high. These compressors are not common, but they may be found in systems that are designed for hotter running refrigeration compressors but are used in an air-conditioning application.

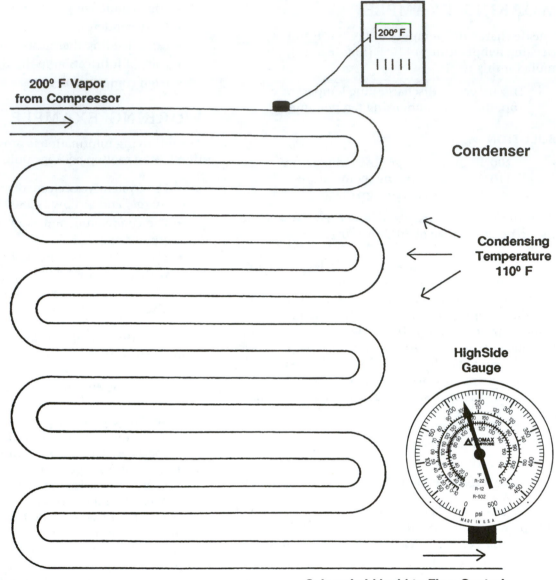

**200° F Vapor
from Compressor**

Condenser

**Condensing
Temperature
110° F**

**HighSide
Gauge**

Subcooled Liquid to Flow Control

200° F – 110° F = 90° Discharge Superheat

Figure 6–13   Measuring discharge superheat.

DISCHARGE-LINE TEMPERATURE

It is important to mention that the discharge-line temperature should not exceed 250°F. This is not discharge superheat but the actual discharge-line temperature. A discharge-line temperature in excess of 250°F means that the internal point of compression inside the compressor may be above 300°F. Operating the system at this extreme temperature will cause the refrigerant lubrication to break down

in the compressor. Lubrication breakdown translates into increased mechanical wear on compressor parts and, ultimately, mechanical seizure. The oil breaks down and may form sludge or acid or cause copperplating on moving mechanical parts. A lack of lubrication may cause mechanical compressor lockup. Mechanical seizure may be reflected as an electrical burnout. A mechanical lockup causes the motor to draw lock-rotor amps, which may burn out the motor windings. Poor lubrication results in the bearing surfaces becoming worn. Worn bearings cause the rotor to drag on the stator. This causes the rotor of the motor rotor to cut into the stator windings. In either case, the motor windings are damaged. The damage will be diagnosed as an open or shorted compressor-motor winding. Even though it was a mechanical problem—lack of lubrication—it will be diagnosed as an electrical failure of the motor windings. An autopsy on the compressor would reveal the real problem.

If the discharge-line temperature is above 250°F, the system has a problem. The problem is a lack of heat removal in the condenser, compressor overheating due to an improper lubrication, or lack of refrigerant cooling. This problem should be corrected before moving on with the discharge superheat check. Incidentally, some compressor manufacturers have maximum discharge temperature recommendations as low as 225° or as high as 275°. The temperature 250° is used as an industry average. Discharge-line temperatures above 200° should be investigated. The temperature at the point of compression will be higher than the discharge-line temperature. The discharged gas cools somewhat by the time it gets to the discharge line.

SUMMARY

- In summary, the second superheat method compares the condensing temperature and the discharge-line temperature. To find discharge superheat, subtract the condensing temperature from the discharge-line temperature. Discharge superheat should be in the range 50°F to 120°F.

- Suction superheat and discharge superheat track each other. High suction superheat will cause high discharge superheat. Low suction superheat will cause low discharge superheat.

- Do not confuse discharge superheat and the importance of knowing the discharge-line temperature. The discharge-line temperature should not exceed 250°F; this is not superheat. When a technician checks discharge superheat, he or she is making two important checks on the air-conditioning system: discharge superheat and the discharge temperature at the point of compression, as indicated by the discharge temperature.

PROBLEMS AND QUESTIONS

1. What is superheat?
2. Describe how to measure suction superheat.
3. Describe how to measure discharge superheat.
4. List the reasons why it is important to know the suction superheat of an air-conditioning system.
5. List the reasons why it is important to know the discharge superheat of an air-conditioning system.
6. What is the correct suction superheat for an air-conditioning system that uses a thermostatic expansion valve (TXV)?
7. A system with a low suction superheat indicates what problem(s)?
8. A system with a high suction superheat indicates what problem(s)?
9. A system with a low discharge superheat indicates what problem(s)?
10. A system with a high discharge superheat indicates what problem(s)?
11. A system with a high discharge temperature indicates what problem(s)?
12. An air-conditioning system with a piston-type metering device shows a suction superheat of 20°F. The technician measures the following temperatures: outdoor, 85°F; indoor dry-bulb, 78°F; a wet-bulb, 65°F. Is the system charged correctly? If not, what is the problem?
13. An air-conditioning system with a capillary tube shows a suction superheat of 15°F. The technician measures the following temperatures: outdoor, 95°F; indoor dry-bulb, 73°F; wet-bulb, 67°F. Is the system charged correctly? If not, what is the problem?
14. An air-conditioning system with a TXV shows a suction superheat of 7°F. The technician measures the following temperatures: outdoor, 100°F; indoor dry-bulb, 70°F; wet-bulb, 67°F. Is the system charge correctly? If not, what is the problem?
15. A technician found that the superheat of an air-conditioning system was correct. What discharge superheat should the technician expect?
16. Why is suction superheat below 5°F dangerous to a compressor?
17. What happens to the suction superheat in a capillary-tube system as the relative humidity drops in a building?
18. What happens to the suction superheat in a piston-type system as the outdoor temperature drops?
19. What is the purpose of a sling psychrometer?
20. A technician recovers refrigerant from an overcharged air conditioning system until the charge is perfect according to the suction superheat method of charging. After 20 min of operation the suction superheat is increased by 6°F over the previous reading. What is the percentage of reduction in the system's charge?

CHAPTER SEVEN

CHARGING PROCEDURES: SUBCOOLING, AMPERAGE DRAW, AND MANUFACTURING RECOMMENDATIONS

AFTER STUDYING THIS CHAPTER, THE STUDENT WILL BE ABLE TO:

- Define subcooling.
- Describe how to measure subcooling.
- State why is it important to know subcooling.
- Determine the correct subcooling temperature.
- Define RLA and LRA.
- Describe how to measure compressor amperage.

- Determine the correct compressor amperage.
- State why is it important to know the compressor amperage.
- Read manufacturers' charging charts.
- Explain the advantages of using a charging chart.
- Describe where to find a charging chart.

SUBCOOLING CHARGING METHOD

SUBCOOLING DEFINED

Subcooling is defined as cooling the liquid refrigerant below its condensing temperature. Subcooling begins at a point in the condenser at which all the refrigerant has been condensed to a liquid. Subcooling occurs near the end of the condenser as the liquid refrigerant makes its final passes through the condenser tubing.

The condenser receives hot, superheated vapor from the compressor. The condenser removes heat from the refrigerant, and the desuperheating process begins. This is a sensible heat process, and the temperature drop of the refrigerant can be measured. The heat from the refrigerant is added to the air or water passing over the condenser. The total amount of heat removed by the condenser equals the amount of heat released by the refrigerant.

Near the center of the condenser the refrigerant is cooled enough that it is converted from a hot,

high-pressure gas into a warm, high-pressure liquid. This condensing process, a latent-heat process, is where most of the heat is removed from the refrigerant. This condensing process is also known as the condensing temperature. The condensing temperature can be found on the inner ring of the high-side gauge (Figure 7–1). The condensing temperature can also be found by taking the condensing pressure and converting it to a temperature on a pressure-temperature chart.

The warm liquid refrigerant proceeds through the condenser, where it is further cooled. Further cooling of this liquid refrigerant is called subcooling, so the liquid is sometimes referred to as a subcooled liquid. The high-pressure, subcooled liquid leaves the condenser through the liquid line. The liquid line feeds the flow control this warm, high-pressure, subcooled liquid refrigerant (Figure 7–2).

Checking subcooling is considered a good charging method. It is valuable because it checks the operation of the condenser and restrictions in the liquid line. Some manufacturers have recommended

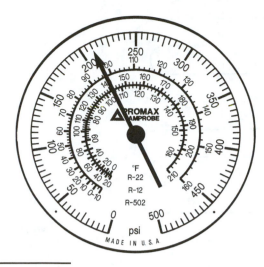

Figure 7–1 Condensing temperature is found on the inner rings for R-22, R-12, and R-502. In this example, using R-22, the pressure indicates 211 psig with a condensing temperature of 104°F. *(Courtesy of Amprobe Promax).*

levels of liquid subcooling set forth in their charging instructions. Having the manufacturer's recommended subcooling temperature helps verify a correct charge. Many manufacturers consider the subcooling method to be an excellent way to verify the system charge after the pressure-temperature conditions are met.

CORRECT SUBCOOLING

Unless specified by the manufacturer, subcooling will fall in the range 10°F to 20°F. Many manufacturers recommend 15°F of liquid subcooling as the ideal subcooling temperature. You will find out that the "ideal" is difficult to achieve in air conditioning and a range of 10° to 20° subcooling may be acceptable. For systems with a TXV, a narrower band of subcooling should be used.

REQUIRED SUBCOOLING FOR TXV SYSTEMS

SEER Rating	Subcooling
10–12	10°F
8.5–9.5	15°F
7–8	20°F

Equipment needs are as follows:

- Refrigeration gauge set
- Accurate digital thermometer
- Pipe insulation
- Tape

HOW TO MEASURE SUBCOOLING

Remember that subcooling is the difference between the condensing temperature and the liquid-line temperature.

1. The equipment should be clean. Clean the condenser coil before taking these measurements. Hook up the refrigeration gauges. Purge air from the hoses. Operate the unit for 15 to 20 min. The coil should be dry if cleaned. Determine refrigerant type from the unit nameplate (normally R-22).

Condensing and evaporating temperatures should be within ±10% of the targeted temperatures, as outlined in the pressure-temperature method of checking a charge. This was discussed in a previous chapter.

2. Measure the temperature of the liquid line at the condenser with a digital thermometer. Do not use a pocket thermometer. Tape the thermometer to the liquid line. Insulate the temperature probe while measuring this temperature. If your liquid line is longer than 25 ft, take the liquid-line temperature near the flow control. With short or long liquid lines, measuring subcooling near the flow control is the most accurate way to obtain this reading.

3. Subtract the difference between the condensing temperature and liquid-line temperature. Condensing temperature is found on the high-pressure gauge. The condensing temperature will be equal to or greater than the liquid-line temperature.

4. *Double-check* all readings.

WORKING EXAMPLE

- R-22 refrigerant (Figure 7–2)
- High-side pressure, 211 psig
- Condensing temperature, 104°F
- Measured liquid-line temperature, 90°F

SOLUTION

Subcooling = condensing temperature −
liquid-line temperature

104°F	condensing temperature from gauge
−90°F	liquid-line temperature from thermometer
14°F	subcooling

Subcooling should be in the range 10°F to 20°F unless otherwise indicated. In this example, the subcooling is OK This subcooling will assure the technician that an adequate amount of liquid refrigerant enters the metering device. It is important to have a 100% stream of liquid refrigerant to the evaporator.

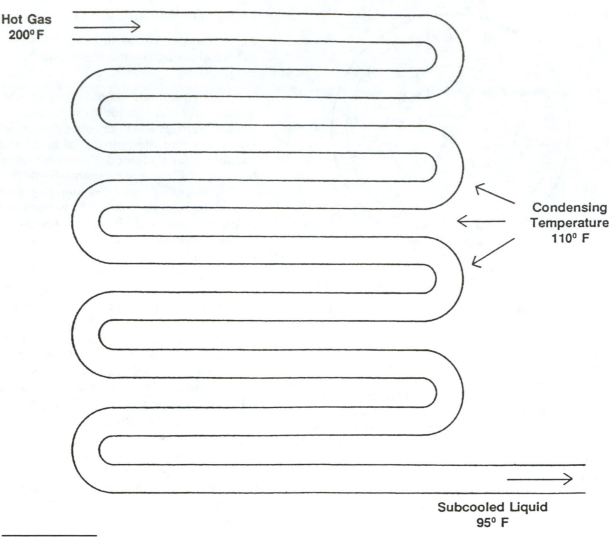

Hot Gas
200° F

Condensing
Temperature
110° F

Subcooled Liquid
95° F

Figure 7–2 Discharge gas temperature, condensing temperature, and liquid temperature. Refrigerant changes state from gas to liquid as heat is removed by the condenser coil.

Figure 7–3 Flowchart for determining subcooling.

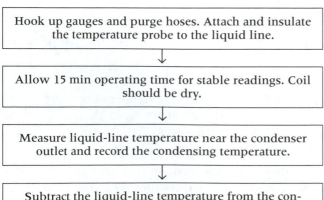

Hook up gauges and purge hoses. Attach and insulate the temperature probe to the liquid line.

↓

Allow 15 min operating time for stable readings. Coil should be dry.

↓

Measure liquid-line temperature near the condenser outlet and record the condensing temperature.

↓

Subtract the liquid-line temperature from the condensing temperature. Adjust the charge to correct the subcooling temperature.

This liquid refrigerant removes most of the BTUs in the evaporator. The cold vapor in the evaporator removes a very small amount of heat from the air or water passing over it. Figure 7–3 shows a flowchart on how to set up and measure subcooling.

WHAT SUBCOOLING TELLS US ABOUT THE SYSTEM

Subcooling tells us how a condenser is functioning. Subcooling over 20°F indicates one of the following:

- An overcharge
- A liquid-line restriction
- A flow-control restriction

- That air or water passing through the coil is too cold

Subcooling below 10°F indicates one of the following:

- An undercharged condition
- A condenser that is not removing enough heat from the refrigerant
- A dirty condenser
- Inadequate condenser air flow or water flow

HIGH-SYSTEM SUBCOOLING

Subcooling increases above 20°F when a system is overcharged or a liquid line is restricted. Liquid begins to back up toward the entrance of the condenser as the refrigerant is overcharged or restricted. When a system is overcharged, the liquid has to be stored somewhere in the refrigeration system, so it is stored in the evaporator and the condenser. In the condenser, the extra liquid takes up space once used for high-pressure gas. There is less space for the discharge gas. The reduced area of refrigerant gas causes an increase in the pressure in the condenser. The condensing pressure and condensing temperature rise. The condensing pressure and condensing temperature stabilize when the amount of refriger-

ant entering the condenser equals the amount of refrigerant being condensed by the condenser. In the case of an overcharge or liquid-line restriction, there is extra liquid in the condenser. The condenser has more coil area for subcooling because excess liquid backs refrigerant into the condenser (Figure 7–4). The longer the liquid is in the condenser, the more it is cooled before leaving; therefore, greater subcooling is obtained.

LOW-SYSTEM SUBCOOLING

The opposite is true of a condenser that is undercharged, is dirty, or has a lack of air or water flow. In the case of an undercharge, there is less liquid refrigerant in the system. Because of the undercharge, there is more refrigerant gas than liquid in the condenser. This reduces the amount of space available for subcooling the liquid refrigerant. In an undercharged condition, the refrigerant condenses closer to the end of the condenser; therefore, it has less coil surface over which subcooling can occur. As stated earlier, the condensing pressure and condensing temperature stabilize when the amount of refrigerant vapor entering the condenser equals the amount of liquid refrigerant being condensed by the condenser. In the case of an undercharge, the pressure is lower, which means that the condensing temperature is

Figure 7–4 Condenser coil with vapor entering from the compressor and gradually becoming a solid stream of liquid refrigerant. In high subcooling, the liquid occurs earlier compared to the low subcooling state.

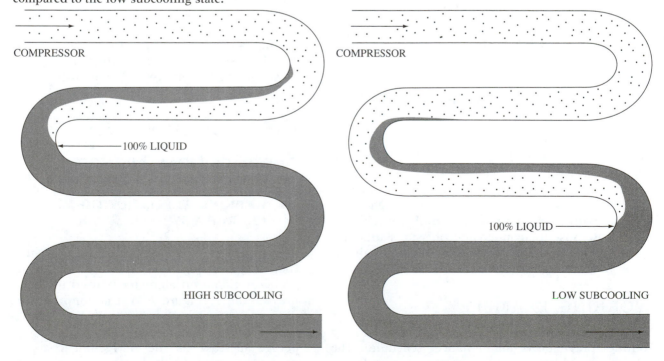

COMPRESSOR

100% LIQUID

HIGH SUBCOOLING

COMPRESSOR

100% LIQUID

LOW SUBCOOLING

lower. Lower condensing temperatures are closer to the liquid-line temperature, which translates into lower system subcooling. There is less heat transferred into the air or water passing through the condenser as the condenser temperature lowers. The closer the temperature difference or delta-T between condenser and air the lower the heat transfer rate.

When a condenser is dirty or fouled or if there is a lack of air- or water flow through the condenser, then the subcooling will be low. The condenser does not have enough heat-removal capability if these conditions occur. The condenser becomes a poor heat exchanger. These conditions do not allow the refrigerant to condense at the normal condensing temperature. The condensing pressure and condensing temperature are higher than normal. The temperature of the liquid refrigerant in the condenser is higher than normal because the heat-removal capabilities have been reduced. Calculated subcooling will be below normal or below 10°F.

WORKING EXAMPLE

An air-conditioning technician is checking the charge on an air-conditioning system. The system was cleaned; the evaporating and condensing temperatures are within range according to the pressure-temperature method. The following information is provided on the unit:

> The system uses R-22 refrigerant.
>
> The high-side pressure is 250°F, which has a corresponding temperature of 117°F.
>
> The liquid-line temperature was measured at 90°F.
>
> Is the subcooling correct?

SOLUTION

$$\text{Subcooling} = \text{condensing temperature} - \text{liquid-line temperature}$$
$$27°F = 117°F - 90°F$$

The measured subcooling is too high. The system could be overcharged or have a restriction anywhere in the liquid line, including the metering device. Checking the superheat and using other charging methods will help determine the problem with the system. Because the condensing pressures and temperatures are good, it is not likely that the air- or water flow through the condenser is too cold. ■

WORKING EXAMPLE

Another air-conditioning system was checked on a routine preventative-maintenance schedule. The system was cleaned and the evaporating and con-

densing temperatures were in range. The following information is provided on the unit:

> The system uses R-22 refrigerant.
>
> The condensing pressure is 230 psig, which corresponds to 111°F.
>
> The liquid-line temperature was measured at 103°F.
>
> Is the subcooling correct?

SOLUTION

$$\text{Subcooling} = \text{condensing temperature} - \text{liquid-line temperature}$$
$$8°F = 111°F - 103°F$$

Subcooling is too low. The system may be undercharged. The air-cooled condenser was cleaned; a dirty condenser is not the problem. A water-cooled condenser may be fouled, thus reducing heat transfer and reducing subcooling. The air- or water flow may not be adequate to produce the correct subcooling. Use other charging methods to verify the results.

■

SUMMARY

In summary, subcooling is a good way to check the system charge; at the same time it conducts a performance check on the condenser. Generally, subcooling should be between 10°F and 20°F. Subcooling for systems with a TXV will vary with the SEER rating. Higher SEER ratings accept lower subcooling temperatures. Some equipment manufacturers have specific subcooling requirements, but they are very close to the subcooling range suggested. Finally, the subcooling charging method can be used to determine the overall performance of the system. Subcooling is used as a diagnostic troubleshooting tool and is discussed again in the chapter on troubleshooting refrigerant-side problems.

AMPERAGE-DRAW METHOD

AMPERAGE-DRAW METHOD OF CHECKING THE CHARGE

> WARNING: THIS IS NOT THE WAY TO CHARGE A SYSTEM.

The amperage-draw method requires a technician to check the amperage draw of a compressor. An Amprobe or clamp-on ammeter is used to check compressor amps. (Figure 7–5) It is important to know the amperage draw of the compressor, but a unit should *not* be charged simply by watching the amperage increase as the refrigerant charge is increased. Charging only by checking the amperage

Figure 7–5 *(Courtesy of Amprobe Promax)*

draw is a poor way to charge an air-conditioning system. Checking the amperage should be used in conjunction with several of the other methods discussed in this textbook.

Before proceeding with this check, you must know the maximum amperage the compressor should draw. This information may be on the condensing unit nameplate or the compressor nameplate. If the compressor has been changed, then the condensing unit amperage may not reflect the amperage draw of the replacement compressor. Many compressors do not have the amperage rating on the compressor nameplate.

RLA AND FLA

The maximum amperage the compressor should draw is stated as

RLA = rated-load amps
or
FLA = full-load amps

In most instances the RLA or FLA should not be exceeded. Exceeding this amperage rating will shorten the life of the compressor motor. The motor may experience an electrical burnout. At a minimum, excess amperage causes the compressor motor to overheat. The overheated motor will temporarily cut out on its overload protective device. The overload de-

vice can be external or internal to the compressor motor windings.

LOCKED-ROTOR AMPERAGE

The compressor has another amperage rating called locked-rotor amps, or LRA. Do not confuse LRA with RLA or FLA. LRA is the amperage a motor will draw when it is mechanically locked up.

LRA is instantaneously drawn when a motor first starts its rotation from the standing position. Many compressor nameplates have the locked-rotor amperage but not the rated-load amperage. Do not select the LRA as the amperage to charge to. On most compressors the LRA is four to six times higher than the rated-load amperage.

Also, check the supply voltage to the condensing unit. Generally, most units accept 208 V through 240 V. The acceptable voltage range may be as wide as 189 V to 264 V. A ±10% supply-voltage variation should not damage the compressor and its assorted electrical components. Measure the voltage while the unit is in operation. Low-voltage conditions will result in higher amperage draw.

MEASURING COMPRESSOR AMPERAGE

Equipment needs are as follows: Clamp-on ammeter (for example, Amprobe). Using an Amprobe, measure the amperage of each of the three wires going into the terminals on a single-phase compressor. Record these amperage readings. Assuming that the compressor is wired for single-phase operation, the highest of the three amperage readings will be the RLA or FLA of the compressor. The compressor amperage will vary with the refrigerant charge in the unit and the operating pressures. The greater the charge, the higher will be the amperage draw. The higher the operating pressures, especially the high-side pressure, the higher will be the amperage draw.

There are three wires going to a single-phase compressor, the common, start, and run wires. The common wire has the highest amperage draw, followed by the run and start wires. The common wire draws the amperage you are interested in documenting (Figure 7–6). This amperage should not exceed the RLA or FLA of the compressor or condenser nameplate rating.

MEASURING AMPERAGE ON THREE-PHASE COMPRESSORS

Measuring amperage on three-phase compressors is a little different when compared to single-phase compressors. Three-phase compressors have three sets of motor windings supplied by three hot power

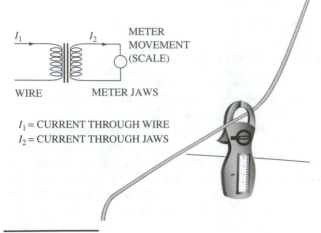

I_1 = CURRENT THROUGH WIRE
I_2 = CURRENT THROUGH JAWS

Figure 7–6   Analog clamp-on Amprobe ammeter measuring motor current. *(Courtesy of Amprobe Promax)*

sources. On a three-phase compressor, all three power wires are individually checked for amperage. The amperage of all three motor windings should be close to the same. A small amount of amperage variation is allowed between the three power wires. This amperage deviation is discussed in the chapter on troubleshooting electrical-side problems. Additionally, the supply voltage should be checked between each of the phases when the compressor is operating. A condition of low or high voltage will affect the amperage draw of the compressor motor.

COMPRESSOR CHANGE OUT

Sometimes a compressor has been changed out in a condensing unit and either the RLA or the FLA is lower or higher than that shown on the condenser nameplate. The amperage rating needs to be obtained from the air-conditioning supply house or the compressor manufacturer.

CONDITIONS THAT AFFECT AMPERAGE

An air conditioner low on charge will draw lower amperage. An air conditioner that is overcharged might draw amperage above the RLA or FLA. An overcharged air conditioning system will *not* always exceed the RLA or FLA rating of the compressor.

Another factor that changes the amperage draw of the compressor is the condensing pressure. The amperage draw will be less when the outdoor temperature is cooler and the condensing pressure is lower. As the temperature rises, so does the amount of amperage the compressor draws. The compressor works harder because of increased compression pressure. It is always better to keep the compressor amperage draw as low as possible. The lower am-

perage draw will translate into longer compressor life and lower utility bills.

The pumping capacity of the compressor as well as the condition of the motor windings also influence the amperage draw. A compressor that has broken internal mechanical parts may still operate, but it will draw lower amperage because it is pumping less refrigerant.

Compressor motor windings may also be defective, causing a higher-than-normal amperage draw. This defect is a winding-to-winding short. The short may not be severe enough to cause the overload to open the common winding. It may be manifested as high-amperage draw.

The compressor has an internal or external overload, over-amperage protection device that will open if the compressor amperage draw is excessive or if the motor windings are too hot. If the compressor cuts out on its overload, the compressor motor will need to cool a few minutes (sometimes a few hours) before this overload device will automatically reset itself and allow the compressor to restart. This overload condition will not affect the outdoor condenser fan motor. In most cases, the fan will continue to run, and it will be difficult to determine if the compressor is operating without the gauges hooked up and the clamp-on ammeter attached to the compressor winding.

The amperage draw method is a poor way to charge an air-conditioning system, but its information will be useful in diagnosing problems.

SUMMARY

Remember: Do not charge the unit up to the RLA or FLA. The only time the compressor should come to this amperage rating is on the hottest day (100°F) or if the compressor is overcharged or defective. Amperage draw in excess of the compressor rating will interrupt the voltage supply to the compressor via an open overload. The overload will reset itself upon cooling. Continuous overloading of the compressor will cause it to overheat. Intervals between resetting of the overload will become greater as the compressor becomes hotter. Sometimes several hours are needed to cool a large compressor, thus allowing the overload time to reset.

Defective or loose wiring can cause high amperage draw. Burned wiring or a loose connection increases the resistance in a circuit, thereby increasing the amperage flow. Inspect the wiring at the compressor, contactor, and in the circuit-breaker panel. Check the contactor for burned contacts. Burned or carbonized contactor points will increase the current flow to the compressor. The resistance across a closed set of contacts should be less than one ohm. Remove power and wiring before conducting resistance checks. The voltage drop across an energized set of compressor contacts should be zero volts. Up to 5 volts is acceptable.

Check the wire connections on the capacitor. Additionally, check the capacitor. A defective run capacitor or improperly sized capacitor will cause high amperage draw and may prevent compressor startup. Many a good compressor has been diagnosed as defective because the run capacitor was bad or weak. Check all starting components before condemning a compressor.

Remember: **Charging a compressor by amperage draw will damage it! Use other charging methods.**

MANUFACTURERS' CHARGING RECOMMENDATIONS

Using the manufacturers' charging recommendations is by far one of the most accurate methods of checking the charge on an air conditioner.

The manufacturer may have a set of instructions, a chart, or graph inside the panel of the condensing unit. The weight of the refrigerant charge is beneficial when charging from a vacuum. If the information is not available on the inside panel of the unit, the technician will need to contact the manufacturer or supplier of the unit for this information. The information is usually included in the installation instructions and is found when installing new equipment. The weight of the total charge may be located on the unit nameplate. When requesting installation instructions, you will need to know the following information:

- Condensing unit name
- Condensing unit model number
- Condensing unit serial number
- Evaporator name
- Evaporator model number

If the manufacturer will not provide this information, contact a local supply house or contractor that handles that line of equipment. Make a file with charging instructions on each new unit installed. This will be a very valuable resource at the time of the installation and as a future reference when servicing the equipment.

What follows are examples of what you might find on a unit (Figures 7–7 and 7–8). Do not use this

Figure 7–7 Sample charging chart for ARI model condensing unit model 123 with evaporator model 456 or model XYZ.

OUTDOOR TEMPERATURE °F	INDOOR TEMPERATURE DB°F	WB°F	PRESSURES (psi) LIQUID	SUCTION
115	75	63	300–331	72–84
		67	305–333	76–78
		71	310–343	81–91
105	75	63	267–300	71–81
		67	272–303	76–88
		71	275–310	79–90
95	75	63	232–265	67–78
		67	237–269	72–83
		71	242–275	76–87
85	75	63	199–230	63–72
		67	202–234	67–78
		71	207–240	70–82
75	75	63	171–204	58–68
		67	175–207	62–72
		71	179–212	67–77
65	75	63	145–175	53–62
		67	148–181	57–67
		71	156–188	60–70

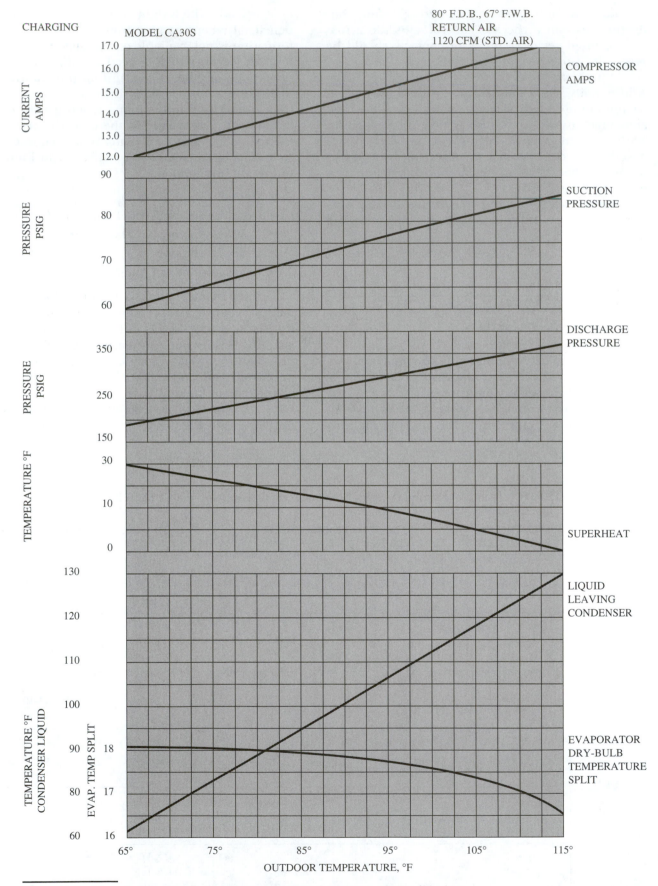

Figure 7–8   Manufacturer's charging chart. *(Courtesy of Armstrong Air Conditioning)*

charging information unless it matches the exact unit model number of the condenser and evaporator that you have.

Remember that the evaporator should be matched with the condensing unit. Mismatched components will invalidate the manufacturer's charging instructions. The filters, evaporator, and condensing coils should be clean prior to checking the charge. Inspect and clean the filter and evaporator if necessary. Clean the condenser coil, even if it seems clean. Finally, the refrigerant lines must be sized properly and the supply airflow sized for 400 cfm per ton. Figure 7–7 shows a chart with manufacturer's charging recommendations. This should be used only on the specific model specified. To use this chart a technician needs to know the following:

- Outdoor temperature entering the condensing unit
- Indoor dry-bulb (DB) temperature
- Indoor wet-bulb (WB) temperatures

From this information, a technician can use the charging chart and determine a range of recommended liquid and suction pressures.

WORKING EXAMPLE

Using the manufacturer's charging chart (Figure 7–7), the following information is measured by the technician:

- Outdoor temperature, 95°F
- Indoor dry-bulb, 75°F
- Indoor wet-bulb, 67°F
- Liquid-line pressure, 260 psig
- Suction-line pressure, 80 psig

According to the charging chart, is the system charged correctly?

SOLUTION: The answer is yes; the unit is charged according to manufacturer recommendations. The range of acceptable liquid-line pressure is 237–269 psig and the range of suction-pressure is 72–83 psig. ∎

Having the benefit of a charging chart is a plus. This information gives a technician confidence that the equipment is properly charged. Other charging methods should be used to verify the charge, even if the manufacturer's charging instructions are used. Figure 7–7 shows a wide range of liquid and suction pressures. An air-conditioning system may not be functioning at peak performance even after complying with this charging chart. The wide range of pressures allows for conditions that may affect good space

cooling. Use other charging methods, such as superheat and subcooling, to verify equipment operation. Compressor amperage should also be considered.

This is a sample charging chart and should be not be used with any equipment. Notice that the return-air indoor dry-bulb conditions must be at 75°F and the wet-bulb temperature must be in a range of 63° to 71°F. Indoor temperatures outside this range invalidate the information provided in the chart.

This chart has limited value to the technician because of the wide pressure ranges that are acceptable and the narrow range of return-air temperatures when using this chart. The next charging chart will be more valuable to the serviceperson and the operation of the air-conditioning system.

A BETTER CHARGING CHART

The more information a charging chart provides, the greater the likelihood the technician will leave the job without being recalled. There is a saying that states, "There is never enough time to do the job correctly the first time, but there is always plenty time to come back and do the job correctly a second time." This holds true in air conditioning, and a good technician will spend the time and do the job properly the first time.

The chart in Figure 7–8 provides a host of good technical information. Starting from the top of the charging chart, the technician can determine these target values:

- Compressor amperage
- Suction pressure
- Discharge pressure
- Suction superheat
- Liquid-line temperature (subcooling)
- Temperature split across the evaporator

WORKING EXAMPLE

This example uses the information provided in Figure 7–8. The technician notices that this charging chart is for Model CA30S. The chart shows that the return-air conditions are

- 80°F dry-bulb
- 67°F wet-bulb
- Airflow, 1120 cfm

Now locate the outdoor-air temperature at the bottom of the chart. For this example, use an outdoor-air temperature of 95°F, as measured at the condenser. On the chart, draw a line from the 95°F outdoor temperature up to the top of the chart. Starting at the top of the charging chart, a technician would expect a compressor amperage reading of 15.2 A or less; a suction pressure of 77 psig; a discharge pressure of 300 psig; a suction superheat of 10°F; a liquid-line temperature of 107°F; and a temperature drop across the evaporator near 18°F. ∎

ANALYSIS OF THESE CHARGING RESULTS

AMPERAGE RESULTS

The amperage recorded on the common winding of this compressor should be less than 15.2 A. There is no minimum amperage reading, but a very low amperage condition indicates a low charge or a compressor mechanical problem.

SUCTION-PRESSURE RESULTS

The suction pressure should be within ±2 psig of the 77-psig target. Lower than normal suction pressure indicates an undercharged condition, low-side, or liquid-line restriction or lack of airflow across the evaporator. Higher than normal suction pressure indicates an overcharged condition, overfeeding TXV, or high ambient or moisture-laden air entering the return-air stream.

DISCHARGE-PRESSURE RESULTS

The discharge pressure should be within ±5 psig of the 300-psig target. Lower than normal discharge pressure indicates an undercharged condition, wet condenser coil, or low ambient conditions. Higher than normal discharge pressure indicates an overcharge, dirty condenser, or lack of proper airflow across the condenser coil.

SUCTION-SUPERHEAT RESULTS

The superheat should be within ±2°F of the 10°F target temperature. Low superheat indicates an overcharged condition, lack of airflow across the evaporator, or overfeeding TXV. High superheat indicates an undercharged condition or liquid-line restriction, including the metering device.

LIQUID-LINE-TEMPERATURE RESULTS

The liquid line should be within ±2°F of the 107°F target temperature. This part of the chart establishes the correct liquid subcooling. Low liquid-line temperatures indicate an undercharged condition or low ambient air entering the condenser coil. High liquid-line temperatures indicate an overcharged condition, dirty condenser coil, or liquid-line restriction. This is not a subcooling temperature.

TEMPERATURE-DROP RESULTS

The temperature drop across the evaporator should be within ±2°F of the 18°F target temperature. Temperature drop below the target indicates a lack of charge or any condition that restricts the flow of refrigerant into the condenser. Higher temperature drops indicate an overcharged condition or lack of airflow across the evaporator. In some instances a lack of charge can create a high temperature drop for a short period of time. This happens because the evaporating temperature is below freezing, thus creating a very cold supply-air temperature. Eventually the coil will freeze over and block the airflow. Cooling capacity will be impaired without adequate refrigerant to the evaporator.

This charging chart gives a technician specific information about the operating characteristics of the system. Information as detailed as this charging chart is a great asset to the technician. The information will allow a system to be charged to peak performance and can be used as a troubleshooting tool as discussed in the chapter on refrigerant-side troubleshooting.

SUMMARY

- Manufacturers' charging recommendations give a technician charging information that has been developed and tested on a specific piece of equipment. The information cannot be generalized and used across model numbers. The information is specific, as stated by the manufacturer. When limited charging information is given, as in Figure 7–7, additional charging methods must be used to verify correct equipment operation.

CHAPTER QUESTIONS

1. What is subcooling?
2. Describe how to measure subcooling.
3. What is the difference between LRA and RLA?
4. What is the purpose of a charging chart?.
5. Explain how to measure rated-load amperage on a single-phase compressor.
6. Explain how to measure rated-load amperage on a three-phase compressor.
7. What is the recommended subcooling temperature range?
8. What are the causes of high subcooling?
9. What are the causes of low subcooling?
10. What are the causes of excessive compressor amperage?
11. What are the causes of lower than normal compressor amperage?

12. A compressor has been changed out in a condensing unit. The condensing unit nameplate shows that the original compressor had an RLA of 25 A and a LRA of 85 A. The replacement compressor has an LRA of 80 A. The RLA is not on the replacement compressor nameplate. Where can the technician find the RLA information?

13. What can happen to a compressor that draws amperage higher than rated-load amperes?

14. A technician needs a charging chart. What information will be needed if the technician calls the manufacturer for an e-mailed copy of the information?

15. Using the chart in Figure 7–7, what is the range of suction- and liquid-line pressures if the outdoor temperature is 105°F and the return-air temperature is 75°F with a wet-bulb of 71°F?

16. Using Figure 7–7, what happens to the liquid-line pressure as the outdoor temperature increases.

17. Using Figure 7–7 what happens to the suction-line pressure as the wet-bulb temperature increases?

18. Using Figure 7–8 list *all* the information that a technician can obtain from an outdoor-temperature reading of 85°F. Be specific.

19. Using Figure 7–8 list *all* the information that a technician can obtain from an outdoor-temperature reading of 105°F. Be specific.

20. Using Figure 7–8 what happens to superheat as the outdoor temperature decreases? Reviewing this charging chart, what type of metering device is used on this unit?

CHAPTER EIGHT

TEMPERATURE-DIFFERENCE, SIGHT-GLASS, AND SWEATBACK METHODS

AFTER STUDYING THIS CHAPTER, THE STUDENT WILL BE ABLE TO:

- Describe the temperature-difference charging method.
- Describe the sight-glass charging method.
- Describe the sweatback charging method.
- Discuss the advantages and disadvantages of each method.
- List the procedures for removing moisture from refrigerant.
- Describe how these charging methods can be used to troubleshoot.

INTRODUCTION

This chapter is a continuation of the chapters on charging procedures. The temperature-difference charging method is useful when checking all air-conditioning charges. The sight-glass method is useful only if the system has a thermostatic expansion-valve-metering device. The sweatback method is a limited charging technique. It is important to be exposed to all the charging methods, even charging methods with limited value.

As with the other charging methods, it is important that the air-conditioning system be clean and maintain the correct airflow of 400 CFM per ton. The condenser and evaporator should have matching capacity and efficiency ratings.

TEMPERATURE-DIFFERENCE METHOD

The temperature-difference, or delta-T or ΔT, method of charging simply compares the supply-air temperature to the return-air temperature. Delta-T is a shortcut term used to express the same meaning. *Delta*, the Greek symbol Δ, means *a difference*. Capital T is an abbreviation for temperature.

It is important to measure the supply air and return air at the proper places in the duct system. The return air should be measured as close to the evaporator as possible. The technician must puncture a hole in the return-air duct system near the inlet to the evaporator (Figure 8–1). The easiest, but not necessarily the most accurate, way to measure the return-air temperature is to place the temperature probe near the return-air grille. Any air leakage in return-air plenum will cause a higher temperature entering the evaporator. It is a good practice to simultaneously measure the temperature at the return-air grille and at the air inlet to the evaporator. The temperature difference between the reading should be less than 2°F. A greater temperature difference indicates warm air is being pulled into the negative pressure of the return-air plenum. In this instance the air entering the evaporator is higher than the air entering the return-air grille. Prior to

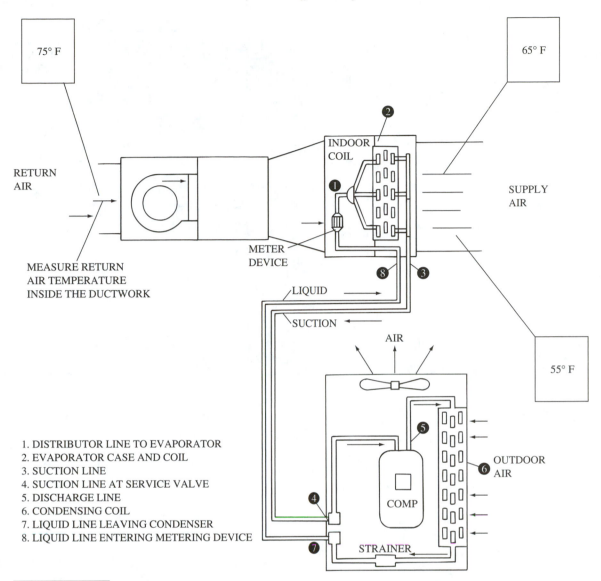

1. DISTRIBUTOR LINE TO EVAPORATOR
2. EVAPORATOR CASE AND COIL
3. SUCTION LINE
4. SUCTION LINE AT SERVICE VALVE
5. DISCHARGE LINE
6. CONDENSING COIL
7. LIQUID LINE LEAVING CONDENSER
8. LIQUID LINE ENTERING METERING DEVICE

Figure 8–1   Measuring temperature difference across the evaporator coil. The return-air (RA) temperature is fairly constant. The supply air temperature near the evaporator coil can vary widely in temperature. It is best *not* to measure the supply air in line of sight of the evaporator outlet.

doing this check, a technician should determine that the readings on the temperature probes register the same temperature.

The supply-air-measuring procedure is also important. Supply air should be measured near the outlet of the evaporator, but not within line of sight of the evaporator. The air coming off the evaporator should have an opportunity to mix in the supply-air plenum and be distributed down the supply-duct system. The airflow coming off any heat exchanger is not the same temperature (Figure 8–2). This is the case with an evaporator coil, gas furnace heat exchanger, or electric heating elements. The temperature of the air streams may vary by as much as 10°F. To obtain an accurate supply-air reading, the air tem-

perature needs to be mixed prior to being sensed by the temperature probe. An average air temperature needs to be measured for this charging procedure.

It is also important that the supply-air duct be measured at a point close to the evaporator outlet but not within line of sight of the coil. Some technicians measure the supply-air temperature at the register outlet. This is a good measurement provided it is from the shortest duct run and that the probe is placed inside the register opening. Measuring the supply-air temperature near, but outside, the register opening will give an inaccurate reading because some of the room air will influence the final reading. Place the temperature probe inside the throat of the duct outlet.

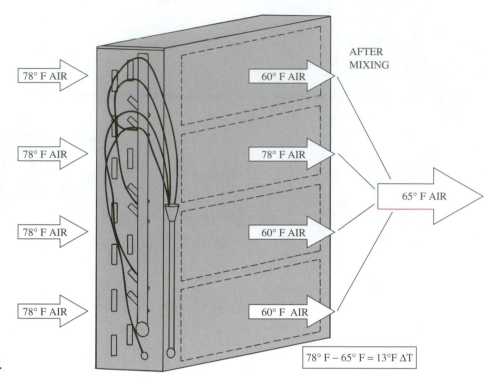

Figure 8–2 Temperatures of air leaving the coil. The average temperature after mixing is 65°F.

THE CORRECT DELTA-T

A technician is looking for a delta-T in a range of 13° to 24°F. Ideally a tighter range of 16° to 18°F is expected under normal operating conditions. Normal operating conditions are defined as a return-air temperature in the range of 72° to 78°F with a relative humidity of 40% to 60%. For an accurate determination of the temperature split, use the chart in Figure 8–3. In order to use this chart, measure the dry-bulb temperature and relative humidity of the return air. Using the chart, plot the dry-bulb temperature on the vertical axis on the left side of the chart. Next, draw a line to the right until it crosses the measured relative humidity. From this point, draw a line straight down to the horizontal axis at the bottom of the chart. The temperature at the bottom of the chart is the target delta-T. The measured delta-T should be within 2°F of this reading.

The chart indicates a greater temperature split as the relative humidity or latent load in the structure decreases. In this instance less of the system's cooling capacity is going to removing moisture, which allows more sensible heat removal or a greater delta-T. A greater temperature split is realized when the structure operates at a higher indoor temperature while maintaining a constant relative humidity.

A temperature difference below 13°F may indicate a low charge or any other condition that would create an inadequate amount of refrigerant in the evaporator. The lack of refrigerant in the evaporator can be caused by a multitude of problems and is covered in detail in the chapter on refrigerant-side troubleshooting.

Figure 8–3 Air-temperature drop for various return-air conditions.

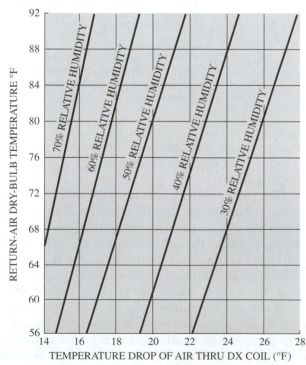

Figure 8–4 Electronic flow hood. *(Courtesy of Shortridge Instruments)*

HIGH AND LOW DELTA-T

Under normal operating conditions, a temperature difference in excess of 20°F does not necessarily indicate an overcharge condition. A system that has low airflow is the most probable cause of a high delta-T. As the airflow slows through the evaporator, more heat is removed from the returning air. The supply air is discharged at a lower temperature, therefore increasing the temperature spread across the evaporator. The problem with this condition is that the evaporating temperature continues to drop as less heat is added to the refrigerant. Less air is flowing across the evaporator. Reduced airflow translates into longer run cycles. If the system runs long enough, the evaporating temperature will drop below freezing. Evaporating temperatures below 32°F will coat the coil with frost and, if the system is allowed to run long enough, will block the airflow. The evaporator will need to be defrosted before reasonably checking the charge. Thick ice can form on the evaporator. It will take time to defrost. Run the air handler and add heat to the return-air stream to reduce defrost time.

Use an airflow hood (Figure 8–4) or the following chart to determine if the airflow is 400 CFM per ton. To use the chart in Figure 8–5, measure the return-air wet-bulb and dry-bulb temperatures and the supply-air dry-bulb temperature. The return-air wet-bulb temperature is plugged into the top horizontal portion of the chart and the return-air dry-bulb temperature is plugged into the left-side vertical axis. The point at which the two temperatures intersect is the target supply-air temperature. If the leaving-air temperature is lower by more than 3°F, then the chart indicates that the airflow through the evaporator is significantly below 400 CFM per ton.

Depending on the equipment design, the delta-T on an overcharged system will peak around 20°F. Some equipment will peak at a lower delta-T and some, at a higher delta-T. Overcharged systems will begin to lose capacity. A technician should expect a lower temperature difference across the evaporator coil as the system is overcharged.

Figure 8–5 Airflow chart that identifies airflow problems.

PROPER AIRFLOW RANGE (COOLING)

	INDOOR ENTERING AIR WET BULB °F																		
	57	58	59	60	61	62	63	64	65	66	67	68	69	70	71	72	73	74	75
PROPER EVAPORATOR COIL LEAVING AIR DRY BULB °F																			
70	51	51	52	52	53	53	54	55	55	56	57	58	59	60	—	—	—	—	—
72	52	52	53	53	54	55	55	56	57	57	58	59	60	61	62	63	—	—	—
74	53	53	53	54	55	55	56	57	58	58	59	60	61	62	63	64	65	—	—
76	54	54	54	55	55	56	57	57	58	59	60	61	62	63	64	65	66	67	—
78	55	55	55	56	56	57	57	58	59	60	61	62	63	64	65	66	67	68	69
80	56	56	56	56	57	58	58	59	60	61	62	63	64	65	66	67	68	69	69
82	57	57	57	57	58	59	60	60	61	62	63	64	65	66	67	68	69	70	71
84	—	58	59	59	60	60	61	61	62	63	63	64	65	66	67	68	69	70	71

(Left vertical axis label: INDOOR ENTERING AIR DRY BULB °F)

NOTE: THIS TABLE IS BASED ON 400 CFM PER TON OF COOLING CAPACITY

A low charge can cause a temporary high temperature difference between the return and supply air. When the undercharged system first begins to operate, the evaporating temperature is low. The low temperature is reflected by very cold air coming off the evaporator. The delta-T will be above 20°F. Initially it will appear that the customer has a super air conditioner with temperature differences much higher that 20°F. The freezing evaporator will eventually be blocked by frost. Frost is caused by humid air contacting the freezing temperature of the evaporator. Frost acts as an insulator and reduces the heat-absorbing capabilities of the evaporator.

WORKING EXAMPLE

A technician measures 78°F dry-bulb at the return air grille at a point 12 in. from the inlet to the evaporator. The relative humidity is calculated to be 55%. The supply air is measured inside a supply-air register on the shortest duct branch from the supply-air plenum at 60°F. Is this delta-T satisfactory?

SOLUTION:

78°F	return-air temperature at a 55% RH
−60°F	supply-air temperature
18°F	temperature drop

According to the chart in Figure 18–3 an 18°F temperature drop is satisfactory because it is within ±2°F of the target delta-T. Other charging methods should be used to verify the charge.   ∎

OTHER PROBLEMS

Low supply-air temperatures can cause other problems. Low supply-air temperatures cause condensation problems. Moisture condenses on the surface of the cold supply-air outlets when the air handler cycles off. This moisture will begin to grow mildew plus attract dust to its moist surface. Additionally, some of the wall or ceiling adjoining the supply-air grille may experience this mildew growth due to contact with the abnormally cold air blowing across its surface. In some instances, drops of moisture may condense on the supply grille during the off-cycle. The water drop will fall to the floor, drain down the wall, or fall on furniture. Metal supply-air grilles may begin to rust. Having the correct charge, airflow, and temperature drop will prevent these problems.

SIGHT-GLASS METHOD

The sight-glass method of checking the charge is one of the easier charging methods. Using only the sight glass as a charge indicator is a poor practice. The sight glass (Figure 8–6) is located in the liquid line. In most installations, the sight glass is located in the liquid line near the condenser outlet. The ideal place for the sight glass is as near as possible to the TXV. The liquid line should feed the metering device a solid stream of subcooled liquid refrigerant; therefore, the sight glass should be located at the metering device. Two sight-glass locations would be best —one near the condenser and one near the metering device. Noncondensables or refrigerant vapor in the

Figure 8–6   Charging by sight glass with a TXV.

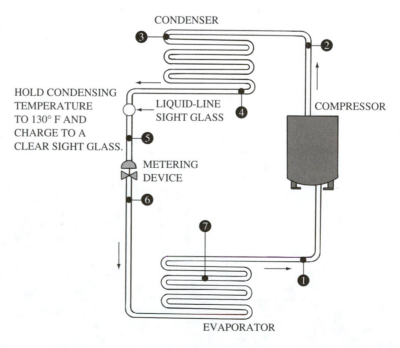

liquid line are seen as bubbles in the sight glass. Noncondensables can be in the form of air, nitrogen, or another type of refrigerant not compatible with the system design. A major restriction in the liquid line or drier upstream from the sight glass also produces bubbles in the sight glass. The restriction acts as a metering device and causes the refrigerant to boil or flash off to a vapor. The vapor is seen as bubbles in the sight glass.

CONDUCTING THE CHARGING PROCEDURE

The sight-glass charging method can be used only on air-conditioning equipment with thermostatic expansion valves. It should be used in conjunction with other charging methods. The system should be clean and the proper airflow established.

In order for a TXV to operate properly, it must have a solid stream of subcooled liquid. Generally, subcooling should be in a range of 10° to 20°F. Remember that the subcooling temperature is found by subtracting the liquid-line temperature from the condensing temperature. Fixed-bore devices such as capillary tubes or piston devices may feed liquid at lower subcooling levels.

To use the sight glass as a charging method, a technician will need to maintain a 130°F condensing temperature. The 130°F condensing temperature represents a charge on a 100°F day. For example, suppose the technician measures the air entering the condenser to be 100°F. Adding the standard rule-of-thumb 30°F to this temperature, the technician would expect a 130°F condensing temperature on a 100°F day.

If the air entering the condenser is not 100°F, the technician will need to simulate this condition. The best way to achieve and control the 130°F condensing temperature is to block the condenser coil with a piece of cardboard or sheet of plastic (Figure 8–7) until the condensing temperature increases to 130°F. At this point in the procedure, the sight glass should be clear with no indication of vapor bubbles. Watch the sight glass a few minutes to be sure that only liquid refrigerant is present. If bubbles are detected, you will need to add refrigerant while maintaining the condensing temperature at 130°F. The condensing temperature will rise as refrigerant is added to the system. If the condensing temperature rises above 130°F, you will need to remove a portion of the condenser covering until 130°F is maintained.

In review, the following apply to the sight-glass method:

- It is useful only with TXV systems.
- A 130°F condensing temperature must be maintained during the charge.
- The system must be charged to a clear sight glass.
- It can indicate moisture contamination of the refrigerant.

ADVANTAGES AND DISADVANTAGES

Of course, the advantage of this charging technique is that it is quick and visual. The setup of the charging procedure is important. Viewing a sight glass under normal operating conditions, a technician may see a few bubbles in the sight glass with a

Figure 8–7 Controlling condensing temperature by blocking condensing coil. *(Courtesy of Lennox Industries)*

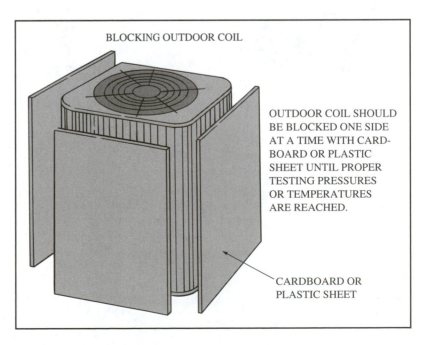

BLOCKING OUTDOOR COIL

OUTDOOR COIL SHOULD BE BLOCKED ONE SIDE AT A TIME WITH CARD-BOARD OR PLASTIC SHEET UNTIL PROPER TESTING PRESSURES OR TEMPERATURES ARE REACHED.

CARDBOARD OR PLASTIC SHEET

perfect charge. Charging up the system until the bubbles disappear may not provide an adequate subcooled charge sufficient for a TXV system. The system may still be undercharged. In an undercharged system, the technician should expect more sight-glass bubbles as the outdoor temperature rises. The condenser is not able to remove as much heat from the refrigerant as the ambient temperature

rises. A marginally charged system will begin to develop bubbles in the sight glass as the temperature rises. A system with the correct charge and adequate subcooling will not show bubbles in the sight glass at the hottest part of the day.

A clear sight glass may indicate a good charge, including the correct amount of subcooling. A clear sight glass may indicate a lack of liquid charge. A sight glass with only vapor will be clear similar to a sight glass with 100% liquid. A clear sight glass can be deceiving, which is why it is important to use other charging measures. Finally, a clear sight glass could indicate a seriously overcharged system.

SIGHT-GLASS MOISTURE INDICATORS

Most liquid-line sight glasses have a moisture indicator (Figure 8–8). The moisture indicator uses a substance that changes color with the content of moisture in the refrigerant. The colors used to indicate a dry and moisture-laden systems vary among sight-glass manufacturers.

On occasion, the moisture indicator will detect moisture when the system is dry. It is best to use another method of moisture detection prior to dehydrating the refrigerant. A moisture and acid test kit (Figure 8–9) can be used to determine the quality of the refrigerant. If the second moisture test indicates abnormal moisture content, then the system needs to be dehydrated. To bring the moisture level to an acceptable level, a technician needs to remove or recycle the refrigerant, change oil, evacuate the system, and change filter driers. Finally, the moisture content must to be checked again. A sight glass with a faulty moisture indicator needs to be changed the next time the system is opened for service.

Figure 8–8 Sight glass with moisture indicator *(Courtesy of Sporlan Valve Company)*

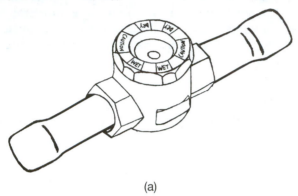

(a)

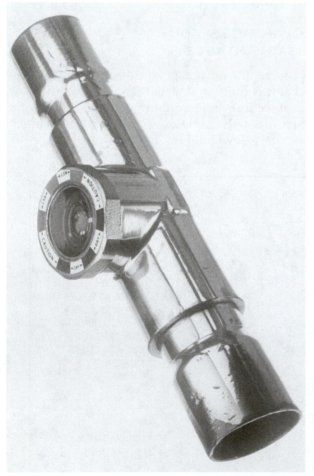

(b)

Figure 8–9 Acid and moisture test kit.

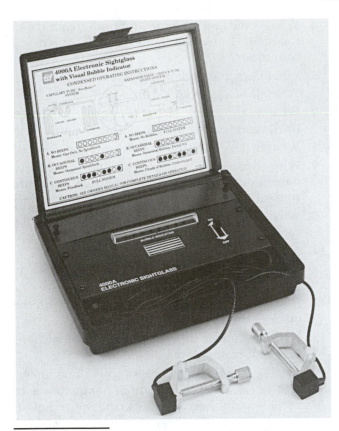

Figure 8–10   Electronic sight glass with both visual and audible bubble detection. *(Courtesy of TIF Instruments)*

ELECTRONIC SIGHT GLASS

Some TXV installations do not have a sight glass. A technician can use a piece of test equipment called an electronic sight glass (Figure 8–10) to detect bubbles. An electronic sight glass has two clamps that attach to the liquid line; it then monitors the condition of the refrigerant in the liquid line. Any bubbles in the liquid line will cause the electronic sight glass to signal with a light or audible sound. The electronic sight glass can be used on the suction line to indicate liquid refrigerant returning to the compressor.

SWEATBACK METHOD

The sweatback method of checking the charge is one of the poorer ways to determine if a system is correctly charged. Even though the procedure is poor, it does have an advantage or two.

In the sweatback method, a technician looks for condensation on the suction line. In some cases, the compressor shell is cold enough to sweat. The theory is that if there is enough cold refrigerant returning to the suction line, it must be properly charged. The colder the suction line, the more condensation or sweat the technician expects. There is a major problem with this theory.

PROBLEMS WITH THE SWEATBACK METHOD

It is important that the suction line be cool to the touch. A warm suction line will translate into high superheat. A suction line at ambient temperature means that the compressor is not operating or that the refrigerant charge is very minimal. A suction line that has a frost coating indicates problems in the refrigeration system including airflow problems. The reasons for this condition are explored in the chapter on refrigerant-side troubleshooting.

The amount of condensation or sweat on a suction line with a correct charge will depend on the temperature of the suction line and the amount of moisture in the outdoor air. Condensation will occur at the dewpoint of the air. Dewpoint is defined as the temperature at which moisture will begin to condense out of the air. In this case, the suction line is cold enough to condense moisture out of the air.

During the hot part of the day, the suction line will be cooler on a system that has a fixed metering device than on a system with a thermostatic expansion valve. The superheat on a fixed metering device varies with the condenser pressure. In the heat of the afternoon, the suction superheat and the suction-line temperature on a capillary-tube system will be lower when compared to a TXV with a fixed superheat between 15° and 20°F. In the cool morning, the suction superheat and suction-line temperature on the same capillary-tube system will be higher when compared to the same TXV system, which maintains a close range of superheat temperatures. The colder the suction lines, the greater the amount of condensation developed. Air that has more moisture content condenses at a higher temperature. Generally, the relative humidity of the air is greater in the morning, because the air has cooled overnight and decreased in volume while maintaining the same moisture content. As the air heats up, it expands and the moisture expands into the additional space and decreases the relative humidity. It this case, a technician would expect to see more condensation on the suction line in the morning, as compared to a suction line at the same temperature in the afternoon.

A cool suction line is important. Using condensation or sweatback as the only indicator of a good charge is not a professional practice.

SUMMARY

- This chapter is the second-to-last chapter on the topic of charging. The information provided here should be used as part of the total charging package. Measuring the temperature drop is important when checking the charge on all equipment. The amount of moisture in the air affects the delta-T. Use the chart provided in this chapter to determine if the system has the correct temperature drop across the evaporator. The amount of air flow across the coil will also determine the delta-T. Use the chart in this chapter to determine if the system has 400 CFM per ton.

- The sight-glass method is valuable only on systems with a thermostatic expansion valve. Maintain a 130°F condensing temperature and charge to a clear sight glass. Once the sight glass is clear, some technicians add a few extra ounces of refrigerant to top off the charge. This can be done with a TXV system, but not with other types of metering devices. A system with a fixed metering device can be charged to a clear sight glass if the outdoor temperature is above 75°F, but the charge *cannot* be determined to be correct using this method.

- A sight glass with a moisture indicator can be used on a system with a fixed metering device. The moisture indicator is the only useful part of this application. The sight glass itself should not be used to determine the refrigerant charge.

- The sweatback method is not a good indicator of the charge, but the suction line should be cool or cold to the touch. The amount of sweat or condensation will depend on the temperature of the suction line and the amount of moisture in the air. Do not use this as significant charging indicator.

PROBLEMS AND QUESTIONS

1. Describe how to check a charge using the sight-glass charging method on a TXV.

2. What is the problem if the sight glass bubbles cannot be cleared using the described charging procedures?

3. Describe how to check the temperature difference across an evaporator.

4. What is the target temperature difference using the delta-T method under normal operating conditions?

5. What does a high delta-T indicate?

6. What does a low delta-T indicate?

7. What is the value of knowing that a suction line is condensing moisture from the air?

8. What is the definition of dewpoint?

9. What procedure would a technician use to dehydrate a refrigerant with high moisture content?

10. What is the advantage of a sight glass on a capillary-tube system?

11. A technician notices that the sight-glass moisture indicator shows a "wet" reading. What is the next step the technician should take?

12. How do these charging methods fit in with the other charging methods in the book?

13. What effect does superheat have on the suction-line temperature?

14. A technician measures a supply-air temperature of 45°F and a return-air temperature of 75°F. Should these readings be expected of an air-conditioning system? If not, what needs to be corrected? What additional reading would be useful to establish the correct delta-T?

15. What precautions should a technician use when setting up the temperature probes for a delta-T charging procedure?

16. What is the purpose of the electronic sight glass?

17. What is the expected delta-T of a coil if the return air measures 76°F dry-bulb and 40% relative humidity? What if the return air is 76°F dry-bulb and 60% RH?

18. What would cause bubbles in a liquid-line sight glass?

19. Using the chart provided in this chapter, give an example of proper airflow through an air-conditioning system.

20. Why is the sweatback method the least valuable charging method discussed in this chapter?

CHAPTER NINE

WEIGH-IN, FROSTBACK, AND APPROACH CHARGING METHODS

AFTER STUDYING THIS CHAPTER, THE STUDENT WILL BE ABLE TO:

• Describe the weigh-in charging method.
• Discuss the equipment needed to weigh in the refrigerant charge.
• Describe the frostback charging method.
• Describe the approach charging method.

• Describe how to use the charging checklist.
• Use the charging checklist.
• Know the value of using a charging checklist.
• Discuss the advantages and disadvantages of each charging method.

INTRODUCTION

This chapter has four major objectives: to introduce each of three new charging methods and describe how to use a charging checklist. The charging methods are

- Weigh-in charging method
- Frostback charging method
- Approach charging method

Using a charging checklist will assist the technician in organizing, documenting, and analyzing the charging procedures. The checklist can be used as a diagnostic tool for troubleshooting refrigerant-side problems. As with all charging procedures, the system should be clean, and the proper airflow should be established.

WEIGH-IN CHARGING METHOD

Weighing in the charge is one of the most accurate charging methods an air-conditioning technician can use. Simply put, the technician weighs the correct amount of refrigerant per manufacturer specifications. The correct charge for weigh-in is readily available for most package and window units. Finding the charging weight for most split systems is a challenge; sometimes such values are not available.

A technician will need most of the following equipment to weigh in the charge:

- Manifold gauge set
- Vacuum pump
- Micron vacuum gauge
- Charging cylinder or digital scale

WEIGH-IN CHARGING PROCEDURE

1. The first step in order to be successful with this method is to find the correct amount of refrigerant that must be weighed into the system. On window units and package units, the information may be on the equipment nameplate (Figure 9–1) or in the installation instructions. If the nameplate is not legible, then try to contact the manufacturer or local distributor for this information.

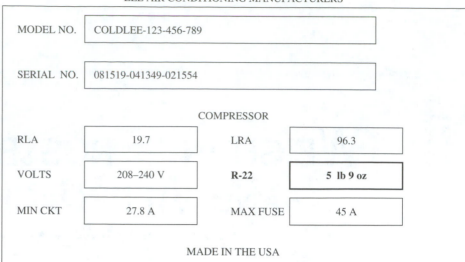

LEE AIR CONDITIONING MANUFACTURERS

MODEL NO.	COLDLEE-123-456-789
SERIAL NO.	081519-041349-021554

COMPRESSOR

RLA	19.7	LRA	96.3
VOLTS	208–240 V	**R-22**	**5 lb 9 oz**
MIN CKT	27.8 A	MAX FUSE	45 A

MADE IN THE USA

Figure 9–1 Condensing unit nameplate with vital information. The unit uses 5 lb 9 oz of R-22.

Finding the correct charging weight is more of a challenge when working with split systems. The charging weight may be on the nameplate or in the installation instructions. Some manufacturers do not make this information available or they recommend some other charging method. Manufacturers of split systems that list the charging weight also stipulate the length of the refrigerant line for the designated charge. For example, a split system may contain 100 oz of R-22 for the condenser and evaporator and 25 ft of interconnecting refrigerant lines. The manufacturer will specify an amount of refrigerant for the liquid line, suction line, and filter drier. Tables 9–1 and 9–2 are provided as a sample of what the manufacturer may provide in the installation instructions. Do not use these charts to charge units, because the requirements for specific units may be different. This is presented as a sample to help you understand how to read and use this information. The information is for Model XYZ, with 100 oz of charge for up to 25 ft of refrigerant lines.

Reviewing Table 9-1 you will notice that the liquid line requires significantly more refrigerant than the suction line. Also, larger line sizes require more refrigerant. Liquid weighs more than vapor; therefore, more ounces of liquid refrigerant are required per foot of refrigerant line. The suction-line charging weight is almost insignificant.

TABLE 9–1

ADDITIONAL REFRIGERANT CHARGE IN OUNCES PER FOOT

Liquid-Line Size		Suction-Line Size	
⅜ in.	½ in.	¾ in.	⅞ in.
0.51 oz	1.1 oz	0.05 oz	0.06 oz

Another factor that is overlooked is the additional refrigerant required by the liquid-line-filter drier. The vapor charge is not significant in the suction-line drier, and many manufacturers do not publish requirements for this drier. Most manufacturers do not recommend a suction-line drier as a permanent installation. The suction-line drier is generally used to trap contaminants after a compressor burnout. Equipment manufacturers expect that the suction-line drier will be removed after a short period of operation in the system. Table 9–2 is provided as a sample of what additional refrigerant the manufacturer recommends based on the type and size of the drier. Do not use this table when weighing in a charge. Consult the equipment or drier manufacturer for exact weight recommendations.

WORKING EXAMPLE

A technician installs a ⅜-in. liquid-line drier manufactured by XYZ company, Model 125. The technician uses Table 9–2 and determines that 4.9 oz of refrigerant will be added to the system to compensate for the volume of the drier.

Reviewing Table 9–2, you will notice that some of the larger-filter driers may require almost 1/2 lb of refrigerant. One-half pound of refrigerant will be important on critically charged units that use a fixed-orifice metering device.

The oil charge may be important to the manufacturer if the refrigerant lines are excessively long. For example, a table like Table 9–3 could be found on split systems where the distance between the condenser and evaporator is in excess of 50 ft. Read the manufacturer's installation instructions to determine the exact additional oil charge required by the installation.

TABLE 9–2

FILTER DRIER CHARGE ALLOWANCE FOR R-22—MANUFACTURER, DRIER MODEL NUMBER, AND CAPACITY

	XYZ		ABC		MNO	
Liquid-Line Size	No.	Oz	No.	Oz	No.	Oz
¼ in.	123	3.8	ABC	3.3	789	4.1
	124	4.0	—	—	790	4.5
⅜ in.	124	4.5	DEF	4.2	791	4.7
	125	4.9	GHI	4.9	792	5.4
½ in.	126	5.9	JKL	5.5	793	7.9
	127	7.4	MNO	7.9	—	—

1. In order to use the weigh-in charging method, the system must be in a vacuum. If the system contains refrigerant, it will need to be recovered to the EPA prescribed level and pulled into a 500-micron vacuum. The system will need to be completely devoid of refrigerant, air, and moisture before weighing in the refrigerant. This is one of the disadvantages of this accurate charging method. It is impossible to determine how much refrigerant is in a working system. The refrigerant in the system must be recovered and the refrigerant weighed into the system for an accurate charge.

2. Select the equipment used to weigh in the charge. The options are the digital charging scale and the charging cylinder. Each device has advantages and disadvantages. As a professional, every technician should have access to both of these devices. ■

DIGITAL SCALE

The digital charging scale (Figure 9–2) can be used to

- Weigh in the charge
- Weigh refrigerant in the recovery process
- Keep refrigerant records as required by the EPA
- Aid in customer billing

TABLE 9–3

ADDITIONAL OIL CHARGE REQUIRED FOR INSTALLATION OF SYSTEM WITH MODEL **XYZ** CONDENSER.

	Model XYZ
Unit Model	Additional ounces of required oil per 5 ft of line in excess of 50 ft*
2 to 3 tons	0.18 oz
3.5 to 6 tons	0.31 oz

*Accessories such as accumulators or receivers will require an additional oil charge. Consult the manufacturer of these accessories for the exact additional oil change recommendation.

The digital charging scale should be a good-quality piece of equipment. The scale increments should be in half (0.5) ounces. The scale should be able to read weight that is being added or removed from a refrigerant cylinder. The zero-adjust control is a normal part of the scale. The instrument should have a digital readout for durability. Analog, or needle, scales are not as accurate as digital scales. The digital scale should be battery operated, which makes it portable and convenient for the technician working around limited power sources. A digital scale that operates on both 120 AC V and battery is the best option for the technician.

Technicians that work in extreme high or low temperatures may need a scale that is analog, or needle-movement type, because a digital readout will blank out in extreme temperature conditions. Most analog scales are not accurate down to the ounce, but they are the instrument of choice when working in temperature extremes. Analog movements are delicate and subject to damage from dropping and rough handling.

Some digital scales have computer-charging features. A technician programs the digital scale to allow charging the system with a specific amount of refrigerant. The refrigerant is routed through the scale and is controlled by the scale computer. At the end of the charging sequence, the computer stops the flow of refrigerant to the system being charged. This allows the technician the opportunity to work on more than one project at a time. Periodically, the technician should monitor the computer charge as with any charging procedure.

When using a digital scale, a technician must take special precautions. The scale should be set to zero at the beginning of each use. To maintain accuracy, the refrigerant cylinder cannot be moved once it is set on the scale-weighing plate and zeroed. The problem with moving the cylinder on the scale-weighing plate is that the cylinder will not be placed on the same spot. Moving the refrigerant cylinder and placing it back on the scale will create

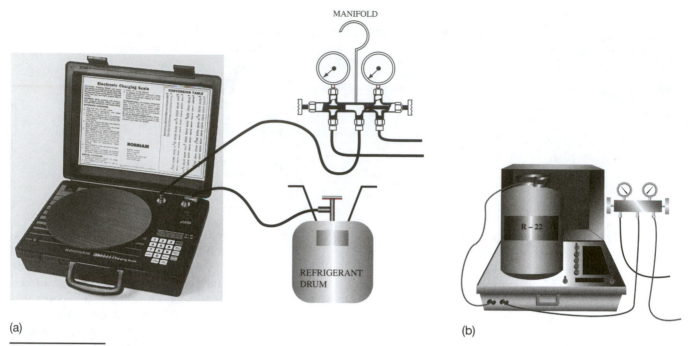

Figure 9–2 (a) Electronic charging scale. (b) Weighing a charge. *(Courtesy of Robinair, SPX Corporation)*

an inaccurate weight reading. Even moving the hoses or manifold gauge set will cause the weight to change. Once the manifold gauge set, hoses, and refrigerant cylinder are in place and zeroed, a technician must not disturb this arrangement if an accurate reading is required. If the hoses or refrigerant cylinder must be moved or adjusted, the technician must take the follow action:

- Record and document the weight of the refrigerant used.
- Make hose or refrigerant cylinder adjustments or changes.
- Zero-adjust the scale a second time after charges are completed.
- When the charging job is finished, add the second charging weight to the first charging weight. The sum of these two weights equals the refrigerant charged into the system.

The technician should purge the air from the refrigeration hoses prior to zeroing the charging scale. This should already be an automatic practice each time the hoses are installed on an air-conditioning system. Finally, two extra ounces should be added to the weighed charge. This additional charge will compensate for the refrigerant loss when removing hoses from Schrader-type access valves. Refrigeration hoses that have automatic or manual check valves do not need to have this extra refrigerant added to the charge. The check valves reduce the refrigerant loss when the hoses are disconnected.

Also, systems with back-seating (King valves) service-valve operation do not require the additional 2-oz charge. Back-seated valves can be considered as a manual check valve at the point of the hose connection; therefore no refrigerant should be lost from the system. The hoses have trapped refrigerant that will be purged to the atmosphere.

WORKING EXAMPLE

A technician has weighed in 55 oz of refrigerant and needs need to switch out a refrigerant cylinder because it is empty or the pressure is so low that the charge is consuming too much time. The system requires 100 oz of refrigerant. What should the technician do?

SOLUTION The technician should record the weight of the refrigerant used. In this case, 55 oz of refrigerant have been charged into the system. Next, the technician should make the appropriate refrigerant cylinder changes and then zero-adjust the scale reading. The technician should then add 45 oz more to complete the charge. A couple extra ounces can be added if non-check valve hoses are used.

55 oz	first charge
45 oz	second charge
100 oz	total charge

OTHER USES

A digital charging scale has other valuable uses. The scale can be used in the recovery process to measure

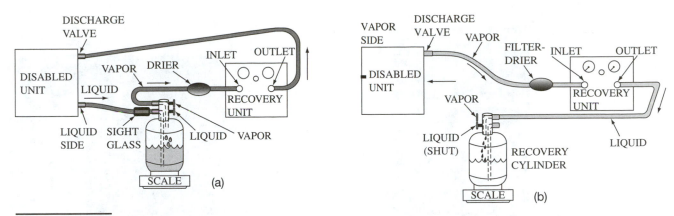

Figure 9–3   (a) Liquid recovery using differential-pressure method. (b) Vapor recovery.

the weight of refrigerant to assure that the recovery cylinder (Figure 9–3) does not exceed 80% of its capacity by weight. The EPA requires that a technician keep records of refrigerant used, recovered, and turned in for reclamation.

A charging scale can be used to measure refrigerant charged into a unit. Many companies bill the customer based on the amount of refrigerant charged into the system. Digital charging scales are an instrument often used by air-conditioning professionals.

SELECTING A QUALITY INSTRUMENT

The following guidelines are recommended when choosing a professional digital charging scale:

1. A sealed LCD digital readout with a minimum character size of 0.4 in.
2. Capacity of 100 lb
3. Accuracy within ±0.5 oz
4. Battery or 120-V power supply
5. Operating range of 32°F to 125°F
6. Keypad with a sealed touchpad style
7. Capable of being factory calibrated for accuracy

► CHARGING CYLINDER

A charging cylinder (Figure 9–4) is another way to weigh in charge. A charging cylinder is constructed of clear, vertical tubes that store refrigerant. The top of the cylinder has a pressure gauge. The top has an access valve for vapor charging and the bottom has a valve for liquid charging. The cylinder has a plastic rotating shroud. The rotating shroud lists the refrigerant types and a pound and ounce scale.

Most charging cylinders have an accuracy of ¼ oz. Charging cylinders have different capacities, ranging from 2½ to 10 lb. It is best to purchase a

charging cylinder to match the largest job. The charging cylinder is designed to meter out the desired amount of refrigerant. Compensation for temperature and pressure variations is accomplished by reading the gauge at the top of the charging cylinder and adjusting the outer rotating weight scale (screen) to match this reading. The refrigerant can be seen in vertical clear tubes that extend vertically behind the rotating weight scale.

The pressure in the charging cylinder will drop as the refrigerant is charged into the air-conditioning system. The refrigerant will be seen bubbling or flashing to a liquid in the clear tubes of the charging cylinder. The charge will slow as the pressure drops in the cylinder and as the pressure rises in the system being charged. The smaller the pressure difference (ΔP), the slower the charge will be. Most charging-cylinder manufacturers offer a refrigerant heater option, which will increase the pressure in the charging cylinder when plugged into a 120-V outlet. This increased pressure will drive the refrigerant into the system and reduce the charging time on the job. The heater is valuable when charging equipment on cold days when the refrigerant pressure will be very low. The heating element may damage the charging cylinder if it is allowed to operate without refrigerant.

The charging cylinder is a fairly rugged instrument, but it should be stored in a carrying case when not in use. The charging cylinder should be selected based on the type of refrigerants used by the technician. With the proliferation of refrigerant types, a technician will need to select a product that will be useful. Older charging cylinders are designed for R-22, R-12, and R-502 refrigerants. Some manufacturers now have an option for R134a refrigerant. Even though the charging cylinder will give a technician years of good service, it is suggested that a repair kit be purchased with the new unit. A repair kit may be needed to correct minor problems such as leaks around the seals or o-rings.

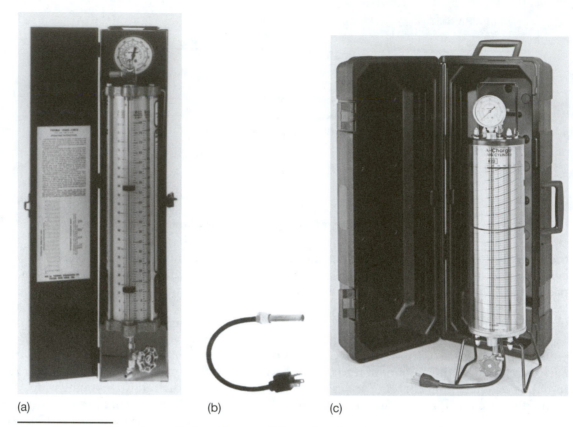

(a)   (b)   (c)

Figure 9–4 (a) Changing cylinder. *(Courtesy of Thermal Engineering Company)* (b) Optional heater. *(Courtesy of Thermal Engineering)* (c) Dial-a-Charge cylinder. Gauges are psi or kgcm²/kPa. *(Courtesy of Robinair, SPX Corporation)*

PREPARING TO USE A CHARGING CYLINDER

The following steps are used the prepare the charging cylinder for use:

- When using a new charging cylinder or empty charging cylinder, pull it into a 500-micron vacuum.
- While the system is under a deep vacuum, hook up a hose from the desired refrigerant to the bottom of the charging cylinder. Purge the connecting hose with refrigerant at the charging cylinder. Next, set up the refrigerant cylinder for a liquid-refrigerant charge. Inverting the refrigerant cylinder may be the only way to obtain liquid refrigerant (Figure 9–5). Many new refrigerant blends have a dip tube from the refrigerant valve to the bottom of the tank. This dip tube allows liquid charging while the cylinder is in the upright position.
- Open the bottom valve on the charging cylinder (Figure 9–5) to allow liquid refrigerant into the charging cylinder. Charge 10 oz extra refrigerant into the cylinder. The extra refrigerant is needed to purge the hoses and for refrigerant loses when removing hoses. The additional re-

frigerant may he useful if a technician wants the option to slightly overcharge the system as a diagnostic tool to aid in troubleshooting. Using a recover unit, remove vapor from the top of the charging cylinder if the liquid level stops rising before the desired level is reached. This action allows more liquid to flow into the charging cylinder. **Caution:** Do not overfill the cylinder. Overfilling the device may cause it to

Figure 9–5 Liquid charging into a charging cylinder. *(Courtesy Robinair, SPX Corporation)*

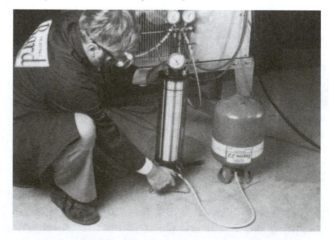

rupture or cause a pressure-relief device to open and dump all or part of the charge. Filling the charging cylinder to maximum capacity and storing it in a warm area will cause the pressure to rise and possibly rupture.

- Rotate the outer plastic shroud to the pressure reading that corresponds to the gauge pressure reading. Align the refrigerant type and pressure under the pressure column on the top gauge (Figure 9–6). When using the optional heater, allow time for the pressure to stabilize before adjusting the outer rotating scale.

- The charging cylinder is ready to use. The center hose of the manifold gauge set is used to receive the refrigerant from the charging cylinder. There are vapor or liquid charging options. The technician can charge refrigerant vapor from the top of cylinder or choose to charge liquid from the bottom of the cylinder.

- Charge the system with the required refrigerant quantity. The pressure gauge on the top of the charging cylinder (Figure 9–7) will change as refrigerant is used. Adjust the rotary scale as the pressure gauge changes. This is an important step for an accurate charge.

WORKING EXAMPLE

A technician reviews the nameplate of a package unit and discovers that the system requires 50 oz of R-22 refrigerant. The technician pulls a 500-micron vacuum on the package unit. After pulling a vacuum on the charging cylinder, the technician charges 60 oz of refrigerant into the evacuated charging cylinder. The technician adds an extra 10 oz to the 50 oz requirement for purging of hoses and for refrigerant that might be lost when removing hoses. On the charging cylinder, the refrigerant is broken down

Figure 9–7 Checking refrigerant pressure during weigh-in charge procedure. The pressure will drop as refrigerant is charged into the system. *(Courtesy of Robinair, SPX Corporation)*

into 1-lb and ½-oz increments (Figure 9–8). Sixty ounces of refrigerant translates into 3 lb 12 oz on the rotating scale at the registered pressure on the top of the gauge. To charge the package unit, 50 oz of refrigerant are removed from the charging cylinder and go into the package unit.

Figure 9–9 depicts the charging setup. The technician charges liquid into the liquid line. A valve-core tool is used to remove the Schrader-valve core, which speeds up completion of the charging process. The charging procedure should be stopped when the refrigerant in the charging cylinder sight glass drops to 10 oz of charge. The technician feels satisfied that the unit is charged correctly but uses other charging methods to be assured that the system is operating as designed. Checking the charge using other charging methods is important. Just because the correct refrigerant is weighed into the system, there is no guarantee that some other performance-reducing situation does not exist, such as a restriction or compressor-pumping problem. Weighing in the charge assures optimal operating characteristics only when the system has no problems.

FROSTBACK METHOD

The frostback method is a way to check the refrigeration circuit operation without tapping into the system. Some equipment does not come with access valves to check the refrigerant charge. This is especially true of small appliances that come precharged from the manufacturer. Access valves must be installed in order to check the refrigerant pressures and condensing and evaporating temperatures, but this is not always necessary.

Figure 9–6 Checking refrigerant type and charging cylinder pressure. *(Courtesy of Robinair, SPX Corporation)*

GAUGE PRESSURE →

SCALE IN
POUNDS AND OUNCES

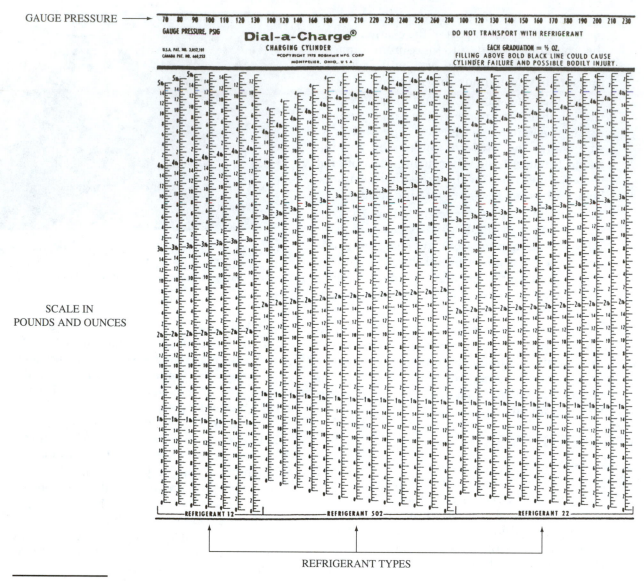

REFRIGERANT TYPES

Figure 9–8 Charging cylinder scale removed from the charging cylinder for review. *(Courtesy of Robinair, SPX Corporation)*

The frostback method of checking a charge can be used only on factory precharged units, such as window units or package units. This method of checking the charge should be considered fair to good. It can be used as a quick check to see if the unit contains an adequate amount of refrigerant without having to install access valves.

Unfortunately, many technicians install line-tap or piercing valves (Figure 9–10) on units without access valves. These line-tap valves are not replaced with a permanent Schrader or service valve. Line-tap valves tend to have a higher leakage rate over a period of time. Line-tap valves should be replaced with more permanent valves to reduce possible leaks. If the line-tap valve is installed on the process tube, it should be pinched off and removed, and the pierced hole should be brazed shut to prevent refrig-

erant leakage. The pinch-off tool must remain on the process tube until the hole is brazed. In this case the system would be left in its original factory-sealed and assembled condition.

USING THE FROSTBACK METHOD

To use the frostback charging method, simply block the airflow through the evaporator while the system is operating in the cooling mode. This can be accomplished by placing plastic sheeting or cardboard over the evaporator inlet. In most window units, the blower suction will keep the covering material against the evaporator. A good seal is important. The same can be done with package units. Do not block the airflow at the air filter on a package unit that is hooked to a duct system. The return-air plenum be-

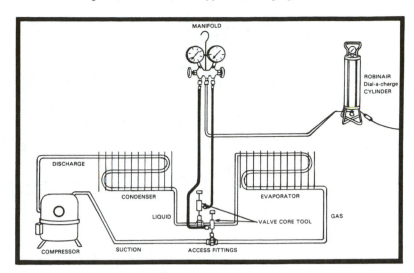

Figure 9–9 Correct hookup for liquid charging from the bottom of the charging cylinder. Liquid should be charged only into the liquid line. Charging liquid into the suction line will dilute the compressor oil and may cause liquid slugging. *(Courtesy of Robinair, SPX Corporation)*

tween the air filter and evaporator may have air leaks. It is best to block the evaporator coil itself. If the package unit has an independent blower, the blower can be disconnected from the power supply. The condenser fan will need to be operated as normal, without airflow restriction.

WHAT TO EXPECT

After a 5- to 10-min running period, a frost line should develop on the suction line (Figure 9–11). The frost line may appear on the housing or stop a few inches away from the compressor shell. Either of these conditions is an indication of a good charge and that the refrigeration circuit is operating as designed. The frostback method does not indicate conclusively that the refrigerant circuit is in perfect condition.

It is difficult to determine if a system is overcharged using this charging method. Reciprocating and scroll compressors may indicate on overcharge if the compressor shell has a large frost pattern on it. If the unit has a rotary compressor, the attached suction-line accumulator (Figure 9–12) should prevent the frost line from touching the compressor shell unless the system is severely overcharged. In the case of a rotary compressor, the frost line will stop at or near the accumulator. A frost line jumping the accumulator will cause damage to the rotary compressor because liquid refrigerant is pulled directly into the compressor chambers (Figure 9–13).

The frostback method will indicate a lack of charge or some other refrigerant-side problem if the frost line does not develop within 10 in. of the compressor in 10 min of running time. Remember, in most cases a factory charge is probably correct. If the frostback procedure develops a frost line on the suction and on a small part of the compressor shell, then

Figure 9–10 Piercing valve. *(Courtesy of Robinair, SPX Corporation)*

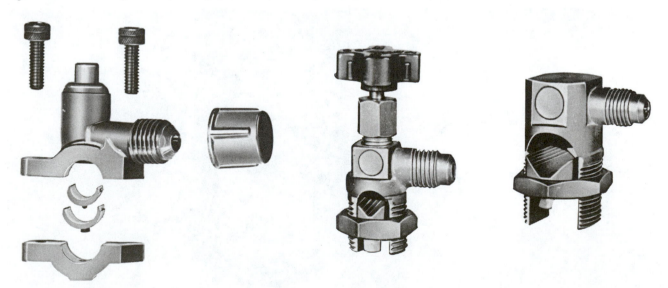

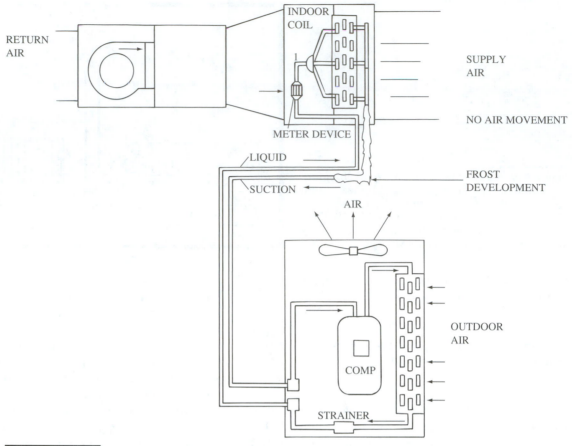

Figure 9–11   Cover or block the airflow to the evaporator and watch for a frost line to develop on the suction line. Frost-back method. Airflow is blocked by removing power to the indoor blower or covering the evaporator with plastic sheeting. A light coating of frost develops on the suction line up to the compressor. This is beneficial when checking package or window units. Results are not conclusive when using the frost-back method on a split system.

Figure 9–12   Rolling-piston rotary compressor. *(Courtesy of Rotorex Company, Inc.)*

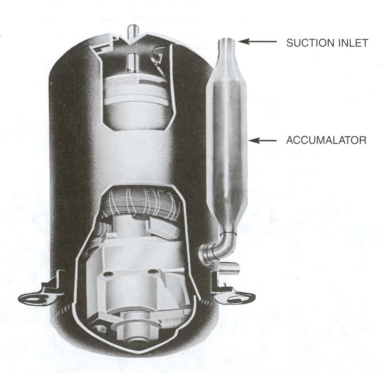

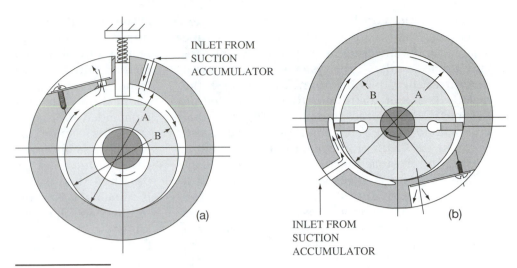

INLET FROM
SUCTION
ACCUMULATOR

(a)

(b)

INLET FROM
SUCTION
ACCUMULATOR

Figure 9–13   Cross-sectional view of rotary compressor. Once the refrigerant passes through the accumulator it passes directly into the compression chamber.

you may assume that the factory charge is adequate. Be suspicious if there are access valves on a factory-charged unit that appears to be field installed. The charge has been adjusted in this instance.

PRECAUTIONS

Operate the system under the frostback condition only as long as necessary to establish frost patterns on the suction line and, in some cases, on the compressor shell. Operating the equipment under the frostback condition may cause droplets of liquid refrigerant to enter the compressor. These liquid droplets will dilute the lubricating qualities of the refrigerant oil and reduce the compressor's mechanical life expectancy. Stop the procedure when readings are completed. Systems with accumulators can stand longer operations without damage to the compressor. The accumulator will allow the liquid refrigerant to boil off prior to entering the compressor. Pressure, temperature, or amperage readings during the frostback procedure will not be valuable because this is not the normal operating condition of the equipment.

PREVENTIVE MAINTENANCE

When conducting preventive maintenance on window units and packaged units, a technician may be tempted to install access valves in order to obtain refrigerant pressure and temperature readings. Customers expect the technician to attach gauges to their systems. This is a normal part of a PM or service routine. There is a disadvantage to this action. Access valves of all types can leak at the depressor core or at the joints where they are connected to the tubing. If a technician does not need an access valve, then one

should not be installed. When going through a preventive maintenance check on a factory-charged unit, first clean the filter, coils, and blower wheel, lubricate bearings, and tighten all connections. Unless there is a complaint of inadequate cooling, the technician should use the frostback method of checking the charge. If there is a complaint of inadequate cooling, the technician must determine if the problem is caused by a dirty system or by some other problem. Cleaning the coils and filter may improve the cooling performance. The amperage draw on the compressor and the delta-T across the evaporator should be checked after the frostback method is successfully completed. When doing PM, install access valves only as a last resort. Figure 9–14 shows a flowchart that summarizes the steps of the frostback method.

TROUBLESHOOTING

Another way the frostback method can be used is to test the individual refrigeration circuits in an evapo-

Figure 9–14   Frostback Flowchart

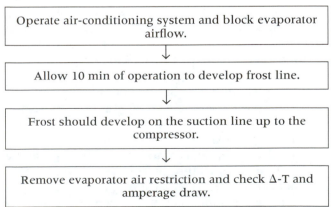

Operate air-conditioning system and block evaporator airflow.
Allow 10 min of operation to develop frost line.
Frost should develop on the suction line up to the compressor.
Remove evaporator air restriction and check Δ-T and amperage draw.

rator. Most evaporators have more than one refrigeration circuit. Each circuit is fed by a distributor tube that attaches the metering device to that evaporator circuit. Some capillary-tube systems feed refrigerant directly from the liquid line to the evaporator. If the metering device or distributor tube is not feeding the evaporator the correct amount of refrigerant, that circuit will not cool and dehumidify as designed. This can be a difficult problem to troubleshoot without using the frostback method. One way to diagnose this problem is to inspect the distributor and evaporator circuits during normal operation. The distributor and evaporator circuits should have the same amount of sweat or condensation as compared to other distributor and evaporator circuits on the coil. If the circuit is totally restricted, the corresponding distributor and portion of the evaporator will be dry. A technician can use the frostback method to diagnose this problem even if the circuit is partially restricted. Stop the airflow through the evaporator. Observe the distributor and evaporator frost pattern. The restricted circuit produces less frost than the unrestricted circuit. The lack of frost is a sure sign that the metering-device feed or the distributor feed has inadequate refrigerant flow. This is a handy way to test for evaporator restrictions in any equipment.

SUMMARY

When doing preventive maintenance or troubleshooting without access valves, a technician should take and record as many readings as possible. A technician should use the *charging checklist,* as discussed at the end of this chapter. Using the frostback method will not tell a technician everything they would like to know about the system, but it is a helpful tool when used on a correctly operating system.

APPROACH CHARGING METHOD

The approach charging method could be classified as a manufacturer's charging method because it is used by a limited number of air-conditioning manufacturers. Lennox Industries is one of biggest manufacturers that recommend this charging procedure on some of their equipment. In the approach method, the outdoor ambient temperature is subtracted from the liquid-line temperature. Depending on the model number of the matched system, the approach charge can range from 1°F to as much as 22°F. The word *approach* is used to indicate how close the liquid line temperature approaches the outdoor ambient-air temperature. Heat exchangers, like condensers and evaporators, never achieve a perfect exchange of heat; therefore, the liquid-line temperature will always be warmer than the outdoor-air temperature blowing through the coil. This charging method will work only on matched systems, and a technician must know the recommended approach temperature. The allowable deviation from the recommended approach temperature is ±2°F.

This chapter discusses the approach method because it is important to know something about all charging methods. Someone hearing about the approach method may not realize that, even though it is valuable, it is limited to certain equipment manufacturers and that the approach temperature must be known in order to be successful. There is no rule of thumb. Additionally, the approach method is not to be confused with the *condenser approach,* which is used to evaluate the efficiency of water-cooled condensers. There are some similarities between the two, but this explanation is limited to air-cooled condensers. Being armed with this knowledge will make you a better technician.

USING THE APPROACH METHOD

The approach method is used only with air-cooled condensers that have a thermostatic expansion valve to feed the evaporator. The outdoor temperature should be above 60°F, condensing pressures should be between 200 and 250 psig, and the indoor dry-bulb temperature must be in a range of 70° to 80°F. The outdoor temperature of 60°F is quite liberal. Many manufacturers recommend outdoor temperatures above 70°F in order to obtain an accurate charge. It may be necessary to restrict the airflow to the condenser in order to reach the liquid pressures in the 200 to 250 psig range. The outdoor coil can be blocked (Figure 9–15) in order to obtain condenser pressure minimums.

The following steps are recommended for finding the system's approach temperature:

1. Connect the gauge manifold. Connect an upright R-22 cylinder to the center port of the gauge manifold.

2. Record the outdoor-air temperature near the air entering the condenser. Do not touch the condenser coil surface with the temperature probe.

Figure 9–15 Controlling condenser pressure by covering the coil with cardboard or plastic. *(Courtesy of Lennox Industries)*

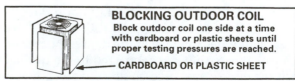

BLOCKING OUTDOOR COIL
Block outdoor coil one side at a time with cardboard or plastic sheets until proper testing pressures are reached.

◄— **CARDBOARD OR PLASTIC SHEET**

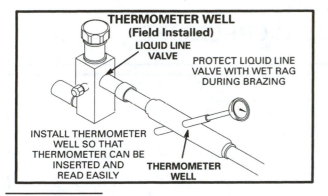

Figure 9–16 Approach method liquid-line temperature check. *(Courtesy of Lennox Industries)*

3. Operate indoor and outdoor units in the cooling mode. Allow the outdoor unit to run until the system pressures stabilize, using 15 min.

4. Make sure that the liquid-line thermometer well is filled with mineral oil before checking the liquid-line temperature (Figure 9–16). If the unit does not have a thermometer well, tape an accurate temperature probe to the liquid line. Tape pipe insulation over the liquid line to prevent the influence of the outdoor temperature.

5. Place the thermometer in the well and read the liquid-line temperature after 10 min. The liquid-line temperature should be warmer than the outdoor air temperature entering the condenser. Use Table 9–4 to determine if the charge is accurate. Add refrigerant to lower the liquid-line temperature. Recover refrigerant to raise the liquid-line temperature. Add refrigerant slowly as the system approaches the correct temperature. This allows refrigerant to stabilize, allowing the correct temperature to be read. (This table is only a sample. This specific approach temperatures need to be determined for an actual system being checked.)

6. Use other charging methods to verify a correct charge.

TABLE 9–4
SAMPLE APPROACH METHOD TABLE

Model Number	Liquid Line Warmer Than Outside Ambient
ABC-024	12°F
XYZ-024	1°F
ABC-036	22°F
XYZ-036	18°F
ABC-048	5°F
XYZ-060	7°F

HVAC CHARGING CHECKLIST

Now that we've learned numerous different charging methods, it is important to be able to organize and use the information gained from each. The information needs to be recorded and documented in some useful format. This is the purpose of the *charging checklist*. The charging checklist (Figure 9–17) is useful when checking charge and as a document for future reference.

When used on the job, the charging checklist will organize a technician's readings, thus allowing the technician to spot problem areas as well as troubleshoot refrigerant-side problems. This important record should be kept by the technician and a copy given to the customer. The air-conditioning company should keep the charging checklist on file for future reference. This record will be useful when future preventive maintenance and troubleshooting problems arise and will become baseline data for future equipment operation. The customer's information should be inserted into a plastic sleeve and attached to the air handler or other convenient area for the use of other technicians. It is not advisable to leave this valuable information in the condensing unit, because even the best-protected documentation will deteriorate when exposed to the outdoor elements.

Completing a charging checklist on a new installation provides an important equipment record for future reference. Identically installed systems have variations in operating characteristics, and the charging checklist will document these differences.

The comprehensive checklist is beneficial if a technician needs to confer with supervisors or manufacturer representatives regarding problems beyond the technician's skill or knowledge level. The information on the checklist will facilitate troubleshooting over the telephone. If all the charging information is collected, the technical telephone representative will be better able to give assistance with the problem.

USING THE HVAC CHARGING CHECKLIST

This section describes line by line how to use the charging checklist. This comprehensive document facilitates charging and is a valuable troubleshooting tool should a refrigerant-side problem arise. It is important that the equipment be clean and lubricated and that the duct system develop 400 CFM per ton of airflow.

The charging checklist heading should be customized to include the contractor's name, address, phone and fax numbers, e-mail address, and state contractor's license number. A professional form is important; it will enhance future return business.

HVAC CHARGING CHECKLIST

Check and Record the Following Conditions:

❶ Date_____

❷ Unit Name_____

❸ OAT_____°F

❹ R/A Temperature_____°F DB ❺ _____°F WB

❻ Metering Device_____

❼ Condensing Temperature_____

❽ Evaporating Temperature_____

❾ Suction Superheat _____ – _____ = _____°F

❿ Discharge Superheat _____ – _____ = _____°F

⓫ Subcooling _____ – _____ = _____°F

⓬ Sweat on Suction Line_____

⓭ Compressor Amperage_____

⓮ Nameplate Amperage_____

⓯ Sight-Glass Condition_____

⓰ Temperature Drop Across Evaporator_____°F

⓱ Manufacturer's Charging Recommendations_____Yes _____No

⓲ Is This Unit Charged OK? _____Yes _____No

⓳ If NO, What Is the Problem? *Be Specific!!*_____

_____

_____

_____

Figure 9–17   Charging checklist.

Line 1 The date and time of service are important to future technicians. The time between service calls will alert technician to how the equipment is maintained. The time of day may influence how the system operates.

Line 2 Condenser and evaporator unit name, model, and serial number aid a technician in determining if the system is of the same manufacturer and matched to operate per manufacturer's design conditions.

Line 3 OAT is an acronym for *outdoor ambient temperature* entering the condenser. The outdoor temperature should be measured as closely as possible to air stream entering the condenser. The temperature probe should not touch the condenser coil, because this will influence the temperature probe and register an improper reading.

Lines 4 and 5 These lines measure the conditions of the air entering the evaporator. A sling or electronic psychrometer can be used to measure the dry-bulb (DB) and the wet-bulb (WB) air temperatures. These temperatures should be measured as closely as possible to the evaporator in case the return-air plenum is leaking air and influencing the air-entering conditions.

Line 6 A technician will need to know the type of metering device installed at or near the evaporator. The type of metering device determines the suction superheat reading. If the metering device is a TXV, expect a superheat in the range of 15° to 20°F at the suction line. If the metering device is a fixed-bore device, a superheat chart will be needed. The suction-line superheat for these devices ranges widely. Use the chart.

Line 7 The condensing temperature is used to determine if a system's high-side temperature (or pressure) is charged within range and is used to determine discharge superheat.

Line 8 The evaporating temperature is used to determine if a system's low-side temperature (or pressure) is within range and is used to determine suction superheat.

Line 9 The suction superheat is calculated by subtracting the evaporating temperature from the suction line temperature. The specific superheat will depend on the type of metering device in use.

Line 10 The discharge superheat is calculated by subtracting the condensing temperature from the dis-charge line temperature. A good discharge superheat is in the range of 50° to 120°F.

Line 11 Subcooling is found by subtracting the liquid line temperature from the condensing temperature. Good subcooling is in the range of 10° to 20°F. Some manufacturers have specific recommendations for the target subcooling, but it is normally near or within the 10° to 20°F temperature range.

Line 12 Sweat, moisture or condensation on the suction line is not a good indicator of a charge. The suction line should be cool to the touch. The amount of condensation, if any, will depend on the suction line temperature and the amount of moisture in the air. It is a general indicator of the operation of the system.

Line 13 On single-phase equipment, measure the amperage on the common winding of the compressor. It is a good practice to measure the amperage on all the windings and make a comparison. The current in the start winding and run winding should nearly equal the current in the common winding. Many manufacturers use a black wire to the common of the compressor, but this is not always the case with all equipment. The rated load amps (RLA) should not be exceeded under normal operating conditions. Extremely high outdoor temperatures may cause the compressor to slightly exceed the RLA rating. Using the amperage draw is not a good indication of the charge. A very low amperage draw may indicate an undercharge or lack of compressor pumping capacity. A high amperage reading may indicate an overcharge, operation in high ambient conditions, or an electrical problem.

When checking three-phase compressors, measure the amperage draw of each of the three phases. The amperage readings between the phases should be almost the same.

Line 14 The nameplate amperage (RLA) is used to determine the maximum rated load amps of the compressor. The locked rotor amperage (LRA) will be useful to the technician if the compressor is tripping out on high amperage.

Line 15 The sight glass is mainly valuable to a system that has a TXV. In order for this device to be useful the condensing temperature should be held near 130°F and the sight glass should be clear. Systems that use fixed bored metering devices find little value in the sight glass charging procedure. A system with a fixed metering device will have a clear sight glass once the outdoor temperature exceeds about 75°F. This does not mean that the system is charged properly, just that there is liquid refrigerant feeding the liquid line. Also, a clear sight glass can be an indication of a lack of charge.

Most sight glasses have a moisture indicator. The color of the moisture indicator should be noted. A moist refrigerant indication may need correction. Check the refrigerant moisture level with an acid and moisture test kit before treating the refrigerant for excess wetness.

Line 16 The temperature drop or delta-T across the evaporator is the temperature difference between the supply air and return air. The supply air should be measured at the shortest duct branch but *not* within "line of sight" of the evaporator. The return air dry-bulb was measured in Line 4. To obtain the delta-T reading, subtract the two temperature readings.

Line 17 Does the system have manufacturer's charging recommendations? A technician may discover recommended charging instructions on the inside of the condensing unit or in the installation instructions. The manufacturer may recommend a specific superheat charge; subcooling temperatures; pressures readings; or the use of a table or chart. When the system is matched, the charging recommendations are what the manufacturer expects of the equipment. Other charging methods, as listed on this checklist, should be used to verify a good charge.

Line 18 A technician must evaluate all the information and determine if the charge is correct.

Line 19 If a system is not charged properly, a technician must again evaluate the information and decide what course of action to take. In some situations the system may not need a charge adjustment but have a restriction, lack of airflow, or a minor or major mechanical malfunction. The system may be cooling and the customer may not want the repair at the time the checklist is conducted. For example, the subcooling may be 5° higher than recommended, but the unit is cooling adequately. High subcooling may be caused by an overcharge or liquid-line restriction. An overcharge can be handled with a simple recovery, but it will require more time to correct a liquid-line restriction problem.

The charging checklist and the billing invoice are good places to document any system problems that the customer needs to know for corrective action. Documenting this information will reduce any misunderstanding between a technician and a customer and helps avoid the loss of customers.

SUMMARY

- This chapter discussed three different charging methods—weigh in, frost-back, and approach methods of charging.

- The charging cylinder is an accurate way to weigh in the correct amount of refrigerant charge. As with the digital charging scale, the charging cylinder must charge refrigerant into a system that is in a vacuum. The technician must know how much refrigerant to weigh in within an ounce or two of the required charge. A charging cylinder can be useful when charging equipment in cold weather. The charging cylinder can be used to build up refrigerant pressure by installing an optional heater at the bottom of the charging assembly. The digital scale has the added advantage of being able to weigh refrigerant removed during the recovery process.

- The frost-back method of checking the charge has the technician blocking the airflow through the evaporator of a package or window unit. The purpose of the frost-back method is to check refrigeration-side operation without having to install access valves. A frost line should develop up to or near the compressor shell. In the case of a rotary compressor the frost line should develop up to the inlet of the accumulator. If the frost line develops as expected the technician can clean up the filters and coils. Next, check the delta-T and amperage on the compressor. If all these reading are in a normal range the system should be operating correctly.

- Finally, the approach method is a manufacturer specific method for assuring a correct charge. Again, as with other charging methods, the approach method must be verified with pressure, temperature, and amperage checks. In an ideal world a heat exchanger would be perfect. A perfect heat exchanger would not have losses. A perfect heat exchanger would transfer all its heat until the temperature of the refrigerant equaled the temperature of the air or water passing over it. The approach method is an easy way to verify the effectiveness of the condenser heat exchanger and an adequate refrigerant charge. A perfect condenser would have an approach of 0° F; however the perfect condenser has not been designed.

PROBLEMS AND QUESTIONS

1. Describe the steps in using the frostback method of charging.
2. What is the advantage of the frostback method?
3. What is the disadvantage of the frostback method?
4. How can the frostback method be used to troubleshoot?
5. Describe how to use the approach charging method.
6. Describe how to weigh in a charge using a charging cylinder.
7. Describe how to weigh in a charge using a digital scale.
8. What are the benefits of using a digital scale?
9. What are the benefits of using a charging cylinder?
10. Why must a technician add extra charge to a charging cylinder?
11. What precautions must be taken when using a charging cylinder?
12. What precautions must be taken when using the frostback charging method?
13. A technician is working on a system with no brand name, and the nameplate is beyond recognition. The technician heard about a new method called the approach method to charging. The technician wants to use this method on this unit. What are you going to tell the technician?
14. A technician is using the frostback method to check the charge on a split system. What advice are you doing to give the technician regarding this method used under these circumstances?
15. A technician is installing a split system. The condenser holds enough charge for 30 ft of refrigerant line. The installation requires 50 ft of refrigerant line. The condenser has a manufacturer-installed liquid-line drier. What recommendations are you going to make regarding this installation?
16. Why is it important to use several different charging methods even if one is highly recommended by an equipment manufacturer?
17. What are the advantages of using a charging checklist?
18. What information would you add to this charging checklist? Why?
19. How can the charging checklist be used to maintain a customer base?
20. How does the charging checklist assist in troubleshooting refrigerant-side problems?

CHAPTER TEN

TROUBLESHOOTING AIR-CONDITIONING SYSTEMS

AFTER STUDYING THIS CHAPTER, THE STUDENT WILL BE ABLE TO:

- List the three areas of troubleshooting.
- Describe how a refrigeration problem can create an electrical problem.
- Understand why it is important to know what caused a mechanical or electrical failure.
- Determine where to start the troubleshooting sequence.
- Describe the foundations of good troubleshooting skill.

- Tell the customer what steps to take while waiting for service.
- Describe the importance of hooking up instruments before starting the equipment.
- Discuss the ways to quickly check the operation of a malfunctioning system.
- Describe what to do when feeling "brain dead."

INTRODUCTION

The following chapters discuss the most important activity that an air-conditioning service technician does as a diagnostic service technician. Troubleshooting is the most difficult, most challenging, and most rewarding aspect of the service technician's job.

Troubleshooting is the process of determining the cause of an equipment malfunction and taking corrective measures to fix the problem. It is normally easy to correct the problem; it is a challenge to correctly determine the malfunction. Troubleshooting is a learned skill and the skill varies between individuals.

These chapters only scratch the surface of the many varied experiences that a technician has in the field. In order to be successful, a technician needs a good fundamental background in troubleshooting. Many technicians learn troubleshooting skills on the job by themselves, from supervisors, or fellow technicians. They will learn good techniques and poor techniques from these experiences. Participating in training that emphasizes troubleshooting skills and/or reading books like this one will advance the level of technicians' troubleshooting skills. The fastest way to improve these skills is by structured learning and lab experiences combined with field experience. This combined effort will propel the technician up the professional ladder. Technicians that can troubleshoot air-conditioning problems are not born; they are cultivated through formal or informal training and hands-on field experience.

TYPES OF SYSTEM TROUBLESHOOTING

There are four types of troubleshooting problems:

- Electrical- or electronic-related problems

- Mechanical- or refrigerant-side-related problems
- Combination of electrical- and mechanical-related problems
- Operator error

Electrical-related problems form the majority of air-conditioning repairs. Electrical-side problems include anything related to the power supply or control-voltage part of an air-conditioning system. Electrical troubleshooting includes, but is not limited to, the power supply feeding the equipment; all wiring; circuit-protection devices; control voltage; electronic components used to operate or control circuits; and the operation of all electrical components, including motors, relays, capacitors, and thermostats.

Refrigerant-side problems are associated with refrigerant flow, airflow or water flow through the air-conditioning system. As can be seen, there is not always a clear demarcation between refrigeration-side and electrical-side problems. An electrical problem can affect the refrigerant circuit, and the refrigerant circuit can affect the electrical operation. For example, too much refrigerant returned to a compressor will cause dilution or washout of the compressor oil. The compressor motor bearing will seize due to a lack of lubrication. Motor windings will burn out if the compressor continually tries to start in the locked-up condition. A technician will find that the compressor motor winding(s) is open. The technician's diagnosis is an electrical problem, even though it was a refrigerant-side problem that caused the motor to burn out.

It is important to determine the cause of a problem. If a technician replaces the compressor without adjusting the charge, the new compressor will be doomed to fail. Troubleshooting includes trying to figure out what caused the component to fail. In many instances the reason for the failure is unknown. A technician simply blames the component manufacturer, voltage problems, or lightning strikes. This may be the case with the basic electrical components, but not a compressor less than 5 years old. Compressors are built tough and are designed to last 15 years and longer. The average compressor lasts 9 to 10 years under normal running conditions. The shorter life expectancy can be blamed on poor installation practices, a lack of equipment maintenance or improper charging practices. When a compressor fails, a technician must decide what caused the failure and act to correct the problem.

On condensing units out of warranty most replacement compressors offer 1-year replacement warranty, not the 5-year warranty found with new condensing units. Manufacturers can design a compressor that will withstand poor replacement practices and operate for 1 year. Compressor manufacturers realize that the problem that caused the original failure may not have been corrected. Sometimes the exact cause of failure is never determined and something in the compressor replacement causes a change that prevents the failure from reoccurring. For example, a compressor is damaged because it was overcharged. A technician may not compare the weight of the refrigerant removed to the weight of the refrigerant charged back into the system. Comparing these weights would alert the technician that the system was overcharged. The overcharge may have been the cause of the compressor failure.

WHERE TO START?

Troubleshooting skills vary between technicians. No particular troubleshooting approach is better than another. The following chapters discuss various troubleshooting procedures. It is up to a technician to develop a troubleshooting plan that is comfortable and successful. Troubleshooting behavior is an individual thought process. The biggest enemy of troubleshooting is time. The customer is hot and uncomfortable. He or she wants the unit fixed now. A quick diagnosis and repair is the goal. The customer does not want to pay extra for the troubleshooting time on the job. A technician also feels pressure from the company to complete the job, satisfy the company, and move on to the next job. Another hot customer is waiting for service. The time and pressure elements can work against a technician in finding the problem. A technician must learn to handle time and pressure constraints to be successful. A technician can react by diagnosing the problem, misdiagnosing the problem, or changing parts until the problem is solved. Having a solid foundation in troubleshooting will lead to success.

UNDERSTANDING THE EQUIPMENT

A technician who has a good understanding of the equipment's sequence of operation will be successful in a rapid diagnosis. Air-conditioning technicians are expected to work on a great variety of equipment. The basic refrigeration and electrical operations of an air-conditioning system are basically the same between manufacturers. Manufacturers have various ideas on how to design equipment. Some refrigerant and control circuits are very simple; some are a little more complex. Residential air-conditioning equipment tends to have a simpler design as compared to commercial and industrial air-conditioning systems. A technician needs to transfer information learned from operating and repairing other equipment. A technician can contact the office, equipment supplier, or manufacturer for technical assistance. These resources are a valuable asset to a technician. A technician should not be afraid to ask for assistance. Even the most experience technician needs advice on occasion.

Technicians need to avail themselves of training sessions on specific equipment. Manufacturers offer training on the operating sequence and troubleshooting of their equipment. Most of this training is offered at little or no cost to the technician. The knowledge acquired can be transferred to other equipment models and manufacturers.

Understanding how a system operates expedites the troubleshooting process. Having a good foundation in understanding how equipment operates helps a technician make the correct diagnostic decision.

REPORTING THE SERVICE CALL

As much information as possible about the problem should be obtained when a customer reports an air-conditioning problem. Is the indoor blower operating? Is the condensing unit operating? Is the outdoor unit fan operating? Are there any unusual noises coming from the inside or outside units? This information should be passed to the technician assigned to the job. The troubleshooting thought process should begin as the technician drives to the job.

The customer should be instructed to turn the unit off at the thermostat (Figure 10–1) until the service technician arrives at the site. If the indoor blower is operating, the customer should be instructed to turn the thermostat to the *off* position and place the fan switch in the *on* position to allow for air circulation in an increasingly warm building.

Allowing the system to continue to operate in the failure mode may do more damage to the equipment. A compressor that is cycling on its overload device will become hotter and hotter as the time intervals between resetting and operation increase. Turning the system off will allow a cool-down time and give the technician a firsthand look at the problem when arriving at the site. Less time will be wasted because the technician will not have to cool an overheated compressor. When the compressor trips out on thermal overload, it will reset much more quickly because it is not as hot as a short-cycling compressor.

In the majority of the service calls, turning off the system will prevent further equipment damage and give the technician a better start when arriving at the customer's location.

Figure 10–1 Advise the customer to turn off the air-conditioning system until the technician arrives on the job. The indoor fan can be set to the *on* position for air circulation.

HOOKING UP THE INSTRUMENTS

One of the first steps a technician should take when arriving at the site is to install the manifold gauge set on the high- and low-pressure sides and clamp the Amprobe around the common wire of the compressor (Figure 10–2). This is very important!

The technician should install these instruments to obtain an immediate reading on what is happening inside the refrigeration system. If the condensing unit is not operating, the technician should take control of its operation by turning off the condensing-unit disconnect or removing one side of the 24-V control to the contactor. Be careful not to allow the 24-V control wire to touch the unit, because it may short out the transformer or burn out the transformer fuse. The recommended way to control the condensing unit is via the supply-voltage disconnect or the control voltage at the outdoor unit. Controlling the system from the thermostat generally means time must elapse between turning the equipment on and going outside to the condensing unit. During this time period the technician may be missing valuable diagnostic information. Be careful; disconnecting and reconnecting the 24-V wiring to stop and start the condensing unit may cause short-cycling of the unit. Short-cycling occurs when a technician does not make a good 24-V connection on

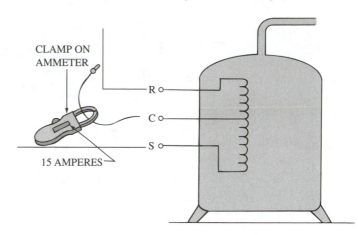

Figure 10–2   Checking compressor amperage on the common winding.

the contactor coil. In trying to reconnect the control voltage, the technician may create a loose connection that will cause the contactor to close and open rapidly. This short-cycling, or "contactor chattering," is not good for the compressor or contactor contacts. A suggested way to prevent short-cycling is to use a set of alligator leads. Attach one end of the alligator clip to the disconnected end of the control-voltage wire. Take the other end of the alligator clip and quickly connect the voltage to the terminal point on the contactor coil. The alligator clip is used as a bridge to control the 24 V at the condensing unit.

Use the disconnect or circuit breaker if it is near the compressor. When closing a disconnect or breaker, stand to the side in case this connection point explodes due to a short circuit in the system (Figure 10–3). This is a safety procedure every technician should follow when closing electrical circuits. Some technicians push the contactor points when they cannot figure out why the condensing unit is not operating. Technicians want to know if the condensing unit will operate by this manual exercise. This is also a dangerous practice. If the system has a short, manually pushing in the contactor moves the short to contactor points, and a dangerous electrical flash or explosion can result.

It is important to get a quick measurement of the gauge pressure (Figure 10–4) and amperage draw of the compressor. The amperage measurement is the most important of these readings. When a compressor is energized, it draws locked-rotor amps (LRAs). Have the Amprobe connected to the common winding of the compressor (Figure 10–2). Watch the meter when the compressor starts. If the compressor draws LRA for several seconds on start-up, the compressor-overload device will open the common power circuit to the compressor. If a technician misses this start-up LRA condition, the compressor may not start again for a considerable amount of time. When a compressor first cuts out on its internal or external overload, it will reset very quickly. The reset time is usually within a few minutes. The time may be longer on commercial units that tie the overload device into a timed-delay circuit. When the compressor is not very hot, the overload cools quickly, closing its contacts and allowing the common circuit to be completed to the compressor. The more often a compressor trips on the overload, the hotter the compressor becomes. The reset time becomes longer and longer. In case of an internal overload, the compressor may take hours to cool before the overload resets itself. A compressor can be cooled by running water slowly down the compressor shell. When doing this operation, the compressor power must be removed, and the wire terminals must be protected from the water. Due to their mass, larger compressors tend to hold heat longer than small-capacity compressors; therefore, larger compressors take longer to reset. Removing it from the compressor shell can cool an external overload-protection device (Figure 10–5). Operating the compressor in this manner can be dangerous because the over-current protection has been reduced. The external overload protector monitors the heat on the compressor shell as well as current draw to assure the

Figure 10–3   Stand to the side when closing or opening a circuit breaker, fuse, or disconnect.

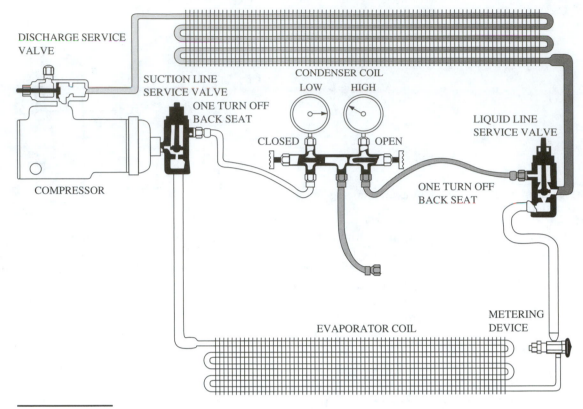

Figure 10–4 Equal pressures indicate that the compressor is not operating.

motor is operating properly. Moving the overload away from the compressor shell removes one of the operating sensing devices. Reattach overload to compressor shell.

Missing the LRA reading may cost a technician extra time and work on the job. Catching the LRA reading will not stop the compressor from cutting out on the compressor overload, but at least a technician can begin the troubleshooting process while the compressor is cooling.

- Are the motor windings measuring correct resistance?
- Are the start components okay?
- Is the supply voltage high enough?
- Does the system have any refrigerant?
- What is the oil level in the compressor?

- Are the connections tight?
- Is there burned wiring?

These are some of the areas a technician can investigate while waiting for a compressor to cool. Again, the technician must take control of the unit. Disconnect the power supply or control voltage. Proceed with a troubleshooting track and control the time when the power should be supplied to the equipment.

Watching the operation of the high- and low-pressure gauges is important once the compressor has established a running condition. A high-pressure or low-pressure condition may cut the compressor off quickly. The pressure safety switch may stop the compressor within a few seconds of operation. Having the Amprobe and manifold gauge set is an im-

Figure 10–5 External overload removed from the compressor shell. *(Courtesy of MARS Components)*

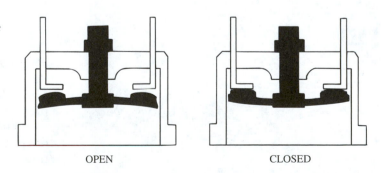

OPEN   CLOSED

portant part of the initial stage of troubleshooting. As a professional technician you are going to hook up these instruments anyway; it is better to hook them up as the first step of the diagnostic process.

QUICK CHECKS

A seasoned technician will use a series of quick checks to determine what is happening in the air-conditioning system. One way to organize your troubleshooting thought patterns is to develop a plan of **act**ion. Use the word *ACT* as a reminder on what to check on an air conditioning system:

> *A* is for airflow.
>
> *C* is for compressor.
>
> *T* is for thermostat.

Let us explore how to use ACT as a quick way to check the vital signs of an air-conditioning system.

A IS FOR AIRFLOW

Check the outdoor unit. Does the condensing unit have airflow? Is the condenser fan operating? In the case of a water-cooled condenser, is the water pump operating? Does the condenser coil look clean? Does the rejected air feel warmer than the outdoor air? Is there a temperature rise across a water-cooled condenser? *Check the indoor unit.* Is there return- and supply-air movement? Do all the supply-air registers have airflow? Is the air filter clean? Is the supply air cold? On a chilled water system, is the water leaving the chiller barrel cold to the touch? Is the chilled water pump operating?

C IS FOR COMPRESSOR

Is the compressor operating? What is the amperage draw of the common winding? Is the suction line cool? Is the discharge line hot? Don't burn your fingers! Is there a significant pressure difference between the high and low side, indicating that the compressor is pumping?

T IS FOR THERMOSTAT

Is the thermostat set to the cooling mode and the temperature set 10°F below the indoor temperature? Does the thermostat control the indoor fan operation by switching it from *fan auto* to *fan on*?

The ACT method (or similar method you devise) will help determine if any of the common problems found in air-conditioning units exist. Narrowing down the problem will direct a technician into diagnosing the problem quickly. Each job is a little different, and a systematic way to start troubleshooting is important. Using the ACT method provides a fast and logical sequence for checking the system and at the same time it familiarizes the technician with the system under scrutiny.

OPERATOR ERROR

Finally, do not assume that the customer knows how to operate the air-conditioning and heating system. In a few instances the customer will have the thermostat set to an incorrect heating or cooling selection. The thermostat may be in the off position. It is not easy to determine if some thermostat switches are set to heating, cooling, or off. In the cooling mode, drop the temperature into the 60°F range to assure that the unit will cut on and not terminate the operation while you are checking out the system. Turn the thermostat up again prior to leaving the job.

Again, it is best to have gauges and instruments hooked up to the outdoor unit prior to checking the thermostat setting. You should remove the control voltage wiring or open the disconnect before checking the thermostat setting.

TROUBLESHOOTING PROCESS

Troubleshooting requires a sequence of logical thought processes. There is no one way that is better than another. The goal of troubleshooting is to diagnose and repair a problem as quickly and accurately as possible. Professional air-conditioning technicians are not just parts' changers. They are mechanical doctors. A technician takes all the vital signs and decides on the correct diagnosis. The mechanical condition may have multiple interrelated problems, which will challenge the technician. Diagnosing and repairing one problem at a time may be the only way to experience victory over the system. At times, even the most seasoned technician will need to request assistance. There will be challenges that make a technician feel "brain dead," and discussing the problem with someone else will clarify the problem. A technician may have the mental tools to diagnosis the problem but may be jammed by unknown brain processes. Getting away from the job for a few minutes to relax is beneficial to the brain in a puzzled condition. Take a break; it may produce the creative thought to solve the problem. In order for technicians to reach this analysis stage, they must have training and experience in troubleshooting skills. The following chapters will bolster these skills.

It is up to individuals to practice what they learns. It is difficult to break troubleshooting behavior that is disorganized and inconsistent. Many technicians learn troubleshooting skills on their own or from coworkers. Unfortunately, many of the on-the-job mentors do not have good troubleshooting skills themselves or they are not able to communicate these skills effectively. Some air-conditioning technicians are not willing to share their information with others for fear of devaluing their knowledge. The thought is that they should keep their knowledge under wraps to affirm their value to their employer. Sharing knowledge and experience will make you an invaluable leader to your company.

SUMMARY

- This chapter marks the beginning of the sections on refrigerant and electrical troubleshooting. There are three major areas of troubleshooting. Electrical problems are the most common, followed by refrigeration-related problems. A combination of electrical and refrigeration problems do also occur. Be aware of operator error.

- It is difficult to teach an individual troubleshooting because everyone learns a little differently and everyone has a different plan for attacking problems. The goal of this textbook is to give readers many different approaches to troubleshooting, thus allowing the opportunity of choosing the best method for themselves. This chapter recommends the ACT method of sorting out a problem. The ACT method is a simple acronym that will remind the technician on what to check. Many problems are very simple and are turned into complex troubleshooting because the obvious was overlooked. Determine what is working and that assists in narrowing down what is not working.

- It is important that the technician take control of the air-conditioning equipment. It is advisable that the system be controlled by the thermostat control wiring at the unit or by a nearby disconnect. Install the gauges and Amprobe prior to starting the equipment. The readings that are obtaining in the first few seconds of operation may be vital in rapid diagnostics.

- Finally, if the technician is confused, take a break or call for technical assistance in a discreet way that does not discredit the trouble-

shooting process in the eyes of the customer. The air conditioning technician will need to not only troubleshoot equipment but also discuss the problem with the customer in an educated manner. This will come with time and experience.

PROBLEMS AND QUESTIONS

1. What are the four types of general troubleshooting problems?

2. What steps should an air-conditioning technician take if he or she becomes temporarily incapable of making a diagnostic decision?

3. Why does a technician need a quick-check strategy?

4. The chapter recommends the ACT method of rapidly checking an air-conditioning system. What method will you use? Detail your personal method. Why did you select this method of troubleshooting logic?

5. Electrical malfunctions make up the majority of air-conditioning problems. List five general areas a technician should investigate when determining electrical problems.

6. Explain how a refrigerant-side problem can be manifested as electrical problem.

7. What is the biggest enemy of troubleshooting? What can a technician do to combat this enemy?

8. What are the biggest assets in troubleshooting an air-conditioning system?

9. What should the customer be asked to do when he or she reports a problem?

10. How should the customer be told to operate the system while waiting for the technician to arrive?

11. Why is it important to hook up the gauges and Amprobe before starting the condensing unit?

12. Describe a personal practice you will use when troubleshooting an air-conditioning system.

13. What is troubleshooting?

14. Why is dangerous to manually push in the contactor points to check the operation of a condensing unit?

15. What are five causes of a compressor tripping out on its thermal overload?

CHAPTER ELEVEN

MECHANICAL TROUBLESHOOTING— PART ONE

AFTER STUDYING THIS CHAPTER, THE STUDENT WILL BE ABLE TO:

•Identify common operating problems associated with residential and small commercial air-conditioning systems.
•Identify useful instruments necessary to troubleshoot air-conditioning systems.
•State air system problems and remedies.
•Explain refrigeration system problems pertaining to refrigerant quantity and refrigerant flow rate.
•Identify expected temperatures for the DX coil, condensing temperature, and refrigerant subcooling temperature.
•Explain the causes and remedies for insufficient, unbalanced, or excessive loads.

•Describe the effects of low and high ambient conditions.
•Identify refrigerant undercharge and overcharge.
•State the conditions that result when there is a liquid-line restriction or plugged capillary tube.
•Describe the effects of a suction-line or hot-gas line restriction.
•Determine the troubleshooting methods for diagnosing an inefficient compressor.
•Analyze the malfunction symptoms to determine the cause of the problem.
•Determine the remedial action necessary to place the system back in proper operating condition.

SAFETY

It is important in troubleshooting that a technician pay utmost attention to safety measures. The following general measures should be strictly followed on every job to provide for personal safety:

1. Wear safety glasses and gloves when handling nitrogen, refrigerants or when brazing.
2. Recover or recycle refrigerant using an approved device.
3. Shut off all power when working on electrical equipment. Verify this condition.
4. If the work must be done while the electrical equipment is energized, remove all watches and rings to reduce the risk of shock.

5. Always read the specific safety recommendations in the manufacturer's installation and service literature.

GENERAL INFORMATION

Technicians servicing air-conditioning equipment find that system performance can be greatly improved not only by good installation practices but also by good maintenance. Careful reference to the manufacturer's installation instructions is important when the system is installed. Good maintenance keeps the equipment running in its original efficient manner.

147

In troubleshooting air-conditioning systems, some of the most useful instruments are:

1. A gauge manifold for measuring suction and discharge pressures at the condensing unit.

2. A minimum of five thermometers or an electronic thermometer with provision for connecting to at least five remote temperature sensors (as discussed later).

3. A sling psychrometer for measuring wet- and dry-bulb temperature to identify the conditions of the supply and return air.

4. A clamp-on ammeter with capability of reading amperes, voltages, and ohms.

The thermometers are used for measuring the following temperatures:

1. Supply air

2. Return air

3. Condenser entering air

4. Suction line at the coil outlet or at the compressor

5. Liquid line at the condenser outlet

By measuring the suction-line temperature, the superheat can be determined. Because the high-side pressure has been measured, the evaporating temperature can be read from a pressure-temperature chart. The superheat is the difference between the suction-line temperature and the evaporating temperature.

By measuring the liquid-line temperature, the subcooling can be determined. Because the discharge pressure has been measured, the condensing temperature can be read from the pressure-temperature chart or guage. The subcooling is the difference between the liquid-line temperature and the condensing temperature.

PROBLEM ANALYSIS

In this chapter the discussion is specifically directed to problems and solutions that apply to the mechanical side of air-conditioning systems. For those solutions that apply to the refrigeration system, please refer to Chapter 8. Electrical troubleshooting is discussed in future chapters.

A typical troubleshooting job begins with a unit that was originally checked out and put into proper operating condition; has been running satisfactorily for a period of time; and has developed a problem. Problems in mechanical air-conditioning system are classified in two categories: *air-system problems* and *refrigeration-system problems*.

AIR-SYSTEM PROBLEMS

The primary problem that can occur in the air-side category is a reduction in quantity of the air flow. Air-handling systems do not suddenly increase in capacity, that is, increase the amount of air across the coil. On the other hand, the refrigeration system does not suddenly increase in heat-transfer ability. The first check is therefore the temperature drop of the air through the DX coil. After measuring the return- and supply-air temperatures and subtracting to get the temperature drop, is it higher or lower than it should be?

This means that "what it should be" has to be determined first. This is done by using the sling psychrometer to measure and determine the return-air wet-bulb temperature and relative humidity. From this the proper temperature drop across the coil can be determined from the chart in Figure 11–1.

Using the required temperature drop as compared to the actual temperature drop, the problem can be classified as either an air problem or a refrigerant problem. If the actual temperature drop is greater than the required temperature drop, the air quantity has been reduced; look for problems in the air-handling system. These could be

1. Dirty air filters

2. Blower motor and drive at reduced speed

3. Unusual restrictions in the duct system

Figure 11–1   Air-temperature drop for various return-air conditions.

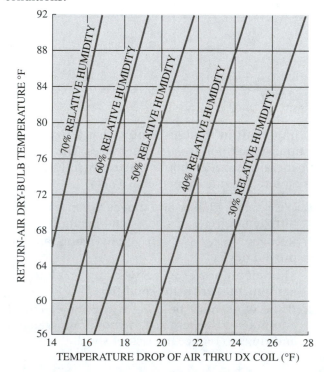

Air Filters

Air filters of the throwaway type should be replaced at least twice each year, at the beginning of both the cooling and heating seasons. In some areas where dust is high, they may have to be replaced as often as every 30 days. In commercial and industrial applications, a regular schedule of maintenance must be worked out for best performance and longest equipment life. Because this is the most common problem of air failure, check the filtering system first. Blower systems that operate continually will need air filter replacement more frequently compared to systems that cycle with the condenser.

Blower Motor and Drive

Check the blower motor and drive in the case of belt-driven blowers to make sure that

1. The blower motor is properly lubricated and operating freely.
2. The blower wheel is clean. The blades could be loaded with dust and dirt or other debris. If the wheel is dirty, it must be removed and cleaned. Do not try brushing only, because a poor cleaning job will cause an imbalance to occur in the wheel. Extreme vibration in the wheel and noise will result. This could cause deterioration of the wheel. Take care not to disturb balancing clips on the blower wheel.
3. On belt-driven blowers, the blower bearing must be lubricated and operating freely.
4. The blower drive belt must be in good condition and properly adjusted. Cracked or heavily glazed belts must be replaced. Heavy glazing can be caused by too much tension on the belt, driving the belt down into the pulleys. Proper adjustment requires the ability to depress the belt midway between the pulleys approximately 1 in. for each 12 in. between the pulley shaft centers.

Unusual Restrictions In Duct Systems

Placing furniture or carpeting over return-air grilles reduces the air available for the blower to handle. Shutting off the air to unused areas will reduce the air over the coil. Covering a return-air grille to reduce the noise from the centrally located furnace or air handler may reduce the objectionable noise, but

ON THE JOB

Problem

Not enough cooling and the conditions are hot and humid. You arrive at the job and find the condenser fan and compressor running. The furnace fan is running and the filter is clean enough. You take a temperature reading across the evaporator coil and the delta-T is 14°F. On a 10 Seer unit the pressures are measured to be 78 psig on the low side and 310 psig on the high side. The outdoor temperature is 80°F, indoor 80°F. What do you do?

Solution

The pressures are high. Check the condenser coil. The surface of the coil is slightly dirty, but a closer look between the fins exposes a buildup of lint inside the coil. A good cleaning with a strong spray from a hose and some coil cleaner caused a drop in discharge pressures from 310 to 295 psig. The system seemed to be cooling better, but the space temperature was not dropping below 80°F after 1 h of continuous running time. The technician suspects problems related to the heat-removing capabilities of the condenser. Upon closer inspection, the technician notices that the condenser coil has a double-row, or back-to-back, design. The technician measures subcooling at 3°F. The technician determines that the condenser coil must have dirt trapped between the double-row coil. The refrigerant is recovered from the condenser and the coil is separated and cleaned. The final check finds the head pressure at 240 psig, subcooling at 16°F, superheat in range, and the unit cycling at a comfortable 72°F. Using the subcooling check early in the procedure would have saved the technician time in determining the exact cause of the problem. Use as many charging methods as possible early in the diagnostic procedure to nail down the problem quickly.

Problem

Not enough cooling. The technician arrives and checks the operation. Temperatures were 85°F outside and 80°F inside. On an even Seer unit the suction pressure was 50 psig and the discharge pressure was 220 psig. Everything was running. The temperature drop across the evaporator was 26°F. The filter was just changed. The customer said they change it once a month. The registers have very little airflow. What do you do?

Solution

The problem seems to be airflow related. The technician inspects the evaporator coil and finds it very dirty. He brushes the coil off with a steel wire brush and cleans it with evaporator coil cleaner. The technician inspects and tests the blower and motor. He lubricates the motor, and it seems to be running at the correct speed. The duct system is in good shape. The system is started. Pressures, superheat, amperage, subcooling and delta-T temperatures are all in range. The customer and technician are satisfied with the cooling output of unit.

it also drastically affects the operation of the system by reducing the air quantity.

Collapse of the return-air duct system will affect the entire duct system performance. Air leaks in the return duct will raise the return-air temperature and reduce the temperature drop across the coil. Check the evaporator coil. It should be clean and free from dirt build-up.

REFRIGERATION-SYSTEM PROBLEMS

When the temperature drop across the coil is less than required, this means that the heat-handling capacity of the system has been reduced.

These problems can be simply divided into two categories: (1) *refrigerant quantity,* and (2) *refrigerant flow rate.* If the system has the correct amount of refrigerant charge and refrigerant is flowing at the desired rate, the system should work properly and deliver rated capacity. Any problems in either category will affect the temperatures and pressures that will occur in the unit when the correct amount of air is supplied over the DX coil for the capacity of the unit. Obviously, if the system is empty of refrigerant, a leak has occurred, and it must be found and repaired. The system must be evacuated thoroughly and recharged with the correct amount of refrigerant. If the system will not operate at all, it is probably an electrical problem that must be found and corrected.

In this chapter the discussion is confined to those problems that affect the operating capacity of the system. The system will start and run but will not produce satisfactory results. This means that the amount of heat picked up in the coil plus the amount of motor heat added and the total rejected from the condenser is not the total heat quantity the unit is designed to handle. To determine the problem, all the information listed in the section "General Information" (begining of chapter) must be measured. These results compared to normal operating results will generally identify the problem. The use of the word *normal* does not imply a fixed set of pressures and temperatures. These vary with each make and model of the system. A few temperatures are fairly consistent throughout the industry and can be used for comparison:

1. DX-coil operating temperature
2. Condensing-unit condensing temperature
3. Refrigerant subcooling

These items must also be modified according to the seasonal energy efficiency ratio (SEER) of the unit. The reason for this is that the amount of evaporation and condensing surface designed into the unit are the main factors in efficiency rating. A larger condensing surface results in a lower condensing temperatures and a higher energy efficiency ratio (SEER). A larger evaporating surface results in a higher suction pressure and a higher SEER. SEER is a measure of efficiency. It is calculated by dividing the seasonal net capacity of the unit in Btu/h by the seasonal watts input. SEER is temporary condition based on the indoor/outdoor temperature, indoor wet-bulb temperature and the system charge. SEER is an indicator of seasonal performance.

DX-COIL OPERATING TEMPERATURE

Normal coil operating temperatures can be found by subtracting the design coil split from the average air temperature going through the coil. The coil split will vary with the system design.

Systems in the SEER range of 7.0 to 10.5 will have design splits in the range 25 to 30°F. Systems in the SEER range of 10.5 to 11.9 will have design splits in the range 20 to 25°. Systems with 12.0+ SEER ratings will have design splits in the range 15 to 20°F. The formula used for determining coil operating temperatures is:

$$COT = \left(\frac{EAT + LAT}{2}\right) - split$$

where

COT = coil operating temperature
EAT = temperature of air entering the coil
LAT = temperature of air leaving the coil

The latter two temperatures added together and divided by 2 will give the average air temperature. This is also referred to as the *mean temperature difference* (MTD).

"Split" is the design split according to the SEER rating. For example, a unit having an entering air condition of 80° DB and a 20°F temperature drop across the evaporator coil, will have an operating coil temperature determined as follows.

WORKING EXAMPLE

For an SEER rating of 7.0 to 10.5,

$$COT = \left(\frac{80 + 60}{2}\right) - 25 \text{ to } 30° = 40 \text{ to } 45°F$$

For an SEER rating of 10.5 to 11.9,

$$COT = \left(\frac{80 + 60}{2}\right) - 20 \text{ to } 25° = 45 \text{ to } 50°F$$

For an SEER rating of 12.0+,

$$COT = \left(\frac{80 + 60}{2}\right) - 15 \text{ to } 20° = 50 \text{ to } 55°F \quad \blacksquare$$

This demonstrates that the operating coil temperature changes with the SEER rating of the unit.

CONDENSING-UNIT CONDENSING TEMPERATURE

The amount of surface in the condenser affects the condensing temperature the unit must develop to operate at rated capacity. The variation in the size of the condenser also affects the production cost and price of the unit. The smaller the condenser, the lower the price, but also the lower the efficiency (SEER) rating. In the same SEER ratings used for the DX coil, at 95°F outside ambient, the 7.0 to 10.5 SEER category will operate in the 25 to 30° condenser split range, the 10.5 to 11.9 SEER category in the 20 to 25° condenser split range, and the 12.0+ SEER category in the 15 to 20° condenser split range.

This means that when the air entering the condenser is at 95°F, the formula for finding the condensing temperature would be:

$$RCT = EAT + split$$

where

RCT = refrigerant condensing temperature
EAT = temperature of the air entering the condenser
split = design temperature difference between the entering air temperature and the condensing temperatures of the hot high-pressure vapor from the compressor.

WORKING EXAMPLE

Using the formula with 95°F EAT, the split for the various SEER systems would be:

For an SEER rating of 7.0 to 10.5,

$$RCT = 95° + 25 \text{ to } 30° = 120 \text{ to } 125°F$$

For an SEER rating of 10.5 to 11.9,

$$RCT = 95° + 20 \text{ to } 25° = 115 \text{ to } 120°F$$

For an SEER rating of 12.0+,

$$RCT = 95° + 15 \text{ to } 20° = 110 \text{ to } 115°F \quad \blacksquare$$

This demonstrates that operating head pressures vary not only from changes in outdoor temperatures but with the different SEER ratings.

REFRIGERANT SUBCOOLING

The amount of subcooling produced in the condenser is determined primarily by the quantity of refrigerant in the system. The temperature of the air entering the condenser and the load in the DX coil will have only a small effect on the amount of subcooling produced. The amount of refrigerant in the system has the predominant effect. Therefore, regardless of SEER ratings, the unit should have, if properly charged, a liquid subcooled to 10 to 20°F. High outdoor temperatures will produce the lower subcooled liquid because of the reduced quantity of refrigerant in the liquid state in the system. More refrigerant will stay in the vapor state to produce the higher pressure and condensing temperatures needed to eject the required amount of heat.

ANALYZING PROBLEMS

Using the information obtained from the two pressure gauges, a minimum of five thermometers, the sling psychrometer, and a clamp-type ammeter, we can analyze the system problems by using the chart in Figure 11–2.

The figure shows that there are 11 probable causes of trouble in an air-conditioning system. After each probable cause is the reaction that the cause would have on the refrigeration system low-side or suction pressure, the DX-coil superheat, the high-side or discharge pressure, the amount of subcooling of the liquid leaving the condenser, and the amperage draw of the condensing unit.

INSUFFICIENT OR UNBALANCED LOAD

Insufficient air over the DX coil would be indicated by a greater-than-desired temperature drop through the coil. An unbalanced load on the DX coil would also give the opposite indication; some of the circuits of the DX coil would be overloaded, while others would be lightly loaded. This would result in a mixture of air off the coil that would cause a reduced temperature drop of the air mixture. The lightly loaded sections of the DX coil would allow liquid refrigerant to leave the coil and enter the suction manifold and suction line.

In TXV valve systems, the liquid refrigerant passing the feeler bulb of the TXV valve would cause the valve to close down. This would reduce the operating temperature and capacity of the DX coil as well as lower the suction pressure. This reduction would be very pronounced. The DX-coil operating superheat would be very low, probably zero, because of the liquid leaving some of the sections of the DX coil.

High-side or discharge pressure would be low due to the reduced load on the compressor, reduced amount of refrigerant vapor pumped, and reduced heat load on the condenser. Condenser liquid subcooling would be on the high side of the normal range because of the reduction in refrigerant demand by the TXV valve. Condensing-unit amperage draw would be down due to the reduced load.

In systems using capillary tubes, the unbalanced load would produce a lower temperature drop of the air through the DX coil because the amount of refrigerant supplied by the capillary tubes would not be reduced; therefore, the system pressure (boiling point) would be approximately the same.

The DX-coil superheat would drop to zero with floodout of the refrigerant into the suction line. Under extreme cases of unbalance, liquid return to the compressor could cause compressor damage. The reduction in heat gathered in the DX coil and the lowering of the refrigerant vapor to the compressor will lower the load on the compressor. The compressor discharge pressure (hot-gas pressure) will be reduced.

The flow rate of the refrigerant will be only slightly reduced because of the lower head pressure. The subcooling of the refrigerant will be in the nor-

---

Figure 11–2   Troubleshooting chart for refrigeration and air-conditioning systems, showing symptoms and probable causes.

Probable Cause	Low-side (Suction) Pressure (psig)	D.X. Coil Superheat (°F)	High-side (Hotgas) Pressure (psig)	Condenser Liquid Subcooling (°F)	Cond. Unit Amperage Draw (Amps)
1 Insufficient or unbalanced load	Low	Low	Low	Normal	Low
2 Excessive load	High	High	High	Normal	High
3 Low ambient (cond. entering air °F)	Low	High	Low	Normal/High	Low
4 High ambient (cond. entering air °F)	High	High	High	Normal	High
5 Refrigerant undercharge	Low	High	Low	Low	Low
6 Refrigerant overcharge	High	Low	High	High	High
7 Liquid line restriction	Low	High	High/normal	High	Low
8 Plugged capillary tube	Low	High	High/normal	High	Low
9 Suction line restriction	Low	High	Low	Normal	Low
10 Hot gas line restriction	High	High	High	Normal	High
11 Inefficient compressor	High	High	Low	Low	Low

Problem

Not enough cooling on a R-22 system. When you arrive you find everything working, including the compressor, the condenser fan, and the furnace fan. The filters are clean. You take a temperature reading on each side of the evaporator coil and hook up your gauges to the service ports. The evaporator delta-T is 9°F, and the pressure on the gauges read 82 psig on the suction side and 160 psig on the discharge side. What should you do next?

Solution

With the high suction pressure and low discharge pressure you suspect a defective compressor. The high side is equalizing with the low side. To validate this diagnosis, pump down the system by operating the system and closing off the liquid line service valve. After 10 min in the pumpdown mode the suction side will pull down only to 5 psig. The target pulldown vacuum is 15 to 20 in. HG. The compressor is replaced and the unit is cooling properly. All pressures, temperatures, and superheat reading are good.

mal range. The amperage draw of the condensing unit will be slightly lower because of the reduced load on the compressor and reduction in head pressure.

EXCESSIVE LOAD

In this case the opposite effect exists. The temperature drop of the air through the coil will be low, so the unit cannot cool the air as much as it should. Air is moving through the coil at too high a velocity. There is the possibility that the temperature of the air entering the coil is higher than the return air from the conditioned area. This could be from leaks in the return-air system drawing air from unconditioned areas.

The excessive load raises the suction pressure. The refrigerant is evaporating at a rate faster than the pumping rate of the compressor. The superheat developed in the coil will be as follows:

1. If the system uses a TXV valve, the superheat will be normal to slightly high. The valve will operate at a higher flow rate to attempt to maintain superheat settings.

2. If the system uses capillary tubes, the superheat will be high. The capillary tubes cannot feed enough increase in refrigerant quantity to keep the DX coil fully active.

The high-side or discharge pressure will be high. The compressor will pump more vapor because of the increase in suction pressure. The condenser must handle more heat and will develop a higher condensing temperature to eject the additional heat. A higher condensing temperature means higher high-side pressure. The quantity of liquid in the system has not changed, nor is the refrigerant flow restricted. The liquid subcooling will be in the normal

range. The amperage draw of the unit will be high because of the additional load on the compressor.

LOW AMBIENT (CONDENSER ENTERING AIR) TEMPERATURE

In this case, the condenser heat-transfer rate is excessive, producing an excessively low discharge pressure. As a result, the suction pressure will be low because the amount of refrigerant through the pressure-reducing device will be reduced. This reduction will reduce the amount of liquid refrigerant supplied to the DX coil. The coil will produce less vapor and the suction pressure drops.

The decrease in the flow rate into the coil reduces the amount of active coil, and a higher superheat results. In addition, the reduced system capacity will decrease the amount of heat removed from the air. There will be higher temperature and relative humidity in the conditioned area and the high-side pressure will be low. This starts a reduction in system capacity. The amount of subcooling of the liquid will be in the normal to high range. The quantity of liquid in the condenser will be higher, but the heat transfer rate of the lower temperatures is less. This will result in a subcooling in the normal to high range. The amperage draw of the condensing unit will be less. The compressor is doing less work.

The amount of drop in the condenser ambient air temperature that the air-conditioning system will tolerate depends on the type of pressure-reducing device in the system. Systems using capillary tubes will have a gradual reduction in capacity as the outside ambient drops from 95°F. This gradual reduction occurs down to 65°F. Below this temperature the capacity loss is drastic, and some means of maintaining head pressure must be employed. The most reliable means is control of air through the condenser via dampers in the airstream, variable-speed condenser fan, or cycling the condenser fan.

Systems that use TXV valves will maintain higher capacity down to an ambient temperature of 35°F. Below this temperature, controls must be used. The control of cfm through the condenser using dampers or the condenser-fan speed control can also be used. In larger TXV-valve systems, liquid quantity in the condenser is used to control head pressure.

HIGH AMBIENT (CONDENSER ENTERING AIR) TEMPERATURE

The higher the temperature of the air entering the condenser, the higher the condensing temperature of the refrigerant vapor to eject the heat in the vapor. The higher the condensing temperature, the higher the head pressure. The suction pressure will be high for two reasons: (1) the pumping efficiency of the compressor will be less; and (2) the higher temperature of the liquid will increase the amount of flash gas in the coil, further reducing the system efficiency.

The amount of superheat produced in the coil will be different in a TXV-valve system and a capillary-tube system. In the TXV-valve system the valve will maintain superheat close to the limits of its adjustment range even though the actual temperatures involved will be higher. In a capillary-tube system, the amount of superheat produced in the coil is the reverse of the temperature of the air through the condenser. The flow rate through the capillary tubes is directly affected by the head pressure. The higher the air temperature, the higher the head pressure and the higher the flow rate. As a result of the higher flow rate, the subcooling is lower.

Figure 11–3 shows the superheat that will be developed in a properly charged air-conditioning system using capillary tubes. Do not attempt to charge a capillary system below 65°F, as system operating characteristics become very erratic. This chart does not consider wet-bulb temperatures entering the evaporator. The entering WB has a big impact on superheat.

The head pressure will be high at the higher ambient temperatures because of the higher condensing temperatures required. The condenser liquid subcooling will be in the lower portion of the normal range. The amount of liquid refrigerant in the condenser will be reduced slightly because more will stay in the vapor state to produce the higher pressure and condensing temperature. The amperage draw of the condensing unit will be high.

REFRIGERANT UNDERCHARGE

A shortage of refrigerant in the system means less liquid refrigerant in the DX coil to pick up heat, and lower suction pressure. The smaller quantity of liquid supplied the DX coil means less active surface in the coil for vaporizing the liquid refrigerant, and more surface to raise vapor temperature. The superheat will be high. There will be less vapor for the compressor to handle and less head for the condenser to reject, lower high-side pressure, and lower condensing temperature.

The amount of subcooling will be below normal to none, depending on the amount of undercharge. The system operation is usually not affected very seriously until the subcooling is zero and hot gas starts to leave the condenser, together with the liquid refrigerant. The amperage draw of the condensing unit will be slightly less than normal.

REFRIGERANT OVERCHARGE

An overcharge of refrigerant will affect the system in different ways, depending on the pressure-reducing device used in the system and the amount of overcharge.

TXV-VALVE SYSTEMS

In systems using a TXV valve, the valve will attempt to control the refrigerant flow into the coil to maintain the superheat setting of the valve. However, the extra refrigerant will back up into the condenser, occupying some of the heat transfer area that would otherwise be available for condensing. As a result, the discharge pressure will be slightly higher than normal, the liquid subcooling will be high, and the unit amperage draw will be high. The suction pressure and DX coil superheat will be normal. Excessive overcharging will cause even higher head pressure, and hunting of the TXV valve. TXV systems can be overcharged to flood an evaporator.

Figure 11–3 The effects of outdoor (ambient) temperature on fixed metering superheat. The chart does not consider the wet-bulb temperature entering the evaporator. This WB is critical in determining correct superheat.

Outdoor Air Temperature Entering Condenser Coil (°F)	Superheat (°F)
65	30
75	25
80	20
85	18
90	15
95	10
105 & above	5

For TXV-valve systems with excessive overcharge:

1. The suction pressure will be high. Not only does the reduction in compressor capacity (due to higher head pressure) raise the suction pressure, but the higher pressure will cause the TXV valve to overfeed on its opening stroke. This will cause a wider range of "hunt" of the valve.

2. The DX coil superheat will be very erratic from the low normal range to liquid out of the coil.

3. The high-side or discharge pressure will be extremely high.

4. Subcooling of the liquid will also be high because of the excessive liquid in the condenser.

5. The condensing unit amperage draw will be higher because of the extreme load on the compressor motor.

CAPILLARY-TUBE SYSTEMS

The amount of refrigerant in the capillary tube system has a direct effect on system performance. An overcharge has a greater effect than an undercharge, but both affect system performance, efficiency (SEER) and operating cost.

Figures 11–4 through 11–6 show how the performance of an air-conditioning system is affected by an incorrect amount of refrigerant charge on a fixed metering devise system.

Shown in Figure 11–4, at 100% of correct charge (55 oz), the unit developed a net capacity of 26,200 Btu/h. When the amount of charge was varied 5% in either direction, the capacity dropped as

the charge varied. Removing 5% (3 oz) of refrigerant reduced the net capacity to 25,000 Btu/h. Another 5% (2.5 oz) reduced the capacity to 22,000 Btu/h. From there on the reduction in capacity became very drastic: 85% (8 oz), 18,000 Btu/h; 80% (11 oz), 13,000 Btu/h; and 75% (14 oz), 8000 Btu/h.

Addition of overcharge had the same effect but at a greater reduction rate. The addition of 3 oz of refrigerant (5%) reduced the next capacity to 24,600 Btu/h; 6 oz added (10%) reduced the capacity to 19,000 Btu/h; and 8 oz added (15%) dropped the capacity to 11,000 Btu/h. This shows that overcharging of a unit has a greater effect per ounce of refrigerant than does undercharging.

Figure 11–5 is a chart showing the amount of electrical energy the unit will demand because of pressure created by the amount of refrigerant in the system, with the only variable being the refrigerant charge. At 100% of charge (55 oz) the unit required 32 kW. As the charge was reduced, the wattage demand also dropped, 29.6 kW at 95% (3 oz), 27.6 kW at 90% (6.5 oz), 25.7 kW at 85% (8 oz), 25 kW at 80% (11 oz), and 22.4 kW at 75% (14 oz short of correct charge). When the unit was overcharged, the wattage required went up. At 3 oz (5% overcharge) the wattage required was 34.2 kW; at 6 oz (10% overcharge), 39.5 kW; and at 8 oz (15% overcharge), 48 kW.

Figure 11–6 shows the efficiency of the unit (SEER rating) based on the Btu/h capacity of the system versus the wattage demand of the condensing unit. At correct charge (55 oz) the efficiency

Figure 11–4 The effect of the refrigerant charge on the capacity of the unit.

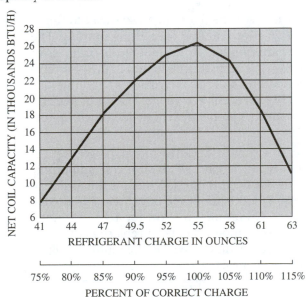

Figure 11–5 The effect of the refrigerant charge on the kW demand of the condensing unit.

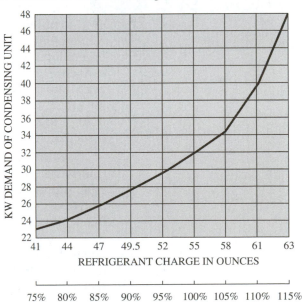

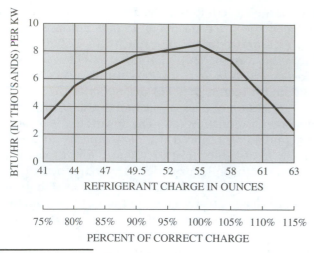

Figure 11–6 The effect of the refrigerant charge on the Btu/h/kW ratio.

(SEER rating) of the unit was 8.49. As the refrigerant was reduced, the SEER rating dropped to 8.22 at 95% of charge, 7.97 at 90%, 7.03 at 85%, 5.2 at 80%, and 3.57 at 75% of full refrigerant charge. When refrigerant was added, adding 5% (3 oz) the SEER rating dropped to 7.19. At 10% (6 oz) the SEER was 4.8, and at 15% overcharge (8 oz) the SEER was 2.29. From these charts the only conclusion is that the capillary-tube systems must be charged to the correct charge with only a −5% tolerance. There charge is critical.

The effect of overcharge produces a high suction pressure because the refrigerant flow to the DX coil increases. Suction superheat will decrease be-

cause of the additional quantity to the DX coil. At approximately 8 to 10% of overcharge, the suction superheat becomes zero and liquid refrigerant will leave the DX coil. This will cause flooding of the compressor and greatly increases the chance of compressor failure. The high-side or discharge pressure will be high because of the extra refrigerant in the condenser. Liquid subcooling will also be high for the same reason. The wattage draw will increase due to the greater amount of vapor pumped as well as the higher compressor discharge pressure.

LIQUID-LINE RESTRICTION

A liquid-line restriction reduces the amount of refrigerant to the pressure-reducing device. Both TXV-valve systems and capillary-tube systems will then operate with reduced refrigerant flow rate to the DX coil. The restriction can be in the form of a blockage in the metering device, drier or tubing. The following observations can be made of liquid-line restrictions:

1. The suction pressure will be low because of the reduced amount of refrigerant to the DX coil.

2. The suction superheat will be high because of the reduced active portion of the coil, allowing more coil surface for increasing the vapor temperature as well as reducing the refrigerant boiling point.

3. The high-side or discharge pressure will be low or near normal because of the reduced load on the compressor.

ON THE JOB

Problem

Not enough cooling or insufficient cooling does not always lead to the refrigeration cycle. You arrive on the job and find everything running. The condenser fan and compressor are running. The temperature drop across the evaporator is 24°F. The pressures are 60 psig suction and 220 psig discharge on a R-22 system. The suction temperature is 39°F. There is a humidifier hooked to the plenum with a bypass to the return air. The humidifier is turned off. It appears to be an airflow problem. Where do you go from here?

Solution

Check everything that relates to airflow. Check the motor, blower squirrel cage, supply- and return-air duct restrictions, evaporator coil, and humidifier bypass damper. The best device for this measurement is the airflow hood. The customer is impressed by the use of the flow hood and the technician is able to determine that CFM rating is too low for a 3-ton system. The technician learns that the blown motor had been replaced last week. It had been wired for medium-speed operation. Wiring the motor to high-speed operation corrects the cooling problems.

4. Liquid subcooling will be high. The liquid refrigerant will accumulate in the condenser. It cannot flow out at the proper rate because of the restriction. As a result, the liquid will cool more than desired.

5. The amperage draw of the condensing unit will be low.

PLUGGED CAPILLARY TUBE OR FEEDER TUBE

Either a plugged capillary tube or plugged feeder tube between the TXV-valve distributor and the coil will cause part of the coil to be inactive. The system will then be operating with an undersized coil, resulting in the following:

1. The suction pressure will be low because the coil capacity has been reduced.

2. The suction superheat will be high in the capillary tube systems. The reduced amount of vapor produced in the coil and resultant reduction in suction pressure will reduce compressor capacity, head pressure, and the flow rate of the remaining active capillary tubes.

3. The high-side or discharge pressure will be low or near normal.

4. Liquid subcooling will be high; the liquid refrigerant will accumulate in the condenser.

5. The unit amperage draw will be low.

In TXV-valve systems, the following will result:

1. A plugged feeder tube reduces the capacity of the coil. The coil cannot provide enough vapor to satisfy the pumping capacity of the compressor and the suction pressure balances out at a low pressure.

2. The superheat, however, will be in the normal range because the valve will adjust to the lower operating conditions and maintain the setting superheat range.

3. The high-side or discharge pressure will be low because of the reduced load on the compressor and condenser.

4. Liquid subcooling will be high because of the liquid refrigerant accumulating in the condenser.

5. The amperage draw of the condensing unit will be low.

SUCTION-LINE RESTRICTION

This could be caused by a plugged compressor suction line strainer, a kink in the suction line, or a solder joint filled with solder. It results in a high pressure drop between the DX coil and the compressor.

1. The suction pressure, if measured at the condensing unit end of the suction line, will be low.

2. The superheat, as measured by suction-line temperature at the DX coil and suction pressure (boiling point) at the condensing unit, will be extremely high.

3. The high-side or discharge pressure will be low because of reduced load on the compressor.

4. The low suction and discharge pressure usually indicate a refrigerant shortage. Warning: The liquid subcooling is normal to slightly above normal. This indicates a surplus of refrigerant in the condenser. Most of the refrigerant is in the coil, where the evaporation rate is low due to the higher operating pressure in the coil.

5. The amperage draw of the condensing unit would be low because of the light load on the compressor.

HOT GAS-LINE RESTRICTION

The high-side or compressor discharge pressure will be high if measured at the compressor outlet or low if measured at the condenser outlet or liquid line. In either case the compressor amperage draw will be high. Therefore:

1. The suction pressure is high due to reduced pumping capacity of the compressor.

2. The DX-coil superheat is high because the suction pressure is high.

3. The high-side pressure is high when measured at the compressor discharge or low when measured at the liquid line.

4. Liquid subcooling is in the high end of normal range.

5. Even with all of this, the compressor amperage draw is above normal. All symptoms point to an extreme restriction in the hot-gas line. This problem is easily found when the discharge pressure is measured at the compressor discharge.

Where the measuring point is the liquid line at the condenser outlet, the facts are easily misinterpreted. High suction pressure and low discharge pressure will usually be interpreted as an inefficient compressor. The amperage draw of the compressor must be measured. The high amperage draw indicates that the compressor is operating against a high discharge pressure. A restriction apparently exists between the outlet of the compressor and the pressure measuring point.

INEFFICIENT COMPRESSOR

This problem is last on the list because it is the least likely to be a problem. When the compressor will not pump the required amount of refrigerant vapor:

1. The suction pressure will balance out higher than normal.

2. The DX-coil superheat will be high.

3. The high-side or discharge pressure will be extremely low.

4. Liquid subcooling will be low because not much heat will be in the condenser. The condensing temperature will therefore be close to the entering air temperature.

5. The amperage draw of the condensing unit will be extremely low, indicating that the compressor is doing very little work.

6. The compressor will not pump down into a 15 inch vacuum.

SUMMARY

- When troubleshooting, the technician should follow the following personal safety measures:

    Wear safety glasses and gloves.

    Recover, recycle, or reclaim refrigerants using approved devices.

    Shut off all power when working on electrical equipment.

    If the work must be done with power applied, remove all watches and rings.

    Always read manufacturer's safety recommendations.

- In troubleshooting air-conditioning equipment, the minimum set of test instruments should include:

    A gauge manifold

    At least five thermometers

    A sling psychrometer

    A clamp-on ammeter

- For analyzing the problem, the following temperature readings should be taken:

    Supply air

    Return air

    Ambient air

    Suction line

    Liquid line

- From these temperature readings and the pressure data obtained using the gauge manifold,

the actual performance of the equipment can be compared to normal operation.

- In testing an operating system to determine the cause of the problem, the following actual readings should be compared with typical (normal) readings:

    Suction pressure

    Evaporator superheat

    Discharge pressure

    Condenser subcooling

    Amperage draw

- Air system problems:

    Air filters

    Blower motor drive

    Unusual restrictions in the duct system

- Refrigeration temperature checks:

    DX-coil operating temperatures

    Condensing-unit condensing temperature

    Refrigerant subcooling

- Symptoms of refrigeration problems

    Insufficient or unbalanced load

    Excessive load

    Low ambient temperature

    High ambient temperature

    Refrigerant undercharge

    Refrigerant overcharge

    Liquid-line restriction

    Plugged capillary tube

    Suction-line restriction

    Hot-gas line restriction

    Inefficient compressor

        Higher-than-normal suction pressure

        Low discharge pressure

        High DX-coil superheat

        Low liquid subcooling

        Low compressor amps

From this comparison of operational data, reference to a troubleshooting chart, such as shown in Figure 11–2, will assist in locating the cause of the problem.

As soon as the cause is verified, in most cases the remedy is self evident. For example, if a wiring connection is loose, it needs to be tightened. When a motor is burned out, it needs to be replaced. Where replacement or service of specialized parts is required, usually the manufacturer provides detailed instructions for performing the work. When maintenance is required, such as cleaning a dirty coil or replacing worn out belts, helpful instructions will be found in the section of this manual on preventive maintenance.

PROBLEMS AND QUESTIONS

1. What instruments are needed to properly diagnose problems in an air-conditioning system?

2. At what locations are air-temperature measurements required?

3. If the air-temperature drop through the coil is higher than when the unit was previously serviced, is the problem in the refrigeration or the air part of the system?

4. Problems in the refrigeration portion of the system can be divided into two categories. What are they?

5. Define SEER.

6. SEER is found by dividing the seasonal _____ by the total _____ used during a cooling season.

7. How many probable causes of trouble are there in an air-conditioning system?

8. If the Btu/h load is increased on the evaporator, will the superheat of the evaporator increase or decrease on a TXV valve coil? A capillary tube coil?

9. What effect does low outside ambient temperature have on the capacity of the air-conditioning unit? Why?

10. What is the minimum outside operating temperature of a capillary-tube system and a TXV-valve system?

11. The minimum outside temperature at which a capillary tube system can be properly charged is _____.

12. What is the easiest way to determine if a unit does not have the proper amount of refrigerant charge?

13. Which has the greatest adverse effect on the capacity of a system, an overcharge or an undercharge of refrigerant?

14. The charge quantity tolerance of a capillary tube system is _____.

15. To properly diagnose trouble in the refrigeration system, five operating characteristics must be known. What are they?

16. A sling psychrometer is used during troubleshooting for measuring the dry-bulb temperature and relative humidity of the supply and return air. True or false?

17. By measuring only the suction-line temperature, the superheat can be determined. True or false?

18. Problems in an air-conditioning system are classified in two categories: air-circuit problems and electrical problems. True or false?

19. Dirty air filters create reduced air flow, which creates increased suction pressure. True or false?

20. Heavy glazing on belts can be caused by too much tension on the belt. True or false?

21. The formula used for determining DX-coil operating temperatures is:
    a. COT = EAT + RT − split
    b. COT = EAT + LAT − split
    c. SEER = RAT + AT − split
    d. None of the above

22. The amount of surface in the condenser affects:
    a. The condensing temperature the unit must develop to operate at rated capacity
    b. The price
    c. The SEER
    d. All the above

23. The unit should have, regardless of SEER ratings, a subcooling temperature of
    a. 5 to 10 °F
    b. 10 to 20 °F
    c. 20 to 30 °F
    d. None of the above

24. During excessive loads, when using the capillary tube, the superheat will be
    a. Higher than normal
    b. Normal operating temperatures
    c. Lower than normal
    d. None of the above

25. The inefficient compressor will have
    a. A higher-than-normal suction pressure
    b. High DX-coil superheat and low compressor amps
    c. A lower-than-normal discharge pressure
    d. All the above

CHAPTER TWELVE

MECHANICAL TROUBLESHOOTING— PART TWO

AFTER STUDYING THIS CHAPTER, THE STUDENT WILL BE ABLE TO:

•Identify problems in the operation of the refrigeration system. Analyze the cause(s) of the problem.
•Interpret manufacturer's service instructions and tables to facilitate troubleshooting.
•Demonstrate efficient means of troubleshooting mechanical problems.
•Identify factors causing shortened compressor life.

•Identify the causes of a failed compressor.
•Identify problems and the troubleshooting process for evaporators and condensers.
•State the problems associated with metering devices.
•Troubleshoot operating problems that specifically apply to hermetically sealed systems.

OVERVIEW

Any operating mechanical/electrical equipment will at some time require service. The repair work necessary to place the equipment back on-line is one of the important functions of the HVAC/R technician. The basic term that describes this type of work is *troubleshooting.* Troubleshooting is the process of determining the cause of an equipment malfunction and performing corrective measures. Depending on the problem, this may require a high degree of knowledge, experience and skill.

Basically there are two types of problems: (1) electrical, and (2) mechanical, although there is much overlapping. Whatever the nature of the problem, it is good practice to follow a logical, structured, systematic approach. In this manner the correct solution is usually found in the shortest possible time.

This chapter focuses on mechanical, or refrigerant-side, problems. Future chapters cover the important aspects of electrical troubleshooting. When conducting the initial phase of troubleshooting, hook up the manifold gauge set and the Amprobe to the common of the compressor. Develop a logical sequence to obtain quick information. The ACT method was recommended in Chapter 10. Each technician needs to develop a comfortable first step that gathers valuable information in the least amount of time.

MECHANICAL REFRIGERATION TROUBLESHOOTING

The most efficient means of troubleshooting mechanical problems in the operation of refrigeration systems is a systematic approach. Shortcuts are possible, depending on the problem, the type of system, and the experience of the technician, but it is usually helpful to follow a step-by-step procedure. Here are the steps:

1. Collect information about the problem.
   a. Collect a description of the problem when the service call was received.

b. Collect direct information about the problem by discussion with the customer.

c. Conduct a preliminary power-off visual system inspection.

d. Conduct a preliminary power-on system inspection.

2. Read and calculate the system's vital signs.

a. Read and record vital signs, including suction and discharge pressures for type of refrigerant being used.

b. Calculate the refrigerant liquid subcooling at the metering device.

c. Calculate the refrigerant gas superheat at the compressor.

3. Compare typical versus actual values.

a. Determine typical values for the conditions and system.

b. Compare typical with actual conditions

4. Consult troubleshooting aids.

a. Perform basic system analysis. Using a basic analysis guide (Figure 12–1), select possible system problems, based on a comparison of the five actual to typical vital values shown in the guide.

b. Using the manufacturer's troubleshooting information (Figure 12–2), perform a detailed analysis. Eliminate possible causes of the problem by test or observation, and select the cause that fits the condition.

In using these steps to diagnose the cause of the problem, the answer may be found in the first two steps, eliminating the need to go further. A difficult problem may require completing all four steps. Proceed through the steps only so far as necessary to find the cause of the problem.

For example, in Step 1c, preliminary power-off inspection, one of the following causes for low capacity may be found:

1. Dirty or missing filters
2. Dirty or loose fan
3. Loose belts
4. Dirty or corroded coil or fins
5. Loose or uninsulated TXV bulb
6. Damaged interconnecting piping

In Step 1d, preliminary power-on inspection, one of the following conditions may indicate the source of the problem:

1. Incorrect fan-rotation direction
2. Insufficient air circulation
3. Noise from a loose pulley wheel
4. Odor from an overheated transformer
5. Hot spots on a bearing

In order to illustrate the use of the step-by-step procedure to locate the problem, follow the example below:

WORKING EXAMPLE

A customer using a split-system air conditioner reports that the compressor is running but that the cooling is inadequate. What should you do?

SOLUTION Proceed with step-by-step analysis and use the basic symptom analysis (Step 4a) to solve the problem. Tests and calculations indicate the following:

1. Discharge pressure is high.
2. Suction pressure is high.
3. Superheat is low.

PRACTICING PROFESSIONAL SERVICES

Learning to troubleshoot refrigeration equipment requires a certain level of mechanical competency, but equally important, it requires patience and perseverance to locate the correct problem. Many times a technician will condemn the first problem without locating the cause of a problem. For example, a compressor has an electrical winding burn out to ground and blows the fuses. A technician may assume an electrical problem caused the electrical failure, but there could be other system problems, which may cause an electrical failure. Possible problems include: (1) a floodback due to poor airflow, (2) a faulty metering device, and (3) high head pressures due to condenser airflow problems causing acidic oil. A refrigerant floodback can cause the oil to be washed out of the crankcase, destroying the compressor bearings. If the bearings do not support the rotor, it will drop onto the stator winding and cause electrical failure.

It is very important to troubleshoot the problem and the cause of the problem. This will improve with time and experience for the novice technician.

System Problem	Discharge Pressure	Suction Pressure	Superheat	Subcooling	Amps
Overcharge	Increase	Increase	Decrease	Increase	Increase
Undercharge	Decrease	Decrease	Increase	Decrease	Decrease
Liquid Restriction	*Decreases or stays the same	Decrease	Increase	**Increase	Decrease
Low Evaporator Airflow	Decrease	Decrease	Decrease	Increase	Decrease
Dirty Condenser	Increase	Increase	***Decrease	Decrease	Increase
Low Outside Ambient Conditions	Decrease	Decrease	***Increase	Increase	Decrease
Inefficient Compressor	Decrease	Increase	Increase or okay	Increase or okay	Decrease
TXV Feeler Bulb Charge Lost	*Decreases or stays the same	Decrease	Increase	**Increase or stays the same	Decrease
TXV Feeler Bulb Loose Mounted	Increases or stays the same	Increase	Decrease	Decrease or stays the same	Increase
Poorly Insulated Feeler Bulb	Increases or stays the same	Increase	Decrease	Decrease or stays the same	Increase

*Condition before charge is added ** Condition after system is overcharged *** Fixed metering device

Figure 12–1 Refrigerant troubleshooting chart.

4. Subcooling is high.
5. Amps are high.

Referring to the chart, this condition could be caused by only one condition: an overcharge of refrigerant. Now that the problem is identified, the manufacturer's charging charts can be used to adjust the charge for the present indoor and outdoor conditions. ∎

TROUBLESHOOTING THE COMPRESSOR

Compressors built today are expected to provide many years of constant, trouble-free, quiet operation. In many applications the compressor is required to run 24 h per day, 365 days a year. Such continuous operation, however, is often not as hard on a compressor as is a cycling operation, where temperatures constantly change and oil is not maintained at a constant viscosity.

The compressor must not only be designed to withstand normal operating conditions, but also occasional abnormal conditions such as liquid slugging and excessive discharge pressure. Compressors

have been designed to take extra punishment and yet function properly. Most compressor failures are caused by system faults and not from operating fatigue. The degree of skill technicians use to install, operate, and maintain the equipment will ultimately

KNOWLEDGE EXTENSION

Many compressor manufacturers such as Copeland, Tecumseh, and Bristol provide operation and maintenance manuals for a minimal fee to a service technician. These manuals have troubleshooting guides, charts, wiring diagrams, and specific operating conditions. They are invaluable to any technician servicing and troubleshooting refrigeration equipment and are available at your local refrigeration supply house or by contacting the manufacturer's representative. Several of these companies also offer specialized educational seminars about their product line for service personnel to learn more about proper operation and maintenance of their product. Contact the manufacturer's training department to find out about a seminar in your area.

General Problem/Symptoms	Possible Causes
Plugged Filter-Drier (Liquid Line)	
• Starved Evaporator Symptoms (See Evaporator Sheet) • Compressor Cycles on Low Pressure Switch	1. Dirty Refrigeration System 2. Improper Evacuation/Dehydration 3. Metal Chips, Scale, etc., in System from Installation
Wet Filter-Drier	
• Moisture-Indicating Sight Glass Shows Wet • Valves Stick Intermittently and System Cycles Off from Internal Ice Blockage • Sealed-Tube Test of Refrigerant Shows Wet	1. System Refrigerant Leak 2. Improper Evacuation/Dehydration 3. Leaking Water-Cooled Condenser Tubes 4. Filter-Drier Exposed to Air Before Installation
Undersized Filter-Drier	
• Low System Capacity • Low Compressor Power Draw (kW) • Low Saturated Suction Temperature • Low Saturated Condensing Temperature • High Discharge Gas Superheat • Flash Gas in Liquid Line Sight Glass • High Liquid Refrigerant Subcooling	1. Bad Design on Field-Piped Systems
Crankcase Heater Inoperative	

• Flooded Start:
  • High Compressor Power Draw (kW)
  • Noisy Operation
  • Excessive Compressor Vibration
  • Overheating of Compressor
• Violent Oil Foaming (Visible in Compressor Sight Glass)
	1. Never Switched On
	2. Heater Element Broken
	3. Control Circuit Problem
	4. Electrical Power or Control Connection Loose or Corroded

Oil Separator Trapping Oil	
• Oil Level Low on Compressor Sight Glass • High Compressor Power Draw (kW) • Compressor Overheating • Compressor Noisy	1. Sludge Blocking Oil Separator Float Valve Orifice 2. Oil Separator Float Assembly Faulty
Oil Separator Float Valve Stuck Open	
• High Saturated Suction Temperature • Hot oil Return Piping • High Compressor Power Draw (kW) • Flooded Start	1. Debris at Oil Separator Orifice Keeps Float Valve from Seating Properly 2. Faulty Float Assembly 3. Liquid Refrigerant Migrates Through Separator to Compressor Oil Sump at Shutdown

Figure 12–2 Troubleshooting chart for refrigeration cycle accessories.

determine the actual life expectancy of the system, particularly the compressor's. It is therefore helpful to review some of the factors that shorten the life of a compressor.

LOSS OF EFFICIENCY

The loss of efficiency of a compressor is usually an indication that the compressor is being subjected to system problems that are wearing some of the component parts. For a reciprocating machine this can result from a number of conditions:

1. If liquid enters the compressor, the efficiency and resulting capacity will be seriously affected. The physical damage reduces the effectiveness of the internal parts.

2. Leaking discharge valves reduce the pumping efficiency and cause the crankcase pressure to rise, increasing the load on the machine.

3. Leaking suction valves seriously affect the compressor efficiency (and capacity) especially at lower temperature applications.

4. Loose pistons cause excessive blow-by and lack of compression.

5. Worn bearings, especially loose connecting rods and wrist pins, prevent the pistons from rising as far as they should on the compression stroke. This has the effect of reducing the clearance volume and results in excessive re-expansion.

6. Belt slippage on belt-driven units.

MOTOR OVERLOADING

When the compressor is not performing satisfactorily, the motor load sometimes provides a clue to the trouble. Either an exceptionally high or exceptionally low motor load is an indication of improper operation. Here are some of the causes of motor overloading:

1. Mechanical problems such as loose pistons, improper suction valve operation, or excessive clearance volume usually lead to reduction in motor load.

2. Another common problem is a restricted suction chamber or inlet screen (caused by system contaminants). The result is much lower actual pressure in the cylinders at the end of the suction stroke than the pressure in the suction line as registered on the suction gauge. If so, an abnormally low motor load will result.

3. Improper discharge valve operation, partially restricting ports in the valve plate (which do not show up on the discharge pressure gauge), and tight pistons will usually be accompanied by high motor load.

4. Abnormally high suction temperatures created by an excess load will cause a high motor load.

5. Abnormally high condensing temperatures, created by problems associated with the condenser, will also lead to high motor load.

6. Low voltage at the compressor, whether the source is the power supply or excessive line loss, will contribute to high motor loading.

NOISY OPERATION

Noisy operation usually indicates that something is wrong. There may be some noisy condition outside the compressor or something defective or badly worn in the compressor itself. Before changing the compressor, a check should be made to determine the cause of the noise. Here are some possible causes outside the compressor:

1. Liquid slugging. Make sure that only the correct superheated vapor enters the compressor.

2. Oil slugging. Possibly oil is being trapped in the evaporator or suction line and is intermittently coming back in slugs to the compressor.

3. Loose flywheel (on belt-driven units).

4. Improperly adjusted compressor mountings. In externally mounted hermetic type compressors, the feet of the compressor may be bumping the studs. The hold-down nuts may not be backed off sufficiently, or springs may be too weak, thus allowing the compressor to bump against the base.

COMPRESSOR NOISES

Noises coming from the inside of the compressor may be one of the following:

1. Insufficient lubrication. The oil level may be too low for adequate lubrication of all bearings. If an oil pump is incorporated, it may not be operating properly, or it may have failed entirely. Oil ports may be plugged by foreign

PREVENTIVE MAINTENANCE

A contractor, concentrating on service work, has many customers who do not buy traditional service contracts or maintenance agreements. They join the company's "Energy Club." The membership certificate is impressive enough to be framed.

Benefits include two service calls (spring and fall), customary discounts, and priority during temperature extremes. Along with service is education on preventive maintenance, such as the importance of changing filters.

Rising costs of replacing modules, repairing complicated electronic systems and cleaning required by environmental contaminants are persuasive reasons for joining the "Club" as an insurance policy. Members are made to feel prestigious by prompt and efficient service people.

matter or oil sludged from moisture and acid in the system.

2. Excessive oil level. The oil level may be high enough to cause excessive oil pumping or slugging.

3. Tight piston or bearing. A tight piston or bearing can cause another bearing to knock—even though it has proper clearance. Sometimes in a new compressor such a condition will "wear in" after a few hours of running. In a compressor that has been in operation for some time, a tight piston or bearing may be due to copper plating, resulting from moisture in the system.

4. Defective internal mounting. In an internally spring-mounted compressor, the mountings may be bent, causing the compressor body to bump against the shell.

5. Loose bearings. A loose connecting rod, wrist pin, or main bearing will naturally create excessive noise. Misalignment of main bearings, shaft to crankpins or eccentrics, main bearings to cylinder walls, etc., can also cause noise and rapid wear.

6. Broken valves. A broken suction or discharge valve may lodge in the top of a piston and hit the valve plate at the end of each compressor stroke. Chips, scale, or any foreign material lying on a piston head can cause the same result.

7. Loose rotor or eccentric. In hermetic compressors a loose rotor on the shaft can cause play between the key and the keyway, resulting in noisy operation. If the shaft and eccentric are not integral, a loose locking device can be the cause of knocking.

8. Vibrating discharge valves. Some compressors, under certain conditions, especially at low suction pressure, have inherent noise which is due to vibration of the discharge reed or disc on the compression stroke. No damage will result, but if the noise is objectionable, some modification of the discharge valve may be available from the compressor manufacturer.

9. Gas pulsation. Under certain conditions noise may be emitted from the evaporator, a condenser, or suction line. It might appear that a knock and/or a whistling noise is being transmitted and amplified through the suction line or discharge tube. Actually, there may be no mechanical knock, but merely a pulsation caused by the intermittent suction and compression stroke, coupled with certain phenomena associated with the size and length of refrigerant lines, the number of bends, and other factors.

ANALYSIS OF A DEAD COMPRESSOR

This exercise assumes that the power system has been checked and that the thermostat is calling for cooling, so that even though the proper power is available at the compressor, it will not start. The compressor may not even "hum" when power is applied to it or it may hum and cycle on overload. In any event, it is the duty of the technician to analyze the problem, locate the cause and provide a remedy. Here are some of the causes of dead compressors:

1. Open control contacts
2. Overload contacts tripped
3. Improper wiring bad connections
4. Overload cutout open
5. Shortage of refrigerant
6. Low voltage
7. Start capacitor defective or wrong one
8. Run capacitor defective or wrong one
9. Start relay defective or wrong one
10. High head pressure
11. Compressor winding burnt out shorted
12. Piping restriction

To locate the cause, the electrical system and, if necessary, the refrigeration system must be thoroughly tested. As can be seen from the preceding information, most problems are electrically related and are not discussed in this chapter.

TROUBLESHOOTING EVAPORATORS AND CONDENSERS

EVAPORATORS

The following is a list of some of the refrigeration system problems associated with evaporators:

1. Low airflow
2. Excessive airflow
3. Uneven airflow over coils
4. Low refrigerant supply
5. Uneven refrigerant distribution to coil circuits
6. Low water flow in cooler (water cooling evaporator for chillers)
7. Uneven water flow through cooler
8. Low refrigerant supply to cooler

The symptoms and possible causes for these conditions are shown in Figure 12–3. The techni-

General Problem/Symptoms	Possible Causes
Low Airflow	
• Compressor Slugging • Low Saturated Suction Temperature • Low Suction Superheat • Low Saturated Condensing Temperature • Low Compressor Power Draw (kW) • Very Cold Supply-Air Temperature • Low System Capacity • Warm Air Temperature In The Space • Iced or Frosted Evaporator • Compressor Liquid Floodback	1. Dirty Evaporator Coil 2. Badly Bent Evaporator Fins 3. Dirty Filter Or No Filter Installed 4. Restrictive-type Air Filter 5. Return-Air Grille Too Small 6. Supply or Return Ducts Undersized 7. Undercharged, Causing Coil Frosting 8. Plenum Or Duct Insulation Restriction
Excessive Airflow	
• Warm Air Temperature In The Space • High Saturated Suction Temperature • Suction Superheat Normal Or Low • High Air Handler Motor Power Draw (kW) • Noisy Grilles or Registers • Noisy Air Handler • Water Dripping from Fan and Supply Ductwork Near Air Handler	1. Fan-Motor Speed Set Too High 2. Wrong Fan Drive Package and/or Setting 3. Undersized Coil (Applied Air Handlers) 4. No Ductwork 5. Greatly Oversized Duct System 6. Condensate Carries Over Air Handler Drain Pan into Fan and Supply Ductwork
Uneven Airflow Over Coil	
• Low Suction Superheat • Low System Capacity • Low Saturated Suction Temperature • Uneven Condensate Coverage Over Coil Surface • Uneven Coil Surface Temperature • Refrigerant Floodback to Compressor • Compressor Slugging	1. Bad Duct Design Near Evaporator Coil 2. Coil Placement Improper 3. Air Turbulence at Coil 4. Lack of Necessary Air Baffling Near Coil 5. Obstruction within Air Handler 6. Obstruction in Duct Work Near Air Handler 7. Mismatched Coil and Air Handler
Low Refrigerant Supply	
• Low System Capacity • Low Saturated Suction Temperature • High Suction Superheat • High Discharge Superheat • Low Subcooling • Low Compressor Power Draw (kW) • Low Saturated Condensing Temperature • Measurable Temperature Drop in Liquid Line • Visible Bubbles in Liquid-Line Sight Glass (TXV) • High Supply-Air Temperature • Frosted or Iced Evaporator	1. System Undercharged 2. Liquid Line Kinked, Crushed, or Restricted 3. Evaporator Tube Crushed (Especially Return Bends) 4. System Refrigerant Leak 5. Restricted Metering Device 6. TXV Power Element Lost Its Charge 7. Undersized Metering Device 8. Undersized Distributor Nozzle 9. Head-Pressure Control Faulty at Low Outdoor-Ambient Temperatures 10. TXV Plugged or Stuck Closed 11. Plugged Distributor Oil Nozzle 12. Free Water in System Forms Ice 13. Filter-Drier Plugged
Uneven Refrigerant Distribution to Coil Circuits or to Cooler Circuits	
• Low System Capacity • Low Saturated Suction Temperature • Low Normal Suction Superheat • TXV Hunts • Compressor Floodback • Compressor Slugging • Uneven Coil Surface Temperature • Uneven Condensate Formation on Evaporator • Frost On Some Areas Of Evaporator But Not On Others	1. Plugged Evaporator Feeder Tube(s) 2. Kinked or Crushed Feeder Tube(s) 3. Partially Blocked Distributor 4. Oversized Distributor 5. Oversized Distributor Nozzle (Applied Air Handlers) 6. Improperly Installed Distributor (Applied Air Handlers) 7. Crushed Evaporator Tube (Especially Return Bends) 8. Plugged Evaporator (or Cooler) Circuit

continued

Figure 12–3   Troubleshooting chart for evaporators.

General Problem/Symptoms, *continued*	Possible Causes, *continued*
Low Water Flow in Chiller	
• Low Saturated Suction Temperature • Low Suction Superheat • Low Saturated Condensing Temperature • Low Compressor Power Draw (kW) • Low Leaving Chilled-Water Temperature • Low System Capacity • High Space Temperature • High Temperature Drop Between Entering and Leaving Chilled Water • Chiller Shuts Down Intermittently (Even Though Thermostat Calls for Cooling) on Low Leaving Water Safety Thermostat	1. Chilled-Water Pump Undersized 2. Faulty Pump Motor 3. Damaged or Blocked Pump Impeller 4. Blocked Chilled-Water Line Valve 5. Water Baffle(s) in D-X Cooler Misplaced Blocking Flow 6. Excessive Water Scaling (Flooded Cooler) 7. Reverse Chilled-Water Pump Rotation 8. Blockage in Chilled Water Piping 9. Water-Flow Control Valve Restricting Flow
Uneven Water Flow Through Chiller	

• Low Suction Superheat
• Low System Capacity
• Low Saturated Suction Temperature
• Compressor Floodback
• Compressor Slugging
• High Leaving Chilled-Water Temperature
• Low Temperature Drop Between Entering and Leaving Chilled Water
	• (D-X Cooler):
	  1. Misplaced or Broken Baffle(s)
	  2. Excess Air in Water System
	  3. Debris Inside Shell of Cooler
	
	• (Flooded Cooler):
	  1. Badly Scaled Water Tube(s)
	  2. Kinked or Crushed Water Tube(s)
	  3. Plugged Water Tube(s) or Water Box

Low Refrigerant Supply to Chiller	

• Low System Capacity
• High Leaving Chilled-Water Temperature
• Low Saturated Suction Temperature
• High Suction Superheat
• Low Compressor Power Draw (kW)
• Low Saturated Condensing Temperature
• Space Temperature Too Warm
• Chiller Compressor Cycles Off Intermittently on Low-Pressure Switch
	1. System Undercharged
	2. Head-Pressure Control Not Working at Low Outdoor Ambient Temperature
	3. Refrigerant System Leak
	4. Flooded Cooler:
	  • Refrigerant Flow from Condenser Blocked
	  • Cooler Refrigerant Supply Valve Stuck
	5. D-X Cooler:
	  • Liquid Line or Accessories Plugged
	  • Liquid Line Kinked or Crushed
	  • TXV Power Element Low on Charge
	  • TXV Plugged or Stuck
	  • Refrigerant Distributor or Nozzle Plugged
	  • Electronic Expansion Valve Faulty or Microprocessor Problem
	  • Frost-Pinched Cooler Tubes

Figure 12–3, *continued* Troubleshooting chart for evaporators.

cian, by process of elimination, uses this information to arrive at the cause of a particular problem. The following information will be helpful in analyzing some of the common problems.

A dirty filter is probably the number one cause of low airflow, particularly if the owner has not provided periodic maintenance of air filters. On larger systems a differential-pressure-sensing meter can be used to indicate the pressure drop through the filters when filter changing is necessary. A responsible person is required to perform the cleaning or replacement necessary periodically. Filters provide a valuable function but do require maintenance.

Excessive airflow can lower the efficiency, interfere with the comfort level and/or produce noisy operation. Excessive airflow is indicated by a low air-temperature drop over the evaporator coil. For proper airflow the temperature drop should normally be between 15 and 22°F. Adjusting the airflow should be a remedial measure in the checkout procedure. Since the power (watts) supplied to the blower motor varies as the cube of the blower rpm, a large savings can be effected on most jobs by supplying the proper air quantity. Too high an airflow reduces the dehumidifying function of an air-conditioning system and therefore interferes with the comfort level. The humidity can be tested by measuring the returning air wet-bulb temperature. The noise factor should be greatly improved by reducing the fan speed.

A low refrigerant supply can often be traced to a leak or a plugged filter-drier. This is a comparatively easy item to check. Due to the restriction in the filter-drier, some vaporization of the refrigerant can occur, lowering the temperature at the outlet. By sensing a temperature drop through the drier, the restriction is located.

A low water flow through a cooler will be evidenced by a low water temperature leaving the chiller. This is usually caused by a restriction in the water line, a water pump with the wrong impeller rotation, or a defective pump. Many times a shutoff valve is found that has been improperly opened and is creating the problem. As soon as the problem is found, a relatively simple solution can often solve the problem.

CONDENSERS

The following is a list of some of the refrigeration problems associated with condensers:

1. High head pressure
2. Refrigerant charge incorrect
3. Low head pressure

The symptoms and possible causes for these conditions are shown in Figure 12–4. A technician, by process of elimination, uses this information to arrive at the cause of a particular problem. The following information will be helpful in analyzing some of the common problems.

A high head pressure can often be traced to noncondensible gases present in the refrigerant. The saturated condensing refrigerant has an equivalent condensing pressure, as shown in the charts. Reading a higher condensing pressure on the equipment is an indication that noncondensibles are present. These noncondensibles, usually air in the system, need to be removed for the system to operate efficiently.

A high head pressure can also be due to a dirty condenser surface. Occasionally an operator will spray an air-cooled condenser coil with water during extremely hot weather to increase its capacity. As a result a deposit is left on the coil that fills the fin space and reduces the capacity of the condenser. This deposit is hard to remove, usually removed with acid and a stiff brush. It can also be so destructive that the coil must be replaced.

If the system has excess refrigerant, depending on the system, too much of the condenser can be

filled with liquid not leaving enough room for condensation. This can cause high head pressure. The overcharge needs to be recovered to have the condenser operate normally.

Low head pressure is often an indication of lack of refrigerant due to a leak. If so, the leak needs to be located with a suitable leak detector (usually electronic), the leak repaired, and the system recharged with the correct amount of refrigerant.

TROUBLESHOOTING METERING DEVICES

The following is a list of some of the refrigeration system problems associated with metering devices:

1. Evaporator overfeeding ("flooding")
2. Evaporator underfeeding ("starving")
3. Thermal expansion valve (TXV) hunting
4. Distributor nozzles unevenly feeding

The symptoms and possible causes for these conditions are shown in Figure 12–5. The technician, by process of elimination, uses this information to arrive at the cause of a particular problem. The following information will be helpful in analyzing some of the common problems.

Evaporator flooding can be caused by a low superheat setting or loose sensing bulb. When this occurs, the application and installation of the valve needs to be carefully checked. Most valves are set at the factory for 10°F superheat. The coil may have a high pressure drop and require an external equalizer to operate properly. The sensing bulb must be tightly held to the suction line and the connection insulated if it is picking up stray heat from other sources.

A starved evaporator may be caused by an undercharged system. If this is the condition look for a refrigerant leak, particularly if the system has been operating satisfactorily for some time.

A hunting TXV is usually a sign that the expansion valve is too large for the application. Most valves operate best for loads down to 50% of their capacity. A valve that is too large will hunt too much of the time. Hunting is the condition where the valve continually opens and closes rather than reaches a stabilized condition.

When the distributor nozzle is furnished with the TXV, it must be sized properly to fit the load and to comply with the installation requirements. The tubes of the coil can be inspected to see that all circuits are being fed equally. If not, a properly sized nozzle needs to be applied.

General Problem/Symptoms	Possible Causes
High Head Pressure	
• Compressor Cycles Off Intermittently on High-Pressure Switch While System Calls for Cooling • Compressor Cycles Off Intermittently on Compressor-Motor Protection Switch • High Saturated Condensing Temperature • High Discharge-Gas Superheat • Compressor Overheats • Compressor Seizure • Compressor Motor Burnout • High Compressor Power Draw (kW) • Low System Capacity • Saturated Suction Temperature Normal to High • Excessive Condenser Water-Flow Rate	1. Faulty Head-Pressure-Control Device 2. Dirty Condenser Coil 3. Faulty Condenser Fan Motor 4. Extensive Fin Damage 5. Condenser Air Recirculation 6. Dirty Condenser Fan 7. Condenser Airflow Blocked 8. Prevailing Winds Prohibit Proper Airflow Across Coil 9. Backward Condenser Fan Rotation 10. Slipping Condenser Fan Belt 11. Bent or Broken Condenser Fan Blade(s) 12. Scaled Water-Cooled Condenser Tubes 13. Faulty Condenser Water Pump 14. Damaged Water Pump Impeller 15. Plugged Condenser Water Lines or Screens 16. Condenser-Water Valve Stuck Closed 17. Cooling-Tower Problems 18. Condenser Vapor Locked by Undersized or Poorly Laid Out Refrigerant Condensate Line, which Prevents Refrigerant from Freely Draining to Receiver 19. System Overcharged 20. Noncondensible Gases Present
Refrigerant Charge Incorrect	
• High Head Pressure • High Liquid Subcooling • Low System Capacity • High Saturated Suction Temperature • High Compressor Power Draw (kW) • Low Discharge Superheat • Low Suction Superheat	1. System Overcharged
• Low Head Pressure • Low Saturated Suction Temperature • Low System Capacity • Low or Nonexistent Liquid Subcooling • Flash Gas at Metering-Device Inlet • High Discharge Superheat • High Suction Superheat	1. System Undercharged
Low Head Pressure	
• Low Saturated Condensing Temperature • Low System Capacity • Low Saturated Suction Temperature • Low Compressor Power Draw (kW)	1. Faulty Head-Pressure-Control Device 2. Refrigerant System Leak 3. Undercharged System 4. Condenser-Water Valve Stuck Open

Figure 12–4 Troubleshooting chart for condensers.

OTHER AREAS FOR TROUBLESHOOTING

In the previous descriptions, troubleshooting certain major components of the system have been suggested. If additional areas need to be examined to find the problem, here are some further areas to troubleshoot:

1. Refrigeration cycle accessories
2. Refrigeration piping
3. The quality of the installation process
4. The quality of the evacuation/dehydration process

General Problem/Symptoms	Possible Causes
Evaporator Flooding	
• High Saturated Suction Temperature • Low Suction-Gas Superheat • Liquid Floodback • Compressor Slugging • Compressor Overheats • High Compressor Power Draw (kW) • Compressor Failure • Compressor Pumps Improperly • TXV Hunts	1. System Overcharge (Fixed Metering Device) 2. Oversized Metering Device 3. TXV Stuck Open 4. TXV Superheat Setting Too Low 5. TXV Type Wrong for Refrigerant in System 6. Uninsulated TXV Sensing Bulb in Warm Areas 7. Loose TXV Sensing Bulb 8. Partial Load Too Low for Metering Device 9. Excess Oil Circulating in System
Evaporator Starving	
• Low System Capacity • Low Saturated Suction Temperature • High Suction-Gas Superheat • Low Compressor Power Draw (kW) • Low Saturated Condensing Temperature • High Discharge-Gas Superheat • High Supply-Air Temperature • Iced or Frosted Evaporator	1. System Undercharged 2. Undersized Metering Device 3. Plugged Metering Device 4. Plugged Distributor or Nozzle 5. Undersized Distributor or Nozzle (TXV Jobs) 6. Kinked or Crushed Capillary Tube 7. TXV Stuck in Closed Position 8. TXV Power Element Low on Charge 9. Wrong TXV for Refrigerant in System 10. Plugged or Crushed TXV External Equalizer Line 11. TXV Superheat Setting Too High 12. Incorrect TXV Sensing Bulb Location 13. Free Water in System Forms Ice and Blocks Refrigerant Flow 14. Low Head Pressure (Fixed Metering Device) 15. Faulty Head Pressure Control Device
TXV Hunting	
• Saturated Suction Temperature Oscillates High and Low, in a Cyclical Fashion • Suction Gas Superheat Oscillates High and Low in a Cyclical Fashion • Compressor Power Draw (kW) Oscillates High and Low in a Cyclical Fashion • Intermittent Floodback and Compressor Slugging • Unstable Supply-Air Temperature • Unstable Evaporator-Surface Temperature	1. Oversized TXV 2. Improper Part-Load Control Operation Loads TXV Too Lightly 3. Very Light Cooling Load 4. Rapid Cooling Load Changes 5. Rapidly Changing High-Side Pressure 6. Intermittent Flashing in Liquid Line 7. Incorrect Evaporator Circuiting Selected (Applied Air Handlers)
Distributor Nozzles (TXV Applications)	
• Evaporator Underfeed (See Symptoms Above)	1. Undersized Distributor Nozzle (Quite Unlikely on Comfort Work)
• Evaporator Unevenly Fed by Refrigerant (See Symptoms on Evaporator Sheet)	1. Oversized Nozzle 2. Nozzle Not Sized for Low Load Stability 3. Faulty Part-Load Controls Sequence for Evaporator Sections

Figure 12–5 Troubleshooting chart for metering devices.

RESTRICTIONS IN THE SYSTEM

In a hermetically sealed system a restriction in the refrigerant circuit sometimes develops. This can be caused by kinked or blocked tubing, a moisture restriction, or a blocked filter-drier. Occasionally a capillary tube is bent and kinked, or contaminants may be in the system that block the opening of the tube, preventing refrigerant flow.

If moisture is in the system and the evaporator is operating below freezing temperatures, ice can form at the metering device, stopping the flow of refrigerant. If the filter-drier is full of waste materials collected from the system, it can cause a restriction in the flow of refrigerant.

A restriction is usually easy to diagnose, since it prevents the normal flow of refrigerant. The dis-

ON THE JOB

Here are some classic quotes from the field that provide insight into professional success.

"Fix the problem, not the blame."

"One employee submitted eight problems that were client-specific—not relating to the work but to things noticed at the client's site. We forwarded these to the client and he was very grateful."

"We rely on a three-point plan: hire the right people, provide training aimed at their advancement, and recognize accomplishment."

"Create your own success by caring about the success of your customers."

"Sometimes the most valuable tool in the truck is your ears. Find out what the customer is really trying to tell you."

"The customer is always right—trite, but everlastingly true."

"Never forget a customer—never let a customer forget you."

"Courtesy is no substitute for efficiency but adds to it enormously."

---

charge pressure, the suction pressure, and the amperage all drop. The superheat and the subcooling both increase.

Normally, restrictions take place in the liquid line. When they do there is a decided temperature drop across the restriction.

Whenever a restriction does occur, its location and cause must be found. Remedial action depends on the nature of the restriction. Kits are available for cleaning out a plugged capillary tube; alternatively, the exact replacement can be installed. A plugged filter-drier can be replaced.

INSTALLATION PROBLEMS

Problems in the installation of hermetically sealed systems can be minimized by following specific installation instructions supplied by the manufacturer. Since there are many types and uses of hermetic systems, only general information can be given here. For example, there is a considerable difference between the proper installation of a room cooler and a domestic refrigerator, although they both contain hermetic systems.

Generally, installation instructions for units with hermetic refrigeration systems include the following topics:

1. Location of equipment
2. Leveling the equipment
3. Electrical and plumbing connections
4. Adding accessories
5. Necessary maintenance

The equipment location must provide the proper clearances to supply ventilation air for an air-cooled condenser, accessibility for service, and suitable support for the unit. High head pressure would be an indication that the outside air was not properly circulating through the air-cooled condenser. This could be caused by improper clearances or a dirty condenser. If necessary, the unit should be relocated. Occasionally the condenser is located outside the building where shrubbery has grown to restrict the air into the condenser. Vegetation should be trimmed away from the unit and this noted for future maintenance.

Units need to be installed in a level position, so that condensate will flow to the drain opening for proper disposal. If the drain pan becomes dirty or the opening for condensate removal is blocked, these areas must be cleaned to permit proper flow after levelling the unit.

The unit needs to be located so that electrical and plumbing connections can be properly applied. Usually a trapped drain is required and one must be provided. Otherwise, under certain conditions the drainage will flow in the reverse direction, causing the unit to flood. Electrical connections must be kept dry and accessible.

Sometimes a unit can be improved by adding certain accessories. In some cases the manufacturer has available an air deflector to direct the moving air over the condenser. Without the accessory, the prevailing wind may pass through the condenser in the wrong direction, not cooling the unit and causing high head pressure.

Another accessory is the access valve. Many smaller systems do not have access valves to check pressures and temperatures. A piercing valve (Figure 12–6) may be a temporary way to check pressures on a system. The piercing valve should be removed or replaced with a permanent valve after serving.

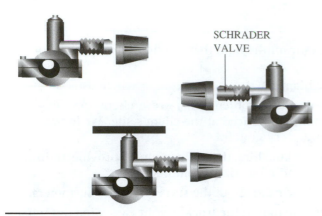

SCHRADER VALVE

Figure 12–6 The use of piercing valves for access to a hermetic system.

Very often, the installation instructions will indicate certain maintenance that must be provided for the equipment to work properly. For example, an air-cooled condenser coil must be cleaned periodically. Maintenance records should be examined to see whether this has been done, what history the problem may have, and what previous actions have been taken.

SUMMARY

- Troubleshooting—The ability of a service technician to determine the cause of an equipment malfunction and to perform corrective measures.
- Troubleshooting requires a high degree of knowledge, experience, and skill.
- Two types of system problems—Electrical and mechanical.
- Key to being a good "troubleshooter" is using a logical, structural, and systematic approach.
- The greatest number of malfunctions are electrical.
- To analyze electrical problems you must know:
    The operating sequence of the unit
    Functions of the equipment
    What is working and what is not
    Electrical test instruments
    What power circuit is driving the system
- Knowledge of operating sequence:
    Usually supplied by the manufacturer in the service and installation instructions.
    Can be determined by the technician by studying the schematic.
- Functions of the equipment:
    Determined by examination and testing
- Required test instruments:

    Voltmeter
    Clamp-on ammeter (Amprobe)
    Temperature testers
    Manifold gauge set
- Figuring things out:
    Use of the ACT method for a quick check.
- Mechanical troubleshooting procedure:
    Collect information about the problem.
    Determine if the problem is electrical or mechanical.
    Read and calculate the system's "vital" signs.
    Compare typical versus actual values.
    Consult troubleshooting aids.
    Perform system analysis.
    Eliminate possible causes of the problem.
    Select the cause that fits the condition.
    Confirm diagnosis.
    Repair or replace fault.
    Rerun test on equipment.
- Compressor malfunctions:
    Loss of efficiency
    Motor overloading
    Noisy operation
    Stuck compressor
    Motor burnout
- Analysis of "dead" compressor:
    Open control contacts
    Overload tripped
    Improper wiring loose connections
    Overload cutout destroyed
    Shortage of refrigerant
    Low voltage
    Start capacitor defective or wrong one
    Run capacitor defective or wrong one
    High head pressure
    Compressor winding burnout
    Piping restriction
- Problems associated with evaporators:
    Low airflow
    Excessive airflow
    Uneven airflow
    Low refrigerant supply
    Uneven refrigerant distribution
    Low water in cooler (chiller)
    Uneven water flow (chiller)
- Problems associated with condensers:
    High head pressure
    Refrigerant charge incorrect

Low head pressure
- Problems associated with metering devices:
  Overfeeding
  Underfeeding
  TXV hunting
  Distributor feeding unevenly
- Sealed system troubleshooting includes:
  Refrigerant leaks
  Restrictions in the refrigerant system
  Installation problems

PROBLEMS AND QUESTIONS

1. Basically, there are two types of problems when troubleshooting. They are ____ and ____ , although there is much overlapping.

2. Does reduced evaporator airflow cause high or low suction pressure?

3. On an evaporator using a TXV valve, will reducing the airflow cause the coil superheat increase, decrease, or stay the same?

4. On an evaporator using a capillary tube, will reducing the airflow cause the coil superheat to increase, decrease, or stay the same?

5. If an expansion valve fails by losing the sensing bulb charge, will the suction pressure rise or fall?

6. Location of the expansion-valve sensing bulb has no effect on the valve. True or False?

7. What is the most common cause of low pressure?

8. What is the most common cause of low suction pressure?

9. What is the easiest way to check for a clogged capillary tube on multitube coils?

10. With the unit off and system at ambient temperature, the pressure in the system should equal the ____.

11. The greatest number of malfunction problems are mechanical. True or False?

12. On a residential air conditioner, when discharge pressure is high, suction is low, and amps are high, this is caused by an overcharge of refrigerant. True or False?

13. Leaking discharge valves
   a. Reduce the pumping efficiency
   b. Cause crankcase pressure to rise
   c. Increase the load on the machine
   d. All the above

14. A starved evaporator may be caused by
   a. An undercharged system
   b. A restricted filter-drier
   c. A refrigerant leak
   d. None of the above

Troubleshoot the following refrigerant-side problems. Using the suction superheat charging chart found in Figure 12–7 will aid in your diagnosis. The system uses a fixed metering device.

15. The outdoor air temperature is 85°F DB. The temperature of the air entering the evaporator is 66°FWB, 75°F DB. The suction pressure is 75 psig. The suction line temperature is 56°F. The system
   a. Is overcharged
   b. Is low on charge
   c. Is charged about right
   d. Has a high-side restriction

16. The outdoor air temperature is 100°F DB. The temperature of the air entering the evaporator is 68°FWB, 75°DB. The suction pressure is 81 psig. The suction line temperature is 53°F. The system
   a. Is overcharged
   b. Is low on charge
   c. Is charged about right
   d. Has a high-side restriction

17. The outdoor air temperature is 70°F DB. The temperature of the air entering the evaporator is 62°FWB, 75°DB. The suction pressure is 63 psig. The suction line temperature is 59°F. The system is
   a. Overcharged
   b. Low on charge
   c. Charged about right
   d. Experiencing low airflow through the condenser

18. The outdoor air temperature is 90°F DB. The temperature of the air entering the evaporator is 64°FWB, 75°DB. The suction pressure is 70 psig. The suction line temperature is 58°F. The system is
   a. Overcharged
   b. Low on charge
   c. Charged about right
   d. Experiencing low airflow through the condenser

COOLING MODE ONLY
CHARGING BY SUPERHEAT METHOD
INDOOR COIL—FIXED ORIFICE METERING

INDOOR CONDITIONS			OUTDOOR <°F> DRY BULB TEMPERATURE									
WB	DB	RH%	65	70	75	80	85	90	95	100	105	110
61	65	80	16	12	8	6						
	70	60	18	14	10	6						
	75	45	20	16	12	8	6					
	80	33	21	17	14	10	6					
	85	23	23	19	16	12	7					
63	70	68	21	17	13	10	8	6				
	75	52	23	19	16	12	9	6				
	80	39	24	20	17	14	10	7				
	85	29	25	21	18	15	12	8	6			
	90	20	26	22	19	16	13	9	7			
65	70	77	24	20	17	13	10	8	6			
	75	59	25	22	19	15	12	10	7			
	80	45	27	24	21	18	14	11	8	6		
	85	33	28	25	22	19	15	12	9	6		
	90	25	29	26	23	20	16	13	10	7		
67	70	86	27	24	21	17	14	11	8	6		
	75	66	28	25	22	18	15	13	10	8	6	
	80	50	30	27	24	21	18	15	12	10	7	
	85	39	31	28	25	22	19	16	13	11	8	
	90	30	32	29	26	23	20	17	14	12	9	6
69	70	95	30	27	25	21	18	15	12	9	7	6
	75	75	31	28	25	22	19	16	13	10	8	6
	80	58	32	29	27	24	21	19	16	14	11	8
	85	45	34	31	28	26	23	20	17	15	12	9
	90	35	35	32	30	27	24	21	19	16	13	11
71	75	82	34	31	29	26	23	21	18	15	12	9
	80	65	35	32	30	28	24	22	20	18	15	13
	85	51	37	34	31	29	26	24	21	19	16	14
	90	39	38	35	32	30	28	25	22	20	17	16
	95	30	39	35	33	30	29	26	23	21	18	17
73	75	92	37	34	32	29	27	24	22	20	17	15
	80	72	37	35	33	30	28	26	24	22	19	17
	85	58	38	36	34	31	29	27	25	23	20	18
	90	45	38	36	34	31	30	28	25	23	21	19
	95	35	38	36	35	32	31	28	26	24	21	19

(For WB 61 upper-right region) CHARGING AT THESE CONDITIONS CAN RESULT IN DAMAGE TO THE COMPRESSOR. CHARGE TO 5° SUPERHEAT AND CHECK SUPERHEAT WHEN CONDITIONS ARE MORE FAVORABLE.

NOTES: SUPERHEAT MEASUREMENTS SHOULD BE TAKEN AT CONDENSING UNIT SERVICE VALVES. CHARGE SYSTEM WITHIN 2° F OF SUPERHEAT INDICATED. RECOMMENDED MINIMUM SUPERHEAT IS 5° F. WHITE AREA IN THE CHART IS THE OPTIMUM WINDOW FOR CHARGING.

DB = DRY BULB TEMPERATURE, DEGREES °F.   WB = WET BULB TEMPERATURE, DEGREES °F.
RH = APPROX. % OF INDOOR RELATIVE HUMIDITY.

Figure 12–7 *(Courtesy Rhearn/Rudd Air Conditioning)*

CHAPTER THIRTEEN

MECHANICAL TROUBLESHOOTING— PART THREE

AFTER STUDYING THIS CHAPTER, THE STUDENT WILL BE ABLE TO:

- Troubleshoot refrigerant-side problems.
- Develop refrigerant-side troubleshooting thought processes.
- Describe how to troubleshoot refrigerant-side problems.
- Ask questions that will lead to the solution of refrigerant-side problems.
- Use the charging checklist to diagnosis problems.
- Know what questions to ask to solve problems over the telephone.

INTRODUCTION

This chapter applies what you have learned in the sections on the various charging methods and the two previous chapters on mechanical, or refrigerant-side, troubleshooting. Electrical troubleshooting is discussed in other chapters. It is extremely important to have a good understanding of charging methods before attempting the exercises in this chapter. The information provided in this chapter will help develop troubleshooting strategies that will be beneficial throughout your professional air-conditioning career. There are many good troubleshooting strategies. Each technician needs to develop organized thought patterns that will be useful for quick and accurate diagnosis. Accuracy is the most important part of this formula. A quick diagnosis is not always a correct diagnosis. Use as many test procedures as possible to verify your final decision.

BASIC REFRIGERANT-SIDE PROBLEMS

This section presents basic refrigerant-side problems. In the first part of the chapter you are working the technician's *help desk* for a major air-conditioning manufacturer. The purpose of the help desk is to provide field technicians with preliminary information about the air-conditioning system your company manufactures. Technicians will be calling for technical assistance on your equipment. You will provide initial diagnostic information on incoming service calls.

The information you received will be limited but adequate enough to make a preliminary decision about the operation of the air-conditioning system. Read the symptoms presented in the telephone call and make a decision based on the limited information provided. More complex field troubleshooting problems will be tackled in the second half of this chapter.

175

HELP DESK TECHNICIAN

A help desk technician answers telephone calls from field technicians. What is your answer to the following troubleshooting problems? Make your decision before reading the solution. Begin to develop troubleshooting thought patterns. What other questions would you ask to determine the answer over the telephone or Internet? Take notes and make your decision. You make the call!

HIGH HEAD PRESSURE AND HIGH SUCTION PRESSURE

Caller 1. The first call of the day is from a technician in Texas. He states that the R-22 unit is not cooling properly. Based on the air temperatures entering the condenser and evaporator you determine that the high-side and low-side pressures (or temperatures) are much higher than normal (Figure 13–1). They are above the 10% range of normal operating pressures.

Technical Assistance For Caller 1 In most cases you will ask the calling technician basic questions before offering a preliminary diagnosis. For example, unless you think it is an electrical problem, each caller should be asked the following questions:

- Is the evaporator clean?
- Is the air filter clean?
- Is the outdoor fan operating?
- Is the indoor fan operating?
- Did you clean the condenser coil?
- What is the metering device?
- Are the system's condenser and evaporator matched?

If the system were clean, then the diagnosis would probably be an overcharged system. The technician would be advised to check the superheat, subcooling, and compressor amperage. In an overcharged system the superheat would be lower than normal, the subcooling would be higher than normal, and the compressor amperage might or might not be in excess of the rated load amperage (RLA). Some systems are overcharged and maintain an acceptable compressor amperage draw.

LOW HEAD PRESSURE AND NORMAL SUCTION PRESSURE

Caller 2 In the second call to the help desk, a field technician states the following information. The R-22 unit is not cooling adequately. During the discussion it is established that all the preliminary cleaning and checks have been conducted by the field technician. The high-side pressure (or temperature) is lower than normal and the low-side pressure (or temperature) is in the middle of a normal range (Figure 13–2).

Technical Assistance For Caller 2 The solution may be to add refrigerant charge to the system. Even though the low-side pressure is normal, the system may be low on charge. In order for an evaporator to perform properly, it must be fed the correct amount of liquid refrigerant and the refrigerant must be at the designated operating temperature. Add more refrigerant until the high side comes into range, while keeping the low-side pressure or temperature within its correct range. When adding refrigerant the high- and low-side pressures and temperatures *do not* change on a one-to-one ratio. There is no set ratio for changes in the air-conditioning system. Adding refrigerant and

Figure 13–1 Low- and high-side gauge pressures (temperatures) are above the normal range for the indoor and outdoor conditions. *(Courtesy of Amprobe Promax)*

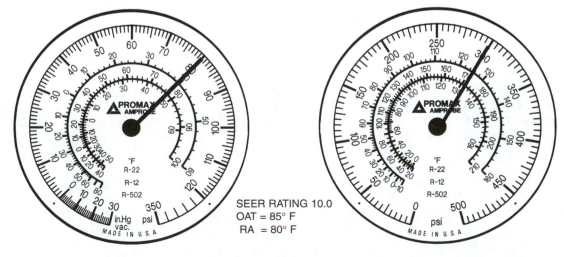

SEER RATING 10.0
OAT = 85° F
RA = 80° F

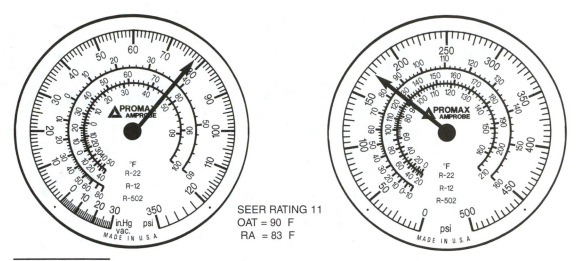

SEER RATING 11
OAT = 90 F
RA = 83 F

Figure 13–2 Normal suction pressure (temperature) and below-normal head pressure (temperature) for the indoor and outdoor conditions. *(Courtesy of Amprobe Promax)*

increasing the high pressure to 10 psig may increase the evaporating pressure 1 or 2 psig (or 1 or 2°F). On other systems, adding 10 psig to increase the condensing pressure may increase the evaporating pressure 5 psig (or 5°F). In this particular case, try adding refrigerant to increase the liquid flow and heat-removing capabilities of the evaporator. As usual, checking superheat, subcooling, air temperatures, and amperage is an important part of checking the charge.

Very Low Discharge Pressure and Higher-Than-Normal Low-Side Pressure

Caller 3 The next call is from a large-volume customer in New England. As usual, the technician is working with a system that is not cooling or cooling is inadequate. The caller reports lower-than-normal discharge pressures and higher-than-normal evaporating pressures for a unit that is an 11 SEER. The technician reported 90°F outdoor air entering the condenser and an indoor return air temperature of 80°F. She also reports 200 psig high-side and 93 psig low-side pressure readings (Figure 13–3).

Technical Assistance For Caller 3 Reviewing the manufacturer's charging recommendations for these conditions, you find the design pressures should be with a 250 psig ±10 high-side pressure and a 72 psig ±3 low-side pressure reading. The technician could use the standard rule of thumb, adding 25°F to the outdoor temperature and subtracting 35°F from the return-air temperature, to determine the rough charging temperatures or pressures. Some technicians prefer to work with the condensing and evaporating temperatures, whereas some technicians prefer to work with the condensing and evaporating pressures. Working

with condensing and evaporating temperatures is best, because you will need to subtract these temperatures when calculating the two superheat readings and the subcooling temperature.

The conclusion is that the compressor is defective. The high-side pressure is equalizing with the low-side pressure. To verify this you ask the technician to measure the rated-load amps (RLA). A very low amperage condition is expected when the pumping capacity is below normal, as found in this case. The technician should do the pumpdown procedure. The compressor should pump down into 15 in. of mercury vacuum in less than 10 min.

These abnormal readings will be caused by anything that exposes the high-pressure side to the low-pressure side. A damaged compressor valve could cause this condition. A blown gasket between the high side and low side will yield similar compressor readings. An open compressor internal relief valve will cause the same pressure readings.

Most compressors have an internal pressure-relief valve. The purpose of the valve is to relieve a high-pressure condition into the shell of the compressor. Under normal operation of the relief valve, the high pressure drops and the low-side pressure rises until the relief valve closes. Sometimes the relief valve does not reseat and causes high pressure to leak into the low side. In either case, the compressor will need to be changed or repaired. A welded hermetic compressor with an open or leaky relief valve would need to be replaced. A serviceable compressor can be field repaired if the valves or gaskets are defective. On some serviceable compressors the internal relief valve will be field replaceable. It will be a challenge to determine where the pressure equalization is occurring in the compressor. Is it the valves, gaskets, or an open or leaky pressure-relief valve?

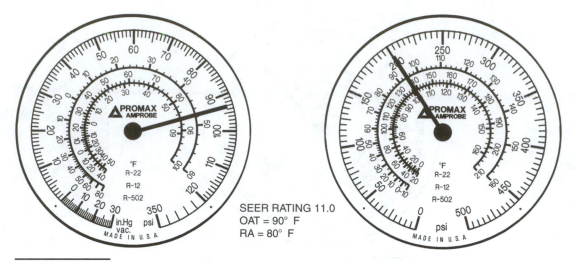

SEER RATING 11.0
OAT = 90° F
RA = 80° F

Figure 13–3   Higher-than-normal suction pressure (temperature) and lower-than-normal high discharge pressure (temperature). *(Courtesy of Amprobe Promax)*

Before leaving the telephone conversation, you remind the caller to check the operation of the manifold gauge set. Open manifold valves or leaking manifold valves will cause the high and low pressures to equalize through the manifold gauge set. Apprentice technicians forget this and confuse themselves with open manifold valves. Open gauge manifold valves are manifested as equalization between the high and low sides in the compressor, as seen in Figure 13–3.

High Superheat and Low Subcooling

Caller 4 The next call is from a technician who has several pieces of valuable information. The technician is not sure what to do with the information. He calls your help desk.

> First, the condensing and evaporating temperatures are in range for the entering temperatures (Figure 13–4) for an R-22 system.
>
> The suction superheat is 25° when using a TXV, which is higher than normal.
>
> The discharge superheat is 160°F, which is higher than normal.
>
> Subcooling is 7°F, which is lower than normal.

Technical Assistance For Caller 4 It is important in any charging procedure to first obtain the correct range of condensing and evaporating temperatures. The technician should be advised to add refrigerant until these temperatures come into range. Adding refrigerant will reduce the suction-superheat and discharge-superheat readings. Charging will also increase the subcooling. You recommend adding charge until the superheat and subcooled temperatures come into range. Remind the technician to check and record the

amperage and delta-T readings. Always recheck the superheat and subcooling temperatures. This equipment model has a charging chart that you fax to the technician's computer.

High Superheat and High Subcooling

Caller 5 The next call is more challenging. The condensing and evaporating temperatures are in range. The technician recorded high suction superheat. The discharge superheat is 140°F, which is also too high. The subcooling is 26°F.

Technical Assistance For Caller 5 All the information provided by the technician leads to a restriction in the liquid line. The condensing and evaporating temperatures are in range, indicating a good flow of refrigerant in the system. There is enough refrigerant flow to complete diagnostic checks, such as superheat and subcooling.

High superheat temperatures indicate a lack of refrigerant flow through the evaporator. There is less liquid refrigerant in the evaporator. The liquid evaporates or boils off to a vapor. The vapor in the evaporator makes more passes in the evaporator coil and picks up more heat as it travels to the suction line. The refrigerant is superheated to a higher temperature as it makes more passes through the evaporator. High suction superheat will also be reflected as high discharge superheat. The compressor, through the process of compression, adds superheat to an already superheated vapor. Discharge superheat should be in the range of 50°F to 125°F. The technician reported a discharge superheat of 140°F.

The subcooling was measured at 26°F. This indicates that there is plenty of refrigerant in the condenser and the condenser is doing its job of removing

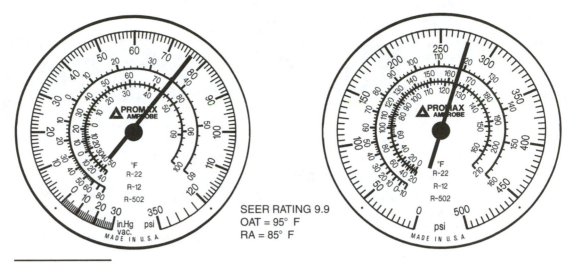

SEER RATING 9.9
OAT = 95° F
RA = 85° F

Figure 13–4 Normal low-side and high-side pressures and temperatures *(Courtesy of Amprobe Promax)*

heat from the refrigerant. A high subcooling temperature means that the condenser is storing extra liquid refrigerant. Extra liquid refrigerant will back up in the condenser. When there is excess liquid in the condenser, the subcooled liquid will form closer to the inlet of the condenser. The longer the liquid is in the condenser, the more subcooling will occur. The temperature of the liquid line will drop the longer it stays in the condenser. A restriction in the liquid line will stack up the refrigerant in the liquid line and back toward the condenser inlet. The restriction can be measured as a higher-than-normal subcooling.

The technician should be advised of the ways a restriction can occur. A liquid-line restriction can be manifested in several forms. The liquid line could be kinked or smashed, which reduces liquid refrigerant flow and holds the liquid longer in the condenser. The liquid line could be undersized and cause a liquid-line restriction. Also, the liquid-line drier could be restricted with contaminants. A pressure drop in excess of 3 psig indicates that the drier is restricted enough to affect the system's performance. The drier could be restricted to the point that it acts as a metering device. In this case, the outlet of the drier will be a cold mixture of liquid and vapor refrigerant. The technician can feel the difference in temperature across a severely restricted drier.

Next, the metering device could be restricted, reducing refrigerant flow to the evaporator while increasing subcooling. A fixed-orifice device can be flushed out or replaced. A high-pressure clean-out pump is available to clear a capillary-tube blockage (Figure 13–5). A thermostatic expansion valve can be opened to improve the refrigerant flow. Contamination on the screen entering the TXV will prevent adequate refrigerant flow even if the valve is adjusted to its most open position.

LOW SUPERHEAT AND HIGH SUBCOOLING

Caller 6 The final call of the day is from a West Coast technician. The unit she is troubleshooting has normal condensing and evaporating temperatures. The suction and discharge superheat are both low. The suction superheat is 3°F. The discharge superheat is 40°F. The subcooling is 27°F.

Technical Assistance for Caller 6 This is a classic case of an overcharged unit. The suction superheat is very low. Low discharge superheat tracks low suction superheat (Figure 13–6) because the refrigerant returning to the compressor is colder than usual and is not heated as much as a normally superheated refrigerant. High subcooling indicates plenty of liquid refrigerant feeding the evaporator. The condenser is

Figure 13–5 Capillary-tube cleaner. *(Courtesy of Thermal Engineering)*

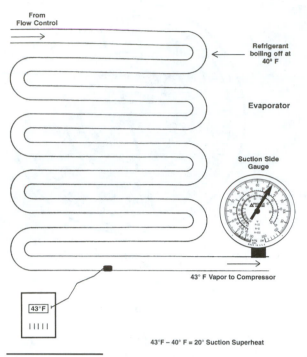

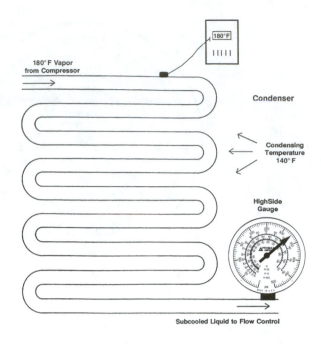

Figure 13–6 Measuring suction and discharge superheat.

doing its job of removing heat and subcooling the refrigerant. The recommendation is to recover refrigerant until the superheat and subcooling fall in line with the correct readings.

SUMMARY

The help desk technician exercises are provided to make you think about the troubleshooting sequence. Unfortunately, in real life the help desk technician may not receive the correct information that is necessary to assist in an accurate diagnosis of the problem. Technicians measure superheat and subcooling incorrectly and end up with readings that confuse more than assist in the decision-making process. The help desk technician must verify that the pressure and temperature readings are properly measured. The caller may be miffed that you need to review the correct setup for measuring superheat and subcooling but this information is important for the outcome of the technical assistance call. When superheat, subcooling, pressures, or temperatures are out of line, it is beneficial to review the test equipment setup.

Some technicians do not realize that their measuring equipment is out of calibration or simply defective. The manifold gauge set should be zero-adjusted before each use. The temperature probes should be checked against each other. Check the probes at the same air temperature and in an ice

bath. An ice bath is a container filled with ice with a little water. Temperature probes in an ice bath should read about 32° to 33°F.

The final problem with a help desk program is receiving feedback from the user. Was valuable information offered to the field technician? Was the equipment repaired? Did the technician figure out some other solution? In most service problems field technicians do not return follow-up calls to the help desk technician to determine if the help desk was of any assistance. The help desk technician could learn from return calls and pass the information along to other callers with the same problem.

ADVANCED TROUBLESHOOTING

This section gives the reader advanced refrigerant-side troubleshooting exercises. The information is provided in more detail than in previous examples. The *charging checklist* should be used to analyze and determine the answer to the problem. Use your critical thinking skills to solve the problems. If necessary, review the chapters on the specific charging methods that will help solve the problem.

All these examples assume that several important steps are taken before finalizing any diagnosis. The condenser coil should be cleaned. The air filter and evaporator coil should be checked and cleaned if necessary. Determine the type of metering device controlling the flow of refrigerant to the evaporator. The metering device will likely be a TXV or fixed-bore flow control. Indoor and outdoor fans and the compressor should be in operation.

A Day in a Life of a Technician

No Cooling, Job 1 You arrive on the job site of a 5-ton package unit that is located on the roof of a resale shop. The complaint is no cooling. The system seems to be working. The indoor temperature is above 80°F. The condenser fan and air-handler blowers are operating. The nameplate reveals the system uses R-22. The rooftop temperature is 80°F. Immediately, you hook up the manifold gauge set and record the following readings:

- High-side pressure is 156 psig.
- Low-side pressure is 150 psig.

From this limited information, what is your preliminary diagnosis? Additional information will need to be gathered to formulate the actual cause of the problem. At this point identify what is operating or not operating. Rule out as much as possible.

Solution 1 There is power to the unit and the thermostat is asking for cooling because the indoor and outdoor fans are operating. Hooking up the Amprobe to the common on the compressor would reveal that the compressor is not operating. This is verified because the high- and low-side pressures are equalized and close enough in pressure to indicate that the compressor is not operating. Conduct electrical troubleshooting to determine the reason for the problem. When the electrical problem is solved, complete the charging checklist to assure that the system has a good charge and that the refrigeration cycle is performing as designed.

Inadequate Cooling, Job 2 The second call of the day brings you to a 3-ton residential split system. The customer complains that it seems that the unit is not cooling well in the afternoon and is running longer than usual. Before checking the coils you elect to take temperature readings and record the following information on the charging checklist:

Refrigerant R-22	Outdoor-Air Temperature, 80°F
Evaporator uses a TXV	Return-Air Temperature, 80°F
Condensing Temperature, 140°F	Evaporating Temperature, 46°F
Suction Superheat, 30°F	Subcooling, 5°F
SEER 9.5	Compressor Amperage, 27 A (RLA 28 A)

Based on this information what would be your diagnosis?

Solution 2 Let us analyze the information and come to some basic conclusions.

1. *What are the refrigerant types and SEER rating?* The system uses R-22 and is not an efficient unit because of the SEER rating, 9.5.

2. *Is the condensing temperature correct?* The outdoor temperature is 80°F, which equates to a 110°F condensing temperature. For systems with SEER ratings 10 or below, add 30°F to the outdoor-air temperature to obtain the target condensing temperature. To determine the condensing temperature, add 30°F to the outdoor-air temperature (OAT). Adding 30°F to 80°F OAT equals a 110°F condensing target temperature. The target condensing temperature should be within ±10% of the 110°F target temperature. In this case the condensing temperature range would be 99°F to 121°F, which is ±11°F. Ten percent of 110°F is calculated by multiplying 0.10 × 110°F = 11°F. The actual recorded condensing temperature is 140°F (337 psig). The condensing temperature is much too high; it should be less than 121°F (110°F + 11°F = 121°F).

3. *Is the evaporating temperature correct?* The indoor return-air temperature is 80°F, which equates to a target 40°F evaporating temperature. For systems with SEER ratings of 10 or below, subtract 40°F from the return-air temperature. The target evaporating temperature should be within ±10% of the 40°F target temperature. In this case the evaporating temperature range is 40°F ± 4°F, or 36°F to 44°F. Ten percent of 40°F is calculated by multiplying 0.10 × 40°F = 4°F. The recorded evaporating temperature is 46°F (78 psig). The evaporating temperature is also too high, leading us to believe that the system is overcharged. Do not jump to conclusions yet. More measurements are necessary to validate our initial decision.

4. *Is the suction superheat correct?* The systems use a thermostatic expansion valve to control the flow of refrigerant to the evaporator. The normal range of suction superheat measured at the compressor is in the range of 15°F to 20°F. Some manufacturers recommend a higher or lower superheat for their equipment. The 15°F to 20°F superheat range is a good standard for air-conditioning equipment unless the manufacturer specifies another range of superheat temperatures.

On this service call, the measured suction superheat of 30°F is very high, indicating a lack of refrigerant in the evaporator. If the discharge superheat were measured, it would also be above the normal temperature range. This information leads us to reevaluate our initial diagnosis of an overcharged system. An overcharged system would have low superheat, not high superheat. More information will help us with our decision.

5. *Is the subcooling correct?* Subcooling should be in a range of 10°F to 20°F. This system measured 5°F of subcooling, which is too low. For every degree below normal subcooling, we lose 1% to 2% of cooling capacity. This unit could be losing as much as 10% of its capacity due to inadequate subcooling. Subcooling is an important factor when it comes to our final evaluation of poor cooling performance.

6. *Is the compressor amperage correct?* The nameplate rated-load amps (RLA) value is 28 A. The compressor is drawing 27 A on its common winding. This information has limited value. Without having special motor curves and capacity charts, the 27-A reading indicates that the compressor is pumping. The amperage draw is higher than what is expected on an 80°F day. You would hope not to exceed 28 A. At 80°F you expect to draw less than 27 A. The compressor has a higher-than-normal pressure differential between the high and low sides; that, along with high superheated vapor, makes it run hotter, thus drawing more amps.

7. *The diagnosis.* Condensing and evaporating temperatures: The condensing and evaporating temperatures are high, which is an indication of enough refrigerant in the system. Before continuing with the remaining charging procedures, the condensing and evaporating temperatures of the system needs to be within the range of its target temperatures (or pressures). Readings above the target are acceptable while gathering system information for analysis purposes. High condensing and evaporating temperatures (or pressures) are an indication of an overcharged system. High condensing temperature or pressure will force up the evaporating temperature (or pressure).

Suction superheat: A high suction superheat indicates a lack of refrigerant in the evaporator. The liquid boils off early in the evaporator and the cool vapor picks up more heat as it travels back to the evaporator. Heat absorbed by the vapor increases the temperature or superheat content of the returning vapor. From this information your troubleshooting thought processes should change from thinking that it is an overcharged system to recognizing it could be a lack of refrigerant flow to the evaporator.

Low subcooling: Low subcooling indicates that the system is undercharged or the condenser is not doing an adequate job of removing heat from the refrigerant. It has been established that there is enough refrigerant at this time. The condensing and evaporating temperatures are high. If the condenser were not removing the right amount of heat, the subcooling would be low and the condensing temperature would be high. This is the case—the condenser is not removing enough heat from the refrigerant.

Compressor amperage: The amp draw is higher than expected on a mild day. The high amperage draw is due to high differential pressures and a lack of cool refrigerant returning to the compressor.

8. *The solution.* After reviewing *all* the information, the final analysis is that something is affecting the heat-removing capabilities of the condenser. The condenser should be cleaned. If the problem persists after the coil is cleaned, the technician should investigate other problems that affect heat removal by the condenser.

- Is the condenser fan located in the shroud, as installed by the manufacturer (Figure 13–7)?
- Has the condenser fan motor been changed lately? The wrong RPM or motor horsepower may be installed. If it is a multispeed motor, it needs to be operating on high speed.
- Is the motor going in reverse direction?
- Was the condenser fan blade replaced recently? The fan blade may look correct, but if it is not the correct replacement, it may not force enough air through the coils. The pitch on the fan blade is important. A few degrees of tilt in the fan blade will affect the airflow. The number of blades must be the same on a replacement fan blade as on the original.
- Is the motor operating on high speed? Some condenser fan motors use a two-speed arrangement. The low speed is used to operate on cool days. The speed-switchover device may be defective, keeping the motor in low-speed operation at warmer ambient temperatures. Low speed translates into low airflow and a reduction in condenser heat-rejecting quality.
- Are motor bearings tight? Spin the fan blade. Does it seem to turn freely or does it drag? Oil the motor if it has lubrication ports.
- Is the condenser coil a double-backed or double-stacked coil? Cleaning the coil will not flush out the dirt trapped between the two coils. The coils need to be separated and cleaned.
- Is the condenser located in a tight area that causes the warm outlet air stream to be recirculated back into the condenser coil?

Any of these conditions could be the solution to the problem. The most common problem is a dirty condenser coil, followed by motor and fan-blade problems.

NOTICE CORRECT
INSTALLATION OF FAN
BLADE ON MOTOR

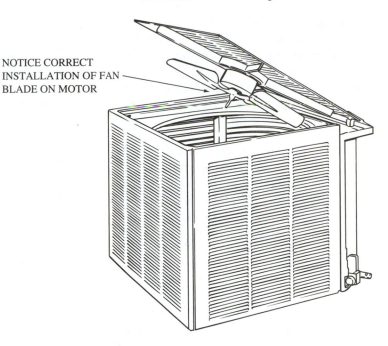

Figure 13–7 Fan-blade location on the motor shaft. *(Courtesy of Rheem/Rudd Corporation)*

Inadequate Cooling, Job 3 This is a return call from the previous day. You were at this school yesterday at 8:30 A.M. responding to a no-cooling service call on a 4-ton split system used to cool an elementary school classroom. Yesterday the system was undercharged. You found and repaired two leaks. The unit was repaired and recharged by noon.

When arriving at the school the teacher using the classroom said that the system cooled satisfactorily yesterday until 2:30 P.M. School was dismissed an hour later. It is about noon when you arrive on the job for the return call. You are wondering what else could be the problem. The Amprobe and manifold gauges are hooked to the system, and it seems to be cooling adequately. Yesterday you were in a big hurry to get to two other jobs and did not check the superheat and subcooling. The condenser was cleaned and the filter and evaporator coil did not require cleaning. The only documentation available is taken from an incomplete charging checklist recorded yesterday. You followed the standard charging procedures on the items that you did check. The previous day's charging checklist provided the following information:

Manufacturer: Be Cool, Model # WEARECOOL-4321
                                SEER Rating: 13.0

OAT: 82°F   Indoor, 77°F

Condensing Temperature,   Evaporating Temperature,
  111°F   38°F

Compressor Amperage,   Delta-T, 17°F
  20A (21 RLA)

Metering Device, TXV

After taking the basic temperature and amperage readings, you find that they are the same as yesterday's readings. The teacher states that the unit was turned off last night and that it had been cooling satisfactorily this morning. What is the problem?

Solution 3 The important fact that the technician overlooked was that the system had a 13.0 SEER rating. Any system with a 12 SEER (Figure 13–8) rating or higher must use the manufacturer's charging recommendations or charging charts developed for that equipment. Using most rules of thumb will not

Figure 13–8 SEER label showing efficiency of a matched air-conditioning system.

Based on standard U.S. Government tests

ENERGYGUIDE

Central Air Conditioner
Cooling Only
Split System

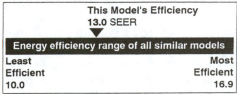

Compare the Energy Efficiency of this Air Conditioner with Others Before You Buy.

This Model's Efficiency 13.0 SEER	
Energy efficiency range of all similar models	
Least Efficient 10.0	Most Efficient 16.9

The **SEER**, Seasonal Energy Efficiency Ratio, is the measure of energy efficiency for central air conditioners.

provide good cooling through the milder and hotter parts of the day. Manufacturers provide specific charging recommendations on systems with SEER ratings higher than 12.

For this service call the technician needs to locate the charging recommendations in the condensing unit or contact the equipment manufacturer or distributor for this specific information. Even with manufacturer's charging instructions, it is important to use other charging methods to check the charge. An air-conditioning system can fall within the charging recommendations and still have problems. For example, the system may be charged to manufacturer's specifications, yet the compressor will cut out on its internal motor overload. Measuring the compressor amperage may reveal an overcurrent problem not addressed by the charging recommendations. Complete the charging checklist and use the manufacturer's recommendations where they apply. Some technicians believe that the manufacturer's recommendations are the only source of information that they need to charge and troubleshoot a system. In most instances this will be true when charging a unit, but this information has limited value for refrigerant-side troubleshooting. When a unit fails to cool properly because of refrigerant-side problems, the completed charging checklist becomes a valuable troubleshooting tool.

Inadequate Cooling, Job 4 The final call of the day finds you troubleshooting a 3-ton split system. You were selected specifically for this job because of your customer-relations and troubleshooting skills. The owner is upset and states that the unit is not cooling properly, especially in the afternoon. Another service company has come out twice in the past week but was not able to solve the problem and satisfy the customer.

Solution 4 From a telephone conversation the office cautions you that the next customer has a warm house and is very upset. Upon arriving on the job you calm down the homeowner by stating that the problem will be diagnosed and repaired today, barring some specialty part that may be difficult to obtain. The fourth call of the day is late in the afternoon; the customer is concerned that you will need a part and that the air-conditioning supply house closes at 5:30 P.M. Your response is that you have a well-stocked service truck; besides this, you tell the customer that the air-conditioning supply house will open up on an emergency purchase basis at any time. You invite the homeowner to follow you around while making the diagnosis. The homeowner likes your suggestion.

Before starting, you request information on the last service call. The homeowner presents the previous two invoices. The two invoices reveal that the first technician repaired one refrigerant leak and recharged the system. The second invoice, dated yesterday, reveals that a different technician added more refrigerant to the system. No other valuable information is provided. It must be noted here that many invoices are inaccurate or incomplete. To use the invoice as a reliable document will cloud your troubleshooting thought processes. On most follow-up service calls it is advisable to review the system operation before coming to any conclusions. Reading another technician's opinion will unduly influence your decision. Make your diagnosis based on facts, not others' actions or previous recommendations. Check it out yourself; you are the professional and in command of the job.

It is always advisable to do a quick operational check. Hook up the manifold gauge set to measure the system operation and clamp the Amprobe to the common of the compressor winding. Does the system operate? Are condensing and evaporating temperatures or pressures close to a normal range? Are the condenser and indoor fans operating? Is the suction line cool and the discharge line hot? Is the supply air cool? It is important to determine if the problem is refrigerant-side or electrical. These quick checks will narrow down the problem and make troubleshooting a little easier.

Together with the homeowner, you check the condition of the air filter and evaporator. They are clean. While checking the evaporator, you discover that the evaporator uses a fixed-piston metering device (Figure 13–9). The condenser coil appeared clean, but flushing it out reveals a fair amount of trapped dirt and debris. The homeowner is surprised by the amount of dirt flushed from a seemingly "clean" condenser. Because the condenser fan motor is exposed, you lubricate it with four drops of nondetergent oil in each lubrication port (Figure 13–10).

Figure 13–9 Fixed metering device. Change the piston to change evaporator capacity.

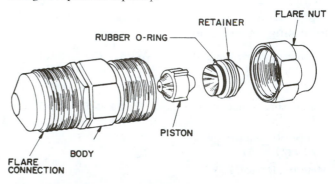

Capped, oiling ports

Figure 13–10 Remove oil-port caps and lubricate and replace caps. Use the recommended quantity and type of lubrication. *(Courtesy of Fasco Motor Company)*

The motor port caps are replaced to keep dirt and moisture from entering the motor bearings.

The system is clean and now you can begin recording information. The system's temperatures and pressures must stabilize before you take a final reading. It is important to allow all the water collected on the condenser to evaporate before the final reading. Water on the fins will keep the condenser coil cool and keep the condensing temperature artificially low. A *dry* condenser should run 15 min before you take final temperature readings. A condenser coil that has just been washed may need as much as 30 min of running time to assure that the water has evaporated off the coil. Condenser coils with moisture on their surfaces will have lower condensing temperatures and pressures. These readings will rise when the coil becomes totally dry. If a system is being charged when the condenser coil is wet, the system will most likely be overcharged. Again, a wet coil keeps the condensing temperature and pressure down and will fool the technician into thinking that the system is undercharged. Taking your time and doing a complete charging checklist will prevent overcharging of the system.

The customer starts the system while you connect temperatures probes to the suction line, discharge line, and liquid line. The manifold gauge set and Amprobe are already connected. The following readings are recorded on the charging checklist:

1. **Date:** *August 15, 3:30 P.M.*
2. **Unit Name:** *LeeStar Condenser Model #AABB+, Evaporator LeeStar #AABB, SEER 10 (R-22)*

3. **OAT:** *80°F*
4. **R/A Temperature:** *80°F DB 67°F WB*
5. **Metering Device:** *Fixed-piston*
6. **Condensing Temperature:** *122°F (266 psig)*
7. **Evaporating Temperature:** *38°F (66 psig)*
8. **Suction Superheat:** 68°F – 38°F = 30°F
9. **Discharge Superheat:** 240°F – 122°F = 118°F
10. **Subcooling:** 122°F – 95°F = 27°F
11. **Sweat on Suction Line:** *None, line feels cool.*
12. **Compressor Amperage:** *24A*
13. **Nameplate Amperage:** *27A*
14. **Sight Glass Condition:** *An occasional bubble. Sight glass moisture indicator is OK.*
15. **Temperature Drop Across Evaporator:** 80°F – 67°F = 13°F
16. **Manufacturer's Charging Recommendations:** *Not available on unit.*
17. **Is This Unit Charge OK?** *No*
18. **If NO what is the problem?** _____

With all the pertinent information gathered, it is time to analyze and act on the data provided in the charging checklist above.

1. Date. The date establishes a record on when and what time of day the work was performed.

2. Unit Name. It is important that the condenser and evaporator match. In our example, the systems are matched per manufacturer's recommendations. The best match is if the condenser and evaporator have the same BTUH capacity; be built by the same manufacturer; and have the same SEER rating. The system should match based on the manufacturer's recommendations. Some manufacturers recommend a ½-ton difference between the size of the condenser and the evaporator. The evaporator will be sized larger.

The second best match involves similar BTUH capacities and like SEER ratings but a different manufacturer of the condenser and evaporator components. The disadvantage of this match is that the customer may not receive the performance and energy savings expected. The components are not a perfect match and have not been tested to performance standards.

It is important to note here that even "matched" systems may not have condenser and evaporator BTUH capacities that are matched according to the model number. Manufacturers "hide" the capacity of their equipment in the model number. For example, manufacturer XYZ may use a sequence of letters and numbers to identify their equipment:

Condenser model DFG-**36**ERT   and evaporator model **36**EWQ-**12**EFG

The **36** is shown in bold numbers to emphasize that the condenser and evaporator have a capacity of 36,000 BTUH. Some manufacturers use a larger evaporator capacity to obtain higher SEER rating. In this example, the evaporator may have a model number of **42EWQ-12EFG**.The manufacturer is using a half-ton larger evaporator to improve the efficiency of the system. The manufacturer will need to recommend this arrangement before it is installed with an expected capacity outcome. This is not a field experiment to be tried by the technician. It is also safer to go with the manufacturer's recommendations in case of a problem. Manufacturers will stand behind their products if they make a design mistake. If technicians decide to deviate from the recommendations, they become liable for the outcome and customer satisfaction.

3. OAT. OAT is the outdoor ambient temperature or outdoor-air temperature entering the condenser. The air temperature should be checked in the airflow stream entering the condenser. Check the temperature as close as possible to the condenser without touching the condenser coil or fins. The coil or fins will be warm and give the technician a false temperature reading.

4. R/A Temperature. The return-air dry-bulb and wet-bulb temperatures are measured at the filter grille or as close as to the evaporator as possible. The best measurement location is at the evaporator. The dry-bulb reading is used to measure the temperature difference across the evaporator and the wet-bulb reading is used to calculate the suction superheat on systems with a fixed-bore metering device.

5. Metering Device. It important to know the type of metering device to determine the correct system suction superheat. This unit has a fixed-piston metering device. A chart is used to determine the correct superheat. Superheat for fixed-flow controls can vary from 5°F to 40°F; the specific superheat is determined from a superheat chart.

6. Condensing Temperature. The R-22 condensing temperature from the inner ring of the gauge set is 122°F, which is equivalent to 266 psig (Figure 13–11). The condensing temperature is outside the target temperature range. To determine the target condensing temperature, add 30°F to the outdoor-air temperature. The target condensing temperature is calculated as follows:

$$80°F + 30°F \pm 10\% = 110°F \pm 10\%$$

$$10\% \text{ of } 110° = 11°F$$

The range of acceptable condensing temperatures is 99°F to 121°F. The condensing temperature from the high-side gauge set is 122°F, which means that the condensing temperature is too high. This is

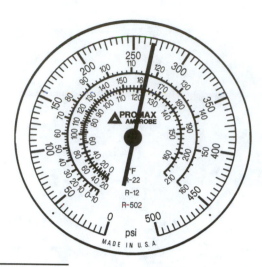

Figure 13–11 Condensing pressure is on the outer ring of numbers. Condensing temperature is located on the inner temperature rings. *(Courtesy of Amprobe Promax)*

the first significant piece of information that will aid in our charging decision.

7. Evaporating Temperature. The evaporating temperature from the inner ring of the low-side gauge is 38°F, which is equivalent to 66 psig (Figure 13–12). The evaporating temperature is in range of the target temperature. To determine the target temperature, subtract 40°F from the return-air temperature: 80°F – 40°F = 40° ± 10%. Ten percent of 40°F is 4°F. The range of acceptable evaporating temperatures is 36°F to 44°F. The evaporating temperature reading from the low-side gauge is 38°F, which means that the evaporating temperature is acceptable. This information tells us that the evaporating temperature is adequate.

Figure 13–12 Evaporating pressure is on the outer ring of numbers. Evaporating temperature is located on the inner temperature rings. *(Courtesy of Amprobe Promax)*

8. Suction Superheat. The system's suction superheat is found by subtracting the evaporating temperature from the suction-line temperature. The evaporating temperature is 38°F, and the suction-line temperature is 68°F. The system's suction superheat is 30°F. Reviewing Figure 13–13, the fixed-orifice metering device chart, the target superheat is 21°F ± 2°F. A superheat range of 19° to 23°F is acceptable. The superheat of 30°F is too high, meaning that there is a lack of liquid refrigerant feeding the evaporator. A lack of liquid refrigerant can be caused by several factors. The system can be undercharged. A restriction in the liquid line, such as a kinked liquid line, partially plugged filter drier, or partially plugged metering device, will reduce the flow of liquid refrigerant. Out of these options, the technician can eliminate an undercharge as being the problem. The condensing temperature is higher than normal and the evaporating temperature is in range; therefore, it can be deduced that the system has enough refrigerant. The problem is that there is *not* enough liquid refrigerant to do the required cooling. Further analysis is necessary to determine and verify the problem.

9. Discharge Superheat. Discharge superheat is calculated by subtracting the condensing temperature from the discharge temperature. The condensing temperature is 122°F and the discharge-line temperature is 240°F. The discharge superheat of 118°F is in range of the 50° to 120°F target superheat, and the discharge temperature does not exceed the 250°F maximum temperature. The information is inconclusive, but it is noted that the discharge superheat is on the high side of normal, which is what we expect when the suction superheat is high. The discharge line temperature of 240°F indicates the compressor motor is not being adequately cooled. Suction superheat directly influences discharge superheat. The two superheat temperatures track each other. When suction superheat is low, discharge superheat will be low. When suction superheat is high, discharge superheat will be high.

10. Subcooling. Subcooling is calculated by subtracting the liquid-line temperature from the condensing temperature. The liquid-line temperature is 95°F and the condensing temperature is 122°F, which gives 27°F. The 27°F subcooling is high. A high subcooling indicates an overcharged condition or a restriction in the liquid line. A restricted liquid line can occur when the liquid line is kinked, the liquid line restricted, or the metering device is restricted. On this service call the system indicates overcharge when evaluating the condensing temperature. Evaluating the superheat, you determined that there was a lack of liquid refrigerant. If the system were overcharged, both the high side and low side would be above the target temperatures or pressures. The high subcooling with high suction superheat points to a liquid-line restriction.

11. Sweat on Suction Line. A lack of condensation or sweat on the suction line is not much of a troubleshooting aid. Outside conditions that create low air moisture dictate there is less likelihood that moisture will condense on a cold suction line. The information provided states that the line feels cool, which means that some cool gas is returning to the compressor. Seeing or feeling moisture on a suction line is subjective and has limited value. Under normal operation the suction line should feel cool or cold to the touch.

12. Compressor Amperage. The compressor amperage is not excessive. The maximum amperage this compressor should draw is 27 A under normal running conditions. The 24 A drawn by the compressor is not excessive and at the same time shows that the compressor is doing work because of the 24 A drawn by the compressor motor.

13. Nameplate Amperage. This is an important reference for the maximum rated-load amps (RLA) drawn by the compressor. Many manufacturers do not include this on their equipment because inexperienced people try to charge the system based on this amperage reading. More manufacturers provide the locked-rotor amps (LRA) than the RLA. Do not confuse the RLA with the LRA. The LRA is four to six times higher than the RLA.

14. Sight-Glass Condition. The liquid-line sight glass is a valuable indicator only with a thermostatic expansion valve. A clear sight glass or bubbles in a sight glass are not used to check the charge of a system that uses a fixed-bore metering device. The moisture indicator in the sight glass does offer some indication of the wetness of the refrigerant. Some moisture indicators give a false moisture indication. It is wise to use a moisture test kit (Figure 13–14) to verify the moisture indication before treating the condition.

15. Temperature Drop across the Evaporator. Also known as delta-T, the technician measures the supply-air temperature and subtracts this temperature from the previously measured return-air temperature. The expected range of temperatures is 13° to 20°F. The measured temperature difference is 13°F, which is on the low-range side of the desired temperatures. The delta-T is in range, but it confirms the suspicion that the system is not cooling adequately.

16. Manufacturer's Charging Recommendations. These recommendations are not available. This is one area manufacturers could improve. Having permanent charging instructions inside the condensing unit would improve the chances that the system is charged correctly. The charging instructions could be used as a troubleshooting tool. A system can be charged according to the manufacturer's

COOLING MODE ONLY
CHARGING BY SUPERHEAT METHOD
INDOOR COIL—FIXED ORIFICE METERING

INDOOR CONDITIONS			OUTDOOR <°F> DRY BULB TEMPERATURE									
WB	DB	RH%	65	70	75	80	85	90	95	100	105	110
61	65	80	16	12	8	6	CHARGING AT THESE CONDITIONS CAN RESULT IN DAMAGE TO THE COMPRESSOR. CHARGE TO 5° SUPERHEAT AND CHECK SUPERHEAT WHEN CONDITIONS ARE MORE FAVORABLE.					
	70	60	18	14	10	6						
	75	45	20	16	12	8	6					
	80	33	21	17	14	10	6					
	85	23	23	19	16	12	7					
63	70	68	21	17	13	10	8	6				
	75	52	23	19	16	12	9	6				
	80	39	24	20	17	14	10	7				
	85	29	25	21	18	15	12	8	6			
	90	20	26	22	19	16	13	9	7			
65	70	77	24	20	17	13	10	8	6			
	75	59	25	22	19	15	12	10	7			
	80	45	27	24	21	18	14	11	8	6		
	85	33	28	25	22	19	15	12	9	6		
	90	25	29	26	23	20	16	13	10	7		
67	70	86	27	24	21	17	14	11	8	6		
	75	66	28	25	22	18	15	13	10	8	6	
	80	50	30	27	24	21	18	15	12	10	7	
	85	39	31	28	25	22	19	16	13	11	8	
	90	30	32	29	26	23	20	17	14	12	9	6
69	70	95	30	27	25	21	18	15	12	9	7	6
	75	75	31	28	25	22	19	16	13	10	8	6
	80	58	32	29	27	24	21	19	16	14	11	8
	85	45	34	31	28	26	23	20	17	15	12	9
	90	35	35	32	30	27	24	21	19	16	13	11
71	75	82	34	31	29	26	23	21	18	15	12	9
	80	65	35	32	30	28	24	22	20	18	15	13
	85	51	37	34	31	29	26	24	21	19	16	14
	90	39	38	35	32	30	28	25	22	20	17	16
	95	30	39	35	33	30	29	26	23	21	18	17
73	75	92	37	34	32	29	27	24	22	20	17	15
	80	72	37	35	33	30	28	26	24	22	19	17
	85	58	38	36	34	31	29	27	25	23	20	18
	90	45	38	36	34	31	30	28	25	23	21	19
	95	35	38	36	35	32	31	28	26	24	21	19

NOTES: SUPERHEAT MEASUREMENTS SHOULD BE TAKEN AT CONDENSING UNIT SERVICE VALVES.
CHARGE SYSTEM WITHIN 2° F OF SUPERHEAT INDICATED. RECOMMENDED
MINIMUM SUPERHEAT IS 5° F
WHITE AREA IN THE CHART IS THE OPTIMUM WINDOW FOR CHARGING.

DB = DRY BULB TEMPERATURE, DEGREES °F.   WB = WET BULB TEMPERATURE, DEGREES °F.
RH = APPROX. % OF INDOOR RELATIVE HUMIDITY.

Figure 13–13 Superheat charging chart. *(Courtesy of Rheem/Rudd Corporation)*

Figure 13–14   Refrigerant testing kit for acid and moisture contamination.

directions and not cool properly. It is advisable to use a charging checklist even if the previous technician followed the manufacturer's charging recommendations.

17. Is the Unit Charge OK? The answer is no. It is not so much that the charge is incorrect as it is a liquid-line restriction that prevents the refrigerant from entering the evaporator at a rate necessary to produce air-conditioning comfort necessary for the space.

SUMMARY

- This chapter provides sample troubleshooting exercises to stimulate thought patterns that will help you diagnosis refrigerant-side problems. The troubleshooting exercises should be solved by information provided by other chapters in this book. Troubleshooting will be the most challenging part of your career. Developing your personal diagnostic habits is important. Every technician has to develop a set of troubleshooting habits that will reduce diagnostic time and improve his or her ability to help fellow technicians. The goal is to be the technician on whom others rely when they can't figure out the problem. This textbook presents the knowledge; it is up to you to formulate the information into useful tools of the trade. Your job will be more interesting and challenging if you are the one the company relies on to troubleshoot problems that others cannot handle. You have the tools; now put them to work as a professional.

PROBLEMS AND QUESTIONS

1. What strategy are you going to use to figure out an air-conditioning problem?
2. As a help desk technician, what are the seven basic questions you should ask the calling technician before diagnosing refrigerant-side problems?
3. What is the purpose of a compressor internal relief valve?
4. What are the causes of very low high-side pressure and very high suction-side pressure?
5. Why is important to develop a personal set of guidelines for troubleshooting air-conditioning problems?
6. Why is a charging checklist important to supplementing a manufacturer's charging recommendation?
7. The accuracy of temperature testers is important. How does a technician check the calibration of a thermometer?
8. What are three causes of high subcooling?
9. Why is it important to dry out a cleaned condenser coil before determining the charge on a system?
10. Why is it important to clean a condenser coil even if it appears clean?
11. How much time do you think it takes to complete a charging checklist on an air-conditioning system?
12. How can you detect a restriction in a liquid-line filter drier?
13. What information do you need to check the charge of a system with a 14.0 SEER rating?
14. What are three causes of high subcooling?
15. What are two causes of low subcooling?

CHAPTER FOURTEEN

THE ELECTRICAL SIDE OF AIR CONDITIONING

AFTER STUDYING THIS CHAPTER, THE STUDENT WILL BE ABLE TO:

•List the basic components of an electrical circuit.
•Describe the differences between series, parallel, and series-parallel circuits.
•Use Ohm's law to solve problems in series and parallel circuits.
•Discuss how Ohm's law will help develop trouble-shooting strategies.

•Use the watts formula to solve power-consumption problems.
•Describe the differences between digital and analog multimeters.
•List the advantages and disadvantages of each multimeter.
•Describe how to use a clamp-on ammeter.

INTRODUCTION

This chapter begins the section on the electrical side of the air-conditioning system. This is a basic chapter on circuit types, electrical symbols, Ohm's law, and meters used in troubleshooting. The information provided in this chapter is essential if you want to develop a solid foundation in electrical trouble-shooting.

COMPLETE ELECTRICAL CIRCUIT

All heating, ventilation, air-conditioning, and refrigeration (HVAC/R) electrical systems are made up of electrical circuits. An electrical circuit has three essential parts and one optional part (Figure 14–1):

1. A source of power (could be a transformer)
2. A load
3. A path for the current to follow
4. A control (optional)

As a result of these electrical components, electrical current is transformed into heat, light, sound, or mechanical motion. Although the control is optional, meaning that the circuit will operate without it, most systems have controls to regulate the supply of power to or remove the power from the load.

 SAFETY TIP Safety is the most important aspect of troubleshooting, installing, and charging air-conditioning equipment. Air-conditioning equipment is somewhat difficult to damage, but human skin, eyes, and digits are less forgiving. Electrical shock can be terminal. The main safety concerns in working with an air-conditioning system are refrigerant burns and electrical shock. These two safety concerns can be reduced by wearing butyl-lined rubber gloves and safety glasses and removing all jewelry and watches when working around electrical equipment.

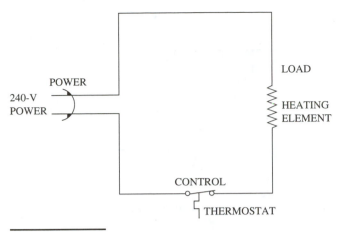

Figure 14–1 A complete circuit must have a power supply, load, and interconnecting wires. This circuit diagram also includes a control thermostat that is optional to the operation.

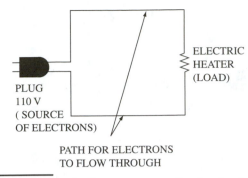

Figure 14–2 Principal components of a simple circuit.

POWER SOURCES

There are two types of power sources:

a. Direct current (DC)
b. Alternating current (AC)

The most common source of direct current is a battery. Batteries are used to supply power to many different types of electrical testing instruments, making them portable and convenient to use. DC power is used for the controls on automotive air-conditioning systems. DC power obtained from AC power supplies is also used on certain solid-state modules for defrost and overcurrent protection.

Alternating current is the most common source of power for most HVAC/R systems. AC power is generated by all the power companies. In residences, 120 V AC is used to power most small appliances. Larger appliances such as electric stoves and residential air-conditioning units use 240 V. The power company supplies residential users with 240 V over the incoming lines. A portion of it is tapped to supply the 120-V requirement. Commercial and industrial customers normally use higher voltages in single-phase and three-phase systems.

Transformers increase or decrease incoming voltages to meet the requirements of the load. For control circuits, it is common to use a transformer to obtain 24 V from line voltages of 120 or 220 V. More detailed explanations of how transformers work and are used will be found in the next chapter.

LOADS

The second condition for an electrical power system is that it must have a load. A *load* is any electrical device that requires power to operate. The most common loads for HVAC/R systems are electric motors. Motors drive the compressors, fans, and pumps. Motors also drive damper motors and zone valves. Many other electrical components require power such as resistance heaters and solenoid valves, to name a few.

ELECTRICAL CIRCUIT: PATH FOR THE CURRENT

The third condition for a power system is that there must be a path for the current to travel (Figure 14–2). Every electric circuit has at least two wires, often indicated as line terminals, L_1 and L_2. In order for there to be a complete circuit, the path of the electrical service (wire connections) flows through one wire of the electric circuit, passes through the load, and returns through the other wire of the electric circuit. In AC systems, the direction of power flow reverses 60 times per second.

CONTROL DEVICE OR SWITCHES

The fourth (optional) condition for the power system is the *control device* or *switch*. The switch is a device to turn the load off and on. It may be manual or automatic, as in the case of a thermostat that turns a unit on and off in response to the surrounding temperature. The switch permits the circuit to be open (Figure 14–3) or closed (Figure 14–4). No current

Figure 14–3 An open circuit: no current flows.

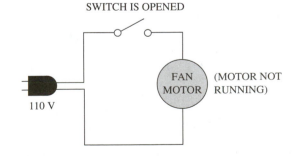

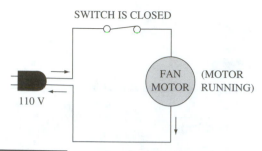

Figure 14–4   A closed circuit, with current flowing through the load.

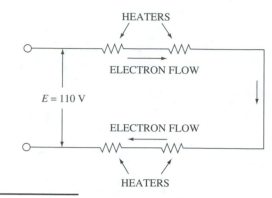

Figure 14–5   A series circuit with four resistances.

flows in the open circuit. When the circuit is closed the load receives power.

TYPES OF CIRCUITS

There are several types of path arrangements for circuits, as follows:

1. The series circuit, which allows only one path for the current to flow;
2. The parallel circuit, which has more than one path;
3. The series-parallel circuit, which is a combination of series and parallel circuits.

SERIES CIRCUITS

In a series circuit, there is only one path for the current to follow. The power must pass through each electrical device in succession in that circuit to go from one side of the power supply to the other. An example of a series circuit is shown in Figure 14–5, where four resistance heaters are placed end-to-end in a single circuit.

Series circuits are common on HVAC/R systems. Usually there is one load controlled by a series of switches, as shown in Figure 14–6. In this diagram the 208-V power-supply terminals are indicated with the symbols L_1 and L_2. The one load is a compressor motor. The switches placed in series with the compressor motor are used to control its operation.

These switches, shown in this diagram, are all safety switches and therefore are all normally closed (NC).

In a series circuit, all switches must be closed in order for current to flow through the circuit. Types of switches include the following:

1. *High-pressure switch,* which senses compressor discharge pressure, opens on a rise in pressure. It is set to cut out at protective high-limit pressure, but remain closed at normal operating pressures. It is also called a high-pressure cutout.

2. *Low-pressure switch,* which senses compressor suction pressure, opens on a drop in pressure. It is set to cut out at a protective low-limit pressure, but remains closed at normal operating pressures. It is also known as a low-pressure cutout.

3. *Compressor internal thermostat,* which senses compressor motor-winding temperature, opens on a rise in temperature. It is set to cut out at a protective high temperature, but remain closed under normal operating conditions.

4. An *operating control* (not shown), also placed in series with the compressor motor, starts and stops the compressor in response to temperature, pressure, humidity or a time clock.

Figure 14–6   Three safety switches in series with a single load.

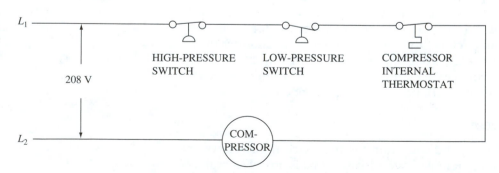

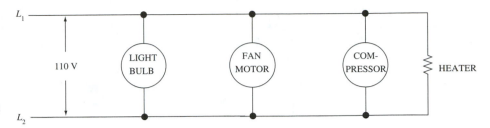

Figure 14–7 Parallel circuit with four different loads.

PARALLEL CIRCUITS

Parallel circuits are used for most HVAC/R equipment wiring. Each load has its own separate path for the current to flow. Most equipment loads are connected directly to the voltage source. The control circuits normally use a lower voltage; 24 V is common. The control system also has load devices, such as relays, which require separate circuits. Control voltage is usually provided by a step-down transformer fed from the power source.

Figure 14–7 illustrates a parallel circuit with four loads. No controls are shown in this diagram, which would be necessary if the circuit were operational. In parallel circuits, each circuit has its own, independent connection to the power source. If the switch is opened in one circuit, the other circuits will continue to operate.

Figure 14–8 shows a number of parallel load circuits, with switches or controls in series with the loads, which could be used for an air-cooled condensing unit. There are three parallel circuits, C_1, C_2, and C_3. Going from top to bottom, they could be described as follows:

Circuit 1. The condenser fan motor 2 has a separate thermostat that turns it on and off.

Circuit 2. The compressor contactor coil (C) is energized when the primary thermostat calls for cooling, provided the two safety switches, LPS and HPS, are closed.

Circuit 3. The compressor contactor has two normally open switches which are in series with the two loads. When the contactor coil in circuit 2 is energized, the two "C" switches in circuit 3 close, supplying power to the compressor motor and the condenser fan 1 at the same time. In effect the two loads, condenser fan motor 1 and the compressor motor, are themselves in parallel and both receive line voltage when the contactor switches close.

In actual practice, a unit may have many parallel circuits for individual loads, all operated in accordance with the design specifications of the control system.

SERIES-PARALLEL CIRCUITS

A series-parallel circuit, as the name implies, combines both a series and a parallel arrangement of electrical loads. A typical diagram is shown in Figures 14–8 and 14–9. It is also known as a combination circuit.

CURRENT ELECTRICITY

Current refers to the flow of electrons through a conductor such as a wire, or through a given space, or past a given point. The flow direction is from negatively to positively charged terminals and occurs because of the potential difference (in the charge) between terminals. The potential difference creates a

Figure 14–8 Condensing-unit wiring diagram showing three separate circuits.

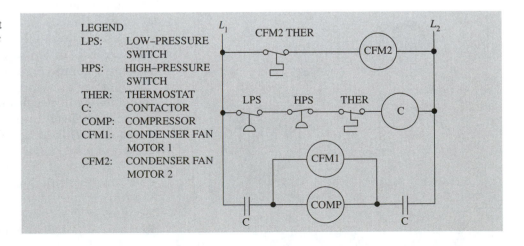

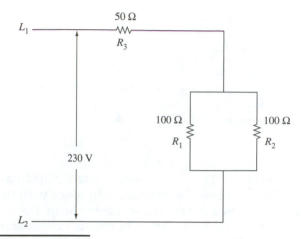

Figure 14–9   Parallel-series circuit showing values of three resistances.

force, called an *electromotive force* (EMF), as shown in Figure 14–10. This force is measured in volts (V). The letter symbol for voltage is *E*, referring to EMF, in circuits.

The unit for electron quantity, the coulomb, is rarely used; however, one definition for a volt is one coulomb of electrons passing a fixed point in the conductor per second.

The *current*, or rate of flow of the charges, is measured by amperes. The rate depends on the amount of voltage applied (the difference of potential between the two ends of the conductor), the size of the conductor, and the material of which the conductor is made. The symbol for the current in amperes (A) is *I*. An electrical system illustrating current flow is shown in Figure 14–11.

Figure 14–10   A simple electric circuit using a battery power source and a lightbulb for the load.

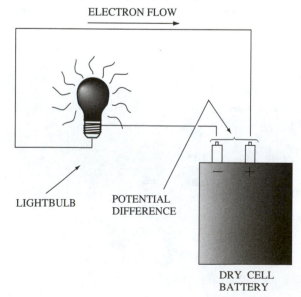

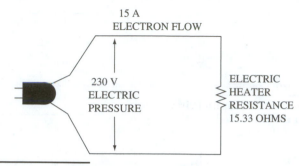

Figure 14–11   Simple electric circuit using an AC power source and an electric heater for the load.

The conductor may permit more or less current to flow because of resistance associated with its physical state. It is identified by applying a known value of voltage and measuring the resulting current. The ratio of voltage to current is the *resistance* and is stated in ohms (Ω). For conductors, this depends on its dimensions and the material of which the conductor is made:

$$R = \frac{\rho l}{A}$$

in which

    R = resistance in ohms
    ρ = (rho), resistivity of material in ohm-
        meters (a constant for a given material at a
        given temperature)
    l = length in meters
    A = cross-sectional area of conductor

The resistivities of conducting materials commonly used are:

Aluminum	2.62
Copper	1.72
Iron	9.71

From this it can be seen why copper is the most frequently used for wires of electrical circuits. For example, a piece of copper wire 0.5 mm (0.02 in.) in diameter and one meter (39.37 in.) long has a resistance of only about 0.09 Ω.

In addition to the resistance of conductors, circuit elements called resistors are also used for various applications (Figure 14–12).

Putting it all together, current in a circuit is found to be directly proportional to the applied voltage and inversely proportional to the resistance of the circuit. This is known as *Ohm's law* and is discussed in more detail later in the chapter.

The flow of electrons through a circuit dissipates energy in the form of heat in passing through the resistances that make up the circuit. Calculated amounts of this energy is called power, *P,* measured in watts (W). A watt of electricity is one ampere (A)

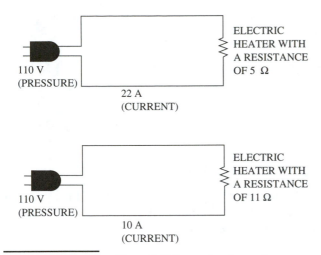

Figure 14–12 The effect of different loads on the current.

of current, *I*, flowing with the force of one volt (V) of voltage, *E*. The power in watts in an electric circuit can be expressed by the equations for singe-phase AC power:

$$P = EI \text{ (PF)}$$

$$P = I^2R$$

$$P = \frac{E^2}{R}$$

where

PF (power factor) = phase angle between *E* and *I*

and

PF = true power/apparent power, as explained later in the chapter.

These equations vary for DC power, single-phase AC power, and three-phase single-phase power, as is explained later on in the chapter.

The amount of power required by electrical devices determines the design of the circuit. The electrical motor is the largest current-consuming device in most heating, cooling, and refrigeration systems. Other units which measure the output of such devices are the horsepower, equal to 746 W (power output), and the British thermal unit (Btu) where 1 W = 3.41 Btu/h (heat output). A more abstract term commonly used by air conditioning manufacturers is the *energy efficiency rate* (EER) which identifies the amount of heat per watt of power consumed by equipment. It is becoming common to specify power ratings of many devices such as refrigeration systems, boilers, etc., in standard kilowatt (kW) terms.

OHM'S LAW

Applying Ohm's law is not difficult, but it does take practice to understand how to apply it. In most cases, you will not use Ohm's law directly but indirectly. Let us examine this statement.

Learning how to use Ohm's law to solve problems is the first step in developing a troubleshooting strategy. As you will see, some Ohm's law problems are simple and some are more challenging. Simple series circuits are the easiest to solve, whereas the parallel and series-parallel circuits take additional calculations. Most of the advance circuits can be solved using alternate methods. Answers should be "proved" by applying more than one calculation to the problem.

Ohm's law is taught in beginning electrical and electronics courses to lay down the basic foundation for understanding circuit operation and planting the seeds that will develop into troubleshooting skills. Think of learning Ohm's law as elementary troubleshooting that will progress in future chapters.

Ohm's law is used strictly with resistive circuits. It cannot be used with inductive circuits such as motors, transformers, and coils, nor can it be used on capacitance circuits. An impedance formula is needed to calculate this information. This is not covered in this material.

Ohm's law, for DC circuits or AC single-phase circuits with purely resistive loads, is as follows:

The relationship between electrical potential, *E*, measured in volts (V); current flow, *I*, measured in amperes (A); and resistance, *R*, measured in ohms (Ω), is expressed in Ohm's law. In simple terms it states that the greater the voltage, the greater the current; and the greater the resistance, the lesser the current flow. Ohm's law is expressed mathematically as "current is equal to electrical potential divided by resistance." Stated in symbols:

$$I = \frac{E}{R}$$

where

I = current
E = electrical potential
R = resistance

The equation can be stated a number of ways. Use is made of whichever one applies. Other versions of Ohm's law are as follows:

$$E = IR \qquad R = \frac{E}{I}$$

This formula is very helpful in analyzing a circuit, because when any two of the terms are known or can be measured, the third value can be calculated using one of the above equations.

It should be noted that Ohm's law was first applied to circuits using direct current. It does not apply, without some modification, to AC circuits because coils of wire produce different effects with alternating current. The flow of current in AC circuits can be influenced by such factors as *inductance* and *inductive reactance* (the "resistance" offered by inductance to the flow of alternating current), and *capacitance* and *capacitive reactance,* which are discussed later in this chapter. If modifications are made to resistance, it will be shown how the general principles contained in Ohm's law do apply to alternating current.

The following examples illustrate the relationships of voltage, current and resistance in a simple electrical circuit. In the examples a DC circuit or a single-phase AC circuit with purely resistive loads is used.

WORKING EXAMPLE

A simple electrical circuit (Figure 14–13) has a power source of 120 V and a resistance load of 10 Ω. How much is the current flow in amperes?

SOLUTION

$$I = \frac{E}{R}$$

$$I = \frac{120 \text{ V}}{10 \text{ }\Omega}$$

$$I = 12 \text{ A} \qquad \blacksquare$$

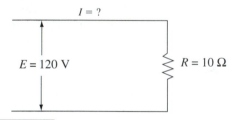

Figure 14–13 Calculating the current using Ohm's law with voltage and resistance known.

WORKING EXAMPLE

Referring to Figure 14–14 what is the resistance of an electric heater, using a power source of 120 V and drawing a current of 10 A?

SOLUTION

$$R = \frac{E}{I}$$

$$R = \frac{120 \text{ V}}{10 \text{ A}}$$

$$R = 12 \text{ }\Omega \qquad \blacksquare$$

Figure 14–14 Calculating the resistance, using Ohm's law, with voltage and current known.

WORKING EXAMPLE

Assuming a circuit as shown in Figure 14–15 with a resistance of 23 Ω and a current flow of 10 A, what is the voltage of the power supply?

SOLUTION

$$E = IR$$

$$E = (10 \text{ A})(23 \text{ }\Omega)$$

$$E = 230 \text{ V} \qquad \blacksquare$$

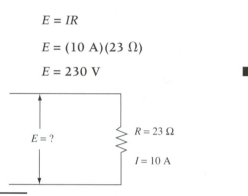

Figure 14–15 Various ways to use Ohm's law.

Note in the preceding examples that by using Ohm's law, if two factors are known, the third can be calculated.

Calculating electric power, for DC circuits or single-phase AC systems with purely resistive loads can be done using the following formula:

$$P = EI$$

where *E* and *I* have the same meaning as used in calculations involving Ohm's law and *P* = power in watts.

In general, for single-phase AC systems, *P* = *EI* (PF), where PF = phase angle between *E* and *I*. For three-phase AC systems, *P* = $\sqrt{3}$ *EI* (PF).

USING THE OHM'S AND WATTS PIE CHARTS

Calculations using Ohm's law or the watts formula (derived from Ohm's law) can be simplified by using a pie chart as shown in Figures 14–16 through 14–24. In the pie diagram in Figure 14–16, *E* (voltage) is always on top of *I* (current), and *R* (resistance). To find voltage, current, or resistance, take the pie chart and cover the value for which you are solving with your finger. The remaining values will represent the formula you will use to find the answer.

For example, if you are looking for voltage, or *E*, cover *E* with your finger and the remaining chart reveals that solution to the problem is to multiply *I* times *R* (Figure 14–17).

To find current, or *I*, cover *I* and use the formula *E* divided by *R*, or *E/R* (Figure 14–18).

To find resistance (ohms), or *R*, cover *R* and use the formula *E* divided by *I*, or *E/I* (Figure 14–19).

WATTS PIE CHART

A pie chart can be used to calculate any information in the watts formula. The watts formula states that

Watts (W) = volts (E) × Amps (I)

Sometimes the word *power* or the letter *P* is used to represent watts.

The pie chart is drawn as shown in Figure 14–20. Just as with the Ohm's law pie chart, you cover the letter that is being solved for in the problem. For example, if you are looking for watts or

power, *W*, cover *W* with your finger. The remaining chart reveals that the solution to the problem is to multiply *E* times *I*, or the voltage by the current. See Figure 14–21.

To find voltage or *E*, cover *E* and use the formula *W* divided by *I*, or *W/E* (Figure 14–22).

To find current or amperage draw, or *I*, cover *I* and use the formula *W* divided by *E*, or *W/E* (Figure 14–23).

Figure 14–24 contains a large pie chart with both Ohm's law and the watts formula combined. Notice that the chart contains additional formulas that will help in solving these types of problems. It would be useful to make a copy of this general pie diagram for future reference and use it when working this type of problem.

Figure 14–20

Figure 14–21 If calculating watts or power, cover *W* and use the formula *W* = *E* × *I*.

Figure 14–22 If calculating *E*, cover *E* and use the formula

$$E = \frac{W}{I}.$$

Figure 14–23 If calculating *I*, cover *I* and use the formula

$$I = \frac{W}{E}.$$

Figure 14–16 Ohm's law pie chart.

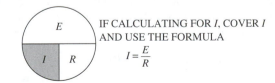

IF CALCULATING FOR *E*, COVER *E* AND USE THE FORMULA

$$E = I \times R$$

Figure 14–17

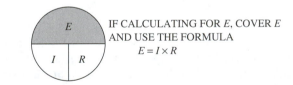

IF CALCULATING FOR *I*, COVER *I* AND USE THE FORMULA

$$I = \frac{E}{R}$$

Figure 14–18

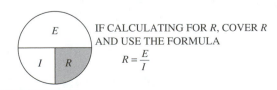

IF CALCULATING FOR *R*, COVER *R* AND USE THE FORMULA

$$R = \frac{E}{I}$$

Figure 14–19

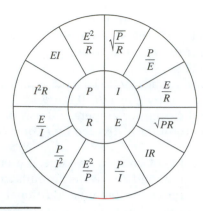

Figure 14–24

VARIATIONS OF THE POWER FORMULA

The power formula can be stated in three different ways with variations for DC, single-phase AC and three-phase AC, shown in the following table.

	DC	Single-Phase AC	Three-Phase AC
Power Formula 1	$P = EI$	$P = EI(PF)$	$P = \sqrt{3}\,EI(PF)$*
Power Formula 2	$P = E^2/R$	$P = E^2/R$	$P = \sqrt{3}\,E^2/R$
Power Formula 3	$P = I^2R$	$P = I^2R$	$P = \sqrt{3}\,I^2/R$

* $\sqrt{3} = 1.732$

The following examples illustrate the use of the power formulas.

WORKING EXAMPLE

(Single-phase AC, assuming a PF = 1.0)
What is the power consumption in a circuit operating with 230 V and 5 A?

SOLUTION

$P = EI$
$P = (230\ V)(5\ A)$
$P = 1150\ W$ ■

WORKING EXAMPLE

(Single-phase AC, assuming a PF = 1.0)
What is the current draw of an 8000-W electric heater operating on a 230-V power supply?

SOLUTION

$I = \dfrac{P}{E}$

$I = \dfrac{8000\ W}{230\ V}$

$I = 34.8\ A$ ■

WORKING EXAMPLE

(DC or single-phase AC)
What is the power consumed by an electric resistance 20 Ω using 6 A?

SOLUTION

$P = I^2R$
$P = (6\ A)(6\ A)(20\ \Omega)$
$P = 720\ W$ ■

WORKING EXAMPLE

(Three-phase AC)
What is the power consumed by a three-phase circuit with a voltage of 230 V, a current of 10 A, and a power factor of 0.80?

SOLUTION

$P = \sqrt{3}\,EI(PF)$
$P = (1.732)(230)(10)(0.80)$
$P = 3186\ W$ ■

CALCULATIONS FOR A SERIES CIRCUIT

For the following calculations a DC circuit or a single-phase AC circuit with purely resistive loads is used.

The current flowing through a series circuit is the same for each load in the circuit. For example, in the circuit shown in Figure 14–25, which has four loads, the current is equal through each load, and expressed in symbols:

$$I_1 = I_2 = I_3 = I_4$$

The total resistance of a series circuit is the sum of all the individual resistances that are placed in series between L_1 and L_2. To state this in symbol form:

$$R_T = R_1 + R_2 + R_3 + R_4$$

The voltage drop across each one of these resistances can be calculated using Ohm's law, and the total of the individual voltage drops should add up to the circuit voltage. Thus,

$$E_T = E_1 + E_2 + E_3 + E_4$$

Figure 14–25   Calculating the current flow through a series circuit.

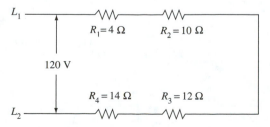

WORKING EXAMPLE

Using Figure 14–25, calculate the individual voltage drops to see if their total equals the circuit voltage.

SOLUTION

Step 1. Determine the total resistance of the circuit.

$$R_T = R_1 + R_2 + R_3 + R_4$$
$$R_T = 4\ \Omega + 10\ \Omega + 12\ \Omega + 14\ \Omega$$
$$R_T = 40\ \Omega$$

Step 2. Determine the current flow in the circuit, using Ohm's law.

$$I = \frac{E}{R}$$

$$I = \frac{120\ V}{40\ \Omega}$$

$$I = 3\ A$$

Step 3. Determine each voltage drop, using Ohm's law. See Figure 14–26.

$$E_1 = I_1 \times R_1 \qquad E_2 = I_2 \times R_2$$
$$E_1 = 3 \times 4 \qquad E_2 = 3 \times 10$$
$$E_1 = 12\ V \qquad E_2 = 30\ V$$

$$E_3 = I_3 \times R_3 \qquad E_4 = I_4 \times R_4$$
$$E_3 = 3 \times 12 \qquad E_4 = 3 \times 14$$
$$E_3 = 36\ V \qquad E_4 = 42\ V$$

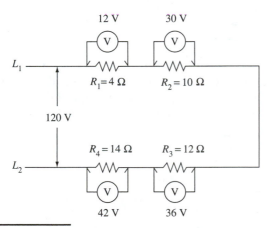

Figure 14–26 Calculating the voltage drop in a series circuit.

Step 4. Check the accuracy of these calculations. Their total should equal the circuit voltage of 120 V. The procedure is as follows:

$$E_T = E_1 + E_2 + E_3 + E_4$$
$$E_T = 12 + 30 + 36 + 42$$
$$E_T = 120\ V \qquad \blacksquare$$

CALCULATIONS USING A PARALLEL CIRCUIT

The following calculations use a DC circuit or a single-phase AC circuit with purely resistive loads. The current draw for a parallel circuit is determined for each of its parts. The current consumed by the entire parallel system is the sum of the individual circuits. The calculation is made using Ohm's law. To obtain the current flowing in the circuit, both the voltage and the resistance of the load(s) must be known. Thus, the total current is calculated as follows:

$$I_T = I_1 + I_2 + I_3 + I_4 + \cdots$$

The resistance of a parallel circuit gets smaller as more resistances are added. If there are only two resistances, the total resistance can be calculated by the following formula:

$$R_T = \frac{R_1 \times R_2}{R_1 + R_2}$$

If there are more than two resistances, use the following formula and solve for R_T.

$$\frac{1}{R_T} = \frac{1}{R_1} + \frac{1}{R_2} + \frac{1}{R_3} + \frac{1}{R_4} + \cdots$$

The voltage drop in a parallel circuit is the line voltage supplied to the loads, or simply stated,

$$E_T = E_1 = E_2 = E_3 = E_4 = \cdots$$

Ohm's law can be used to calculate voltage, amperage, or resistance, if the other two values are known.

WORKING EXAMPLE

From the information given in Figure 14–27, calculate the current draw for each of the individual circuits and the total current draw.

Figure 14–27 Calculating the current through parallel circuits.

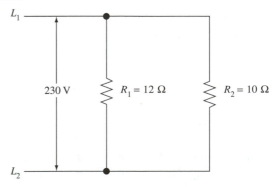

SOLUTION

$$I = \frac{E}{R}$$

$$I_1 = \frac{120 \text{ V}}{12 \text{ }\Omega}$$

$$I_1 = 10 \text{ A}$$

$$I_2 = \frac{120 \text{ V}}{10 \text{ }\Omega}$$

$$I_2 = 12 \text{ A}$$

$$I_T = I_1 + I_2$$

$$I_T = 12 \text{ A} + 10 \text{ A}$$

$$I_T = 22 \text{ A} \qquad \blacksquare$$

WORKING EXAMPLE

Find the total resistance of the complete parallel circuit shown in Figure 14–27.

SOLUTION

$$R_T = \frac{R_1 \times R_2}{R_1 + R_2}$$

$$R_T = \frac{12 \text{ }\Omega \times 10 \text{ }\Omega}{12 \text{ }\Omega + 10 \text{ }\Omega}$$

$$R_T = \frac{120}{22}$$

$$R_T = 5.4 \text{ }\Omega \qquad \blacksquare$$

WORKING EXAMPLE

What is the resistance of parallel electric heating elements with a resistance of 3 Ω, 4 Ω, and 5 Ω? See Figure 14–28.

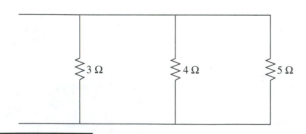

Figure 14–28 Parallel circuits with electric heat strips.

SOLUTION:

$$\frac{1}{R_T} = \frac{1}{R_1} + \frac{1}{R_2} + \frac{1}{R_3}$$

$$\frac{1}{R_T} = \frac{1}{3} + \frac{1}{4} + \frac{1}{5}$$

Write each fraction with a common denominator:

$$\frac{1}{R_T} = \frac{20}{60} + \frac{15}{60} + \frac{12}{60}$$

$$\frac{1}{R_T} = \frac{47}{60}$$

Invert both sides of the equation to obtain R_T.

$$R_T = \frac{60}{47}$$

$$R_T = 1.28 \text{ }\Omega$$

ALTERNATIVE METHOD OF CALCULATION You can also solve using decimals. Solution using a decimal solution:

$$\frac{1}{R_T} = \frac{1}{R_1} + \frac{1}{R_2} + \frac{1}{R_3}$$

$$\frac{1}{R_T} = \frac{1}{3} + \frac{1}{4} + \frac{1}{5}$$

$$\frac{1}{R_T} = 0.33 + 0.25 + 0.20 = 0.78$$

Invert both sides of the equation to obtain R_T.

$$R_T = \frac{1}{.78}$$

$$R_T = 1.28 \text{ }\Omega \qquad \blacksquare$$

CALCULATION RESISTANCE, AMPERAGE, AND VOLTAGE IN A SERIES-PARALLEL CIRCUIT

As in previous calculations, a DC circuit or single-phase AC circuit with purely resistive loads is used.

Both the voltage of the circuit and the values of all resistances are known (Figure 14–29). With two

Figure 14–29 Parallel-series circuit showing values of three resistances.

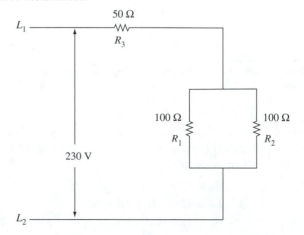

of the factors provided, the third (current flow) can be determined using Ohm's law from the following steps in the working example that follows:

WORKING EXAMPLE:

Calculate the resistance through the parallel circuit consisting of R_1 and R_2 (Figure 14–30).

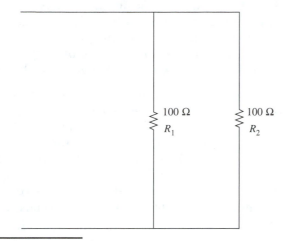

Figure 14–30 Determine the combined resistance of the two parallel resistances.

SOLUTION: Step 1:

$$R_{1,2} = \frac{R_1 \times R_2}{R_1 + R_2}$$

$$R_{1,2} = \frac{100 \times 100}{100 + 100}$$

$$R_{1,2} = \frac{10,000}{200}$$

$$R_{1,2} = 50 \ \Omega$$

Therefore, 50 Ω can be substituted for the parallel resistances. The main circuit has now become a strictly series circuit (see Figure 14–31).

Step 2. Calculate the current flow though the revised main circuit (Figure 14–31).

$$I_T = \frac{E}{R_{1,2} + R_3}$$

$$I_T = \frac{230 \text{ V}}{50 \ \Omega + 50 \ \Omega}$$

$$I_T = \frac{230}{100}$$

$$I_T = 2.3 \text{ A}$$

Step 3. Calculate the current flow through R_1 and R_2. Since R_3 is in series with $R_{1,2}$, in Figure 14–29, we calculate the voltage drop across R_3 and $R_{1,2}$ (Figure 14–30).

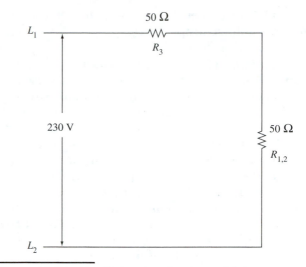

Figure 14–31 Parallel-series circuit combining two parallel resistances to form a simple series circuit.

$$E_3 = I \times R_3$$
$$E_3 = 2.3 \text{ A} \times 50 \ \Omega$$
$$E_3 = 115 \text{ V}$$

$$E_{1,2} = E - E_3$$
$$E_{1,2} = 230 - 115$$
$$E_{1,2} = 115 \text{ V}$$

Step 4. Calculate the current through R_1 and R_2.

$$I_1 = \frac{E_{1,2}}{R_2}$$

$$I_1 = \frac{115 \text{ V}}{100 \ \Omega}$$

$$I_1 = 1.15 \text{ A}$$

$$I_2 = I_1$$
$$I_2 = 1.15 \text{ A}$$

$$I_{1,2} = I_1 + I_2$$
$$I_{1,2} = 1.15 \text{ A} + 1.15 \text{ A}$$
$$I_{1,2} = 2.3 \text{ A}$$

Thus, the combined current through R_1 and R_2 is the same as through R_3, which is correct for a series circuit, and the answer is verified. ∎

SUMMARY

Knowing how to solve Ohm's law and watts formula problems is the first step in understanding circuits and learning elementary troubleshooting.

Series Circuits The current draw in a series circuit is the same throughout the circuit. The total resistance is found by adding all the resistances in the circuit. The total voltage is equal to the voltage drop across each load. The greater the resistance of the load, the higher the voltage drop.

Parallel Circuits The total current draw in a parallel circuit is equal to the sum of the amperages used in each branch. The total resistance is found by using special formulas. *Hints:* The total resistance will always be lower than the resistance of the lowest resistor. Parallel circuits have more paths for current flow; therefore, the total circuit resistance will be lower. If the resistance in a parallel circuit is the same in each circuit, the total resistance can be found by dividing the number of resistors into the resistance of one branch. For example: Four 12-Ω electric heating elements are wired in parallel. The resistance can be calculated by dividing 12 Ω by 4, which is equal to 3 Ω of total resistance. The voltage drop across each parallel branch is the same.

Series-Parallel Circuits When calculating *E, I,* and *R* in series-parallel circuits, treat each circuit as a series or parallel circuits with the rules that apply to these circuits.

▼ ELECTRIC METERS

The three electric meters that have the greatest use for installers and service personnel are the *voltmeter, ammeter,* and *ohmmeter.* They can be purchased as separate meters, or most commonly, they are all combined in a single meter called a multimeter.

There are two basic types of meters, the *analog* and the *digital.* These terms are familiar since they also apply to watches. The digital meter is solid-state and gives a direct numerical readout of the measured value. The analog meter has a needle that points to the measured value. Each is shown in Figure 14–32.

ANALOG METERS

All analog meters operate on the same principle. When the current flows through a conductor, it pro-

duces a magnetic field around the conductor. If a magnetic needle is placed close to the current, the needle will attempt to line up with the field, as shown in Figure 14–33. The analog meter movement is controlled by a magnetic field. The movement is attached to a needle that indicates a meter reading.

In order to conserve space, the current-carrying wire is coiled and a scale is provided to indicate the position of the needle. The mechanism is so constructed that the greater the current flow, the greater the deflection of the needle on the scale.

Three important characteristics of analog meters need to be considered by those who use them:

1. The most accurate reading is at or near the midpoint of the scale. This is because the spring that opposes the deflection of the meter does not exert constant pressure across the scale. The meter may be inaccurate at either end of the scale. So whenever the operator has a choice of scales, the one selected should place the pointer in the most favorable (central) position.

2. Analog meters periodically need to recalibrated. Most meters include some type of adjustment and instructions for calibration.

3. The small coil of wire that forms part of the meter movement is sensitive to excessive current. The meter may be made completely inoperable if subjected to excessive current. In using a multiple-scale meter, always use the higher scale first and move down to the scale required.

Analog meter accuracy is normally specified as a percent of full scale reading, so readings should be taken in the upper two-thirds of the scale for testing accuracy.

More expensive (and more accurate) analog meters are often furnished with a mirror scale to enable more accurate readings by avoiding *parallax,* an apparent difference in readings taken from different perspectives. The pointer and scale are aligned so the mirror image of the pointer disappears behind the pointer. This gives the most accurate reading.

Quality analog meters will no doubt become more expensive as digital meters become more popular. Many service technicians will want to have both. Just like analog watches, analog meters have advantages such as ease of reading changes and variations in the measurements.

DIGITAL METERS

Digital meters offer a number of advantages, although they are usually more expensive. Rugged versions are available and recommended. Because of the expense of digital meters, it is important that the technician be thoroughly familiar with the use

Figure 14–32   Two types of electrical meters.

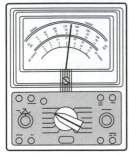

ANALOG   DIGITAL

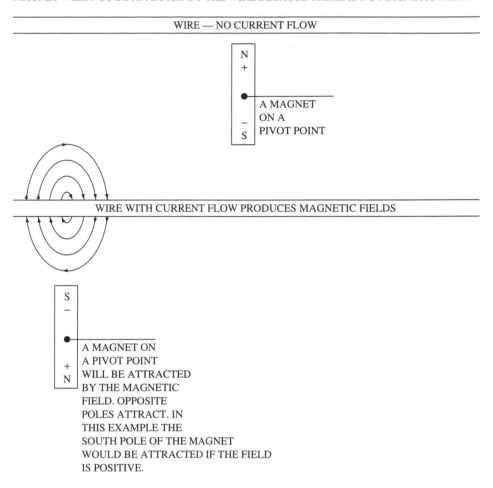

A WIRE WITH NO CURRENT FLOW PRODUCES NO MAGNETISM IN THE WIRE, A PERMANENT MAGNET WILL NOT BE AFFECTED BY THE WIRE BECAUSE THERE IS NO MAGNETIC FIELD.

WIRE — NO CURRENT FLOW

N
+

–
S

A MAGNET
ON A
PIVOT POINT

WIRE WITH CURRENT FLOW PRODUCES MAGNETIC FIELDS

S
–

+
N

A MAGNET ON
A PIVOT POINT
WILL BE ATTRACTED
BY THE MAGNETIC
FIELD. OPPOSITE
POLES ATTRACT. IN
THIS EXAMPLE THE
SOUTH POLE OF THE MAGNET
WOULD BE ATTRACTED IF THE FIELD
IS POSITIVE.

Figure 14-33 The effect of the magnetic field on the position of the meter needle. A magnetic field is used to rotate a movement that is attached to a needle indicator on an analog meter.

of both types. Some of the advantages of digital meters are:

1. They are direct reading. There is no need to interpret the scale.
2. Digital meters can be obtained that will give accurate readings to three decimal places
3. They have no moving parts and are less likely to fail or get out of calibration than analog meters.
4. They often have automatic scaling features.

Some precautions must be observed when using digital multimeters.

1. Digital meters can read "ghost voltages." Ghost voltage readings are caused by an induced or bleed voltage found in the circuit being read by the digital meter. The current is so low that there is not enough power to operate the circuit, but it is adequate to be detected by a digital voltmeter. The digital meter is a very sensitive device and will not load down electronic circuits that use very low current to operate solid-state devices. Using a digital meter will confuse a technician with voltage readings that do not

have capacity to operate the components. In some cases the voltage will be much different than the expected voltage readings. For example, a 120-V circuit may measure 100 V. In comparison, an analog meter has enough resistance in its internal circuitry to load down the circuit and eliminate false and confusing ghost voltage readings.

2. Digital readouts are not stable. The digital readout reflects the true circuit operation. Voltage from a power supply does vary, sometimes as much as several volts. The variation in the digital readout is usually to the right of the decimal point. Better-quality meters have a circuit damper to prevent the reading from jumping around so much to the right of the decimal point. Reading fluctuations are characteristic of most digital multimeters. The user should get accustomed to this condition.

3. Digital meters always require a battery or external power source. An analog meter requires batteries for resistance reading only. Have spare batteries available at all times.

4. Digital meters have temperature limits before the readout is blanked out. Digital meters work well

in temperatures above freezing to about 120°F. Working in walk-in freezers or in hot attics will temporarily impair the readout of most digital displays.

As with other test equipment, it is advisable to purchase a good-quality instrument. Purchasing a digital meter with a built-in capacitor checker will give you the certainty that the capacitor is near the value listed. It is difficult to check a capacitor using the resistance scale of the digital meter. The capacitor (microfarad) checking capabilities are worth the additional cost of the instrument.

Finally, digital multimeters can lose calibration or accuracy. Like other instruments, they should have an annual inspection and checkout. Periodically, compare the voltage, resistance and current reading to another digital or analog meter. Every professional technician has two or more multimeters for these comparative purposes as well as for backup in case a meter is in the shop for repairs or calibration.

AMMETERS

The clamp-on ammeter is one of the most useful of the electric meters for HVAC/R technicians. It is used to measure current flow through a single wire by enclosing the wire within the jaws of the instrument, as shown in Figure 14–34.

This instrument functions like a transformer. The primary coil is the test wire enclosed by the jaws of the instrument. The secondary is a coil of wire within the instrument that is connected to the current-indicating mechanism. The current in the primary

---

Figure 14–34 Construction and use of a clamp-on ammeter, commonly referred to as an Amprobe. *(Courtesy of Amprobe Promax)*

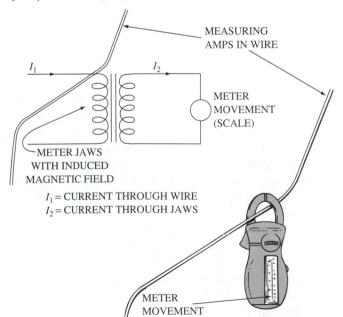

wire induces a flow of current in the secondary winding, measuring the current flow. The greater the current flow through the test wire, the greater the induced current and the greater the deflection of the needle reading on the scale. It is not necessary to disconnect or make contact with any wires to obtain a reading. This is very convenient.

CARE OF THE INSTRUMENT

More accurate tests and longer life for the clamp-on ammeter are obtained by proper care of the instrument. Some of the ways the instrument should be treated are:

1. Keep the jaws clean and aligned.
2. When taking a reading always start on the highest possible scale and then work down to the most appropriate scale.
3. Do not cycle a motor off and on while taking readings unless the meter is first set to the highest scale.
4. Never put the clamp around two wires at the same time. If the current is flowing in opposite directions, the meter will read the difference between the two. If the current is moving in the same direction in both wires, the meter will add the two.

READING SMALL AMOUNTS OF CURRENT

The clamp-on ammeter is useful in reading small amounts of current. The procedure is to loop the wire around the jaws, as shown in Figure 14–35. Passing the wire twice through the jaws doubles the strength of the magnetic field. It is therefore necessary to divide the meter reading by two to determine the actual current. For two passes, the actual current is ½ the reading; for three passes the actual current is ⅓ the reading, etc.

One application of the clamp-on ammeter is to adjust the anticipator setting of a thermostat. The *anticipator* supplies false heat to the thermostat, causing it to shut off the heat to the space being heated before the temperature in the room reaches the set point. This prevents the residual heat in the furnace, supplied by the fan after the burners are off, from overheating the room.

For the anticipator adjustment, 10 passes of wire are wrapped around the jaws of the instrument. To obtain the actual current flow through the anticipator, the scale reading is divided by 10. The diagram for the anticipator circuit is shown in Figure 14–36. The location of the ammeter with its coil of wire is shown in Figure 14–37. The wire that passes through the jaws of the ammeter is connected to the "W" terminal of the thermostat.

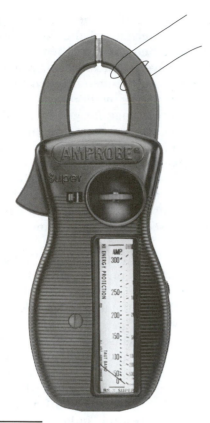

Figure 14–35 Low current can be measured by a clamp-on ammeter. Divide the number of wire wrapped around jaws by the number registered on the scale. The Amprobe will be on the lowest scale. Amperage less than 3 Amps can be accurately measured by using this procedure. *(Courtesy of Amprobe Promax)*

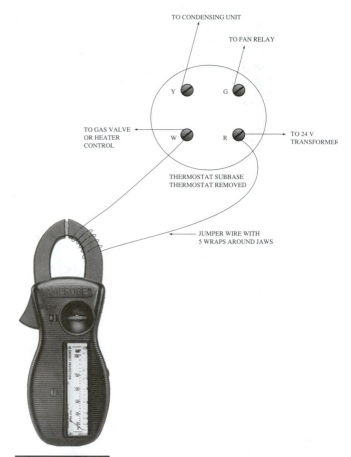

Figure 14–37 Set to lowest scale. Divide the reading by 5 to obtain the correct current flow in the heating-control circuit. Set the heat anticipator to this setting. *(Courtesy of Amprobe Promax)*

CLAMP-ON METER OPTIONS

Clamp-on ammeters, such as the Amprobe, have inexpensive options to extend the usefulness of the instrument. The Amprobe has a carrying case and probes that can be used to measure voltage and ohms. The accuracy of the voltage option is good, but the resistance option is limited to measuring low resistance which is useful for most HVAC/R components and circuits. The clamp-on accessories can convert the instrument into a backup multimeter.

THE IN-LINE AMMETER

Occasionally it is desirable to use an in-line ammeter. The proper location for its connections in a circuit are shown in Figure 14–38. Note that it is connected in series with the circuit being tested. In DC circuits, verify the correct polarity of the ammeter used before energizing the circuit.

Never connect an ammeter across a load. It will be destroyed by line current because there is no load to limit it!

Figure 14–36 The heat anticipator in the thermostat is wired in series with the 24-V control circuit. It is a variable resistor that should be set to the current draw in the secondary of the transformer. The amperage will usually be less than 1 A.

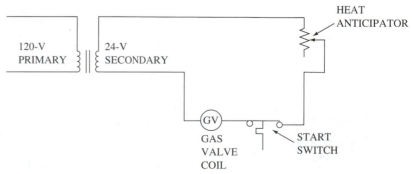

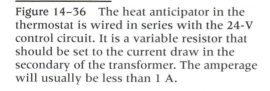

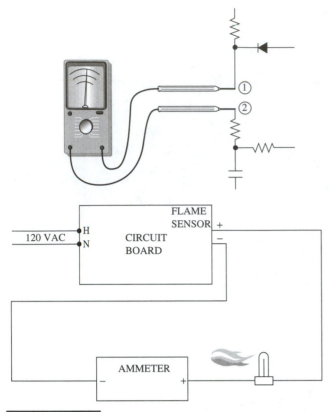

Figure 14-38 The ammeter must be in series with the circuit for proper amperage measurements. Placing the ammeter in parallel with the circuit will damage the meter because there is no resistance in the ammeter movement. *(Courtesy of Amprobe Promax)*

VOLTMETERS

Two leads from the voltmeter are connected to the circuit being tested, as shown in Figure 14–39. Voltmeters are connected in parallel with the load to read the voltage drop.

Figure 14–40 shows a voltmeter that has multiple scales. Each scale has a different resistance in the

Figure 14-39 Step-down transformer for a 230-V, three-wire, single-phase power supply. Check at the breaker or fuse panel.

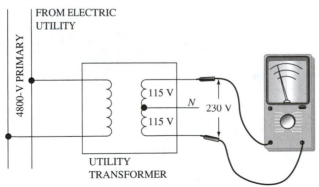

meter placed in series with the circuit being tested. A knob in the center face of the meter adjusts the meter to the scale being used. When using a multi-range meter, always start to measure voltage using the highest range on the meter. When the approximate voltage is read, the meter range can then be reduced to the proper range for greater reading accuracy. Also, using the meter with a higher voltage than the range of the meter could cause burnout or otherwise damage the meter. Some digital voltmeters provide an auto-scaling function and the reading automatically adjusts to the correct range.

In DC circuits, verify the correct polarity of the probes that are used before connecting the meter to the DC circuit.

In testing a motor to see that it has proper voltage, one lead goes on each side of the load, as shown in Figure 14–41. The illustration is a AC circuit in which the polarity of the leads do not have to be observed.

The voltmeter can also be used to determine if a hidden switch is open or closed. This is very helpful in troubleshooting. If there is power in the circuit, and the leads of the voltmeter are placed on each side of the switch, a voltage reading indicates the switch is open and a zero reading indicates the switch is closed (if no other open switches are in the circuit).

OHMMETERS

The ohmmeter is different from the other two popular meters in that it uses a battery as a power supply. The battery furnishes the current needed for resistance measurements The wiring for a typical meter is shown in Figure 14–42.

It is a direct application of Ohm's law. The higher the resistance, the lower the current flow and the less the meter deflection. The resting place for the needle is on the left of the scale. For a high resistance the deflection is small. For a small resistance the deflection is large.

One thing that is extremely important is the power to the circuit being tested must be turned off. Further, if there are any capacitors in the circuit, they must be discharged before the meter is used. There can be only one source of power to the meter and that must be from the battery within the meter itself or the meter may be damaged.

Using the meter to check for open circuits is called continuity testing. Figure 14–43 shows three diagrams representing the three possible responses that the meter can give.

1. In the diagram at the left, the meter is measuring the resistance through a good fuse and it registers zero. This indicates current flow or zero ohms. Measure a fuse or switch on RX1.

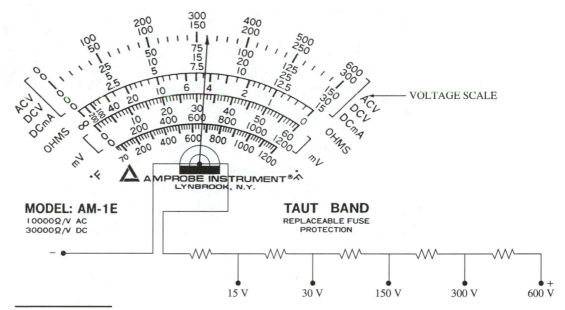

Figure 14-40 Electrical diagram of an analog multimeter. Resistance is added in series with the meter movement as the voltage measured increases. *(Courtesy of Amprobe Promax)*

2. In the diagram in the center, the meter is measuring the resistance of a transformer coil of wire, which has a measurable resistance, which is read on the meter. The range is set for RX1.

3. In the diagram on the right, the meter is measuring the resistance of an open circuit, which is read on the meter as infinity. Infinity means that the resistance is so large that it cannot be measured. It means that at this point there is a lack of continuity or no current flow. The range is set for RX10.

An ohmmeter must be able to read resistances from a few ohms to tens of millions of ohms (megaohms). In order to do this, more than one battery is used, as shown in Figure 14–44. The higher amount of power is required for higher resistances.

One very handy feature on an analog ohmmeter is the zero ohm adjustment, Figure 14–45. This knob makes possible a quick and easy method of calibrating the instrument each time it is used. To test the adjustment, the two leads are touched together and if the reading is not zero, it is adjusted to zero before the instrument is used and when switching ranges.

Unlike the voltmeter, one scale is used to read all resistance ranges. To determine the resistance value, multiply the meter reading by the number shown next to the selector switch setting. For example, in Figure 14–46, the reading 3.2 is multiplied by the selector setting (×1), giving a resistance of 3.2 Ω.

Some ohmmeters have a selector position of R × 100,000, which is used in measuring very high resistances such as motor windings to ground.

Care must be taken to prevent errors in reading resistance when two or more circuits are connected in parallel, as shown in Figure 14–47. The meter in the illustration is actually reading the combined resistance of two parallel resistances, Comp and OFM.

In order to read only one resistance, one side of the component being tested is disconnected, as shown in Figure 14–48.

One caution that needs to be followed: do not use an ohmmeter to test a solid-state circuit unless the manufacturer specifically requires it. The internal battery voltage of the ohmmeter can damage the solid-state circuit Figure 14–49.

Figure 14-41 *(Courtesy of Amprobe Promax)*

VOLTAGE IS ALWAYS MEASURED ACROSS THE COMPONENT BEING CHECKED.
METER IS BEING USED TO MEASURE THE VOLTAGE ACROSS RUN WINDING.

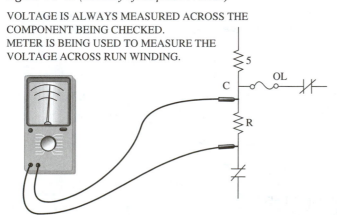

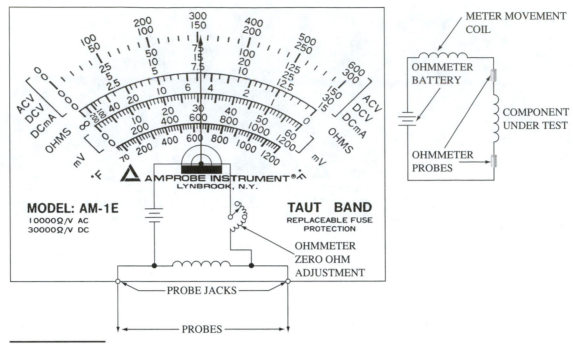

Figure 14–42 *(Courtesy of Amprobe Promax)*

Figure 14–43   Use of ohmmeter for measuring circuits. *(Bottom picture courtesy of Amprobe Promax)*

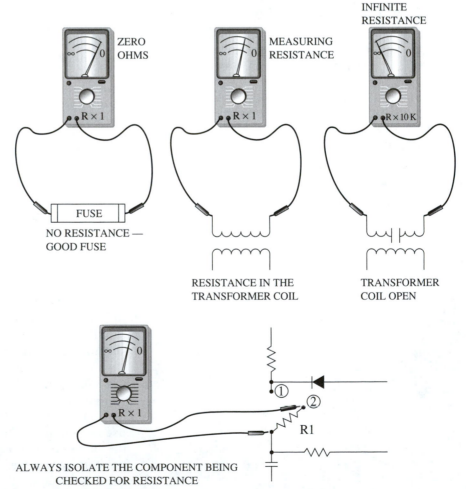

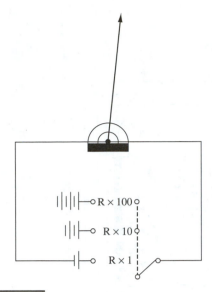

Figure 14-44 Use increased battery voltage to increase resistance range.

Figure 14-45 Amprobe Instrument Model AM-3E industrial analog multimeter. *(Courtesy of Amprobe Promax)*

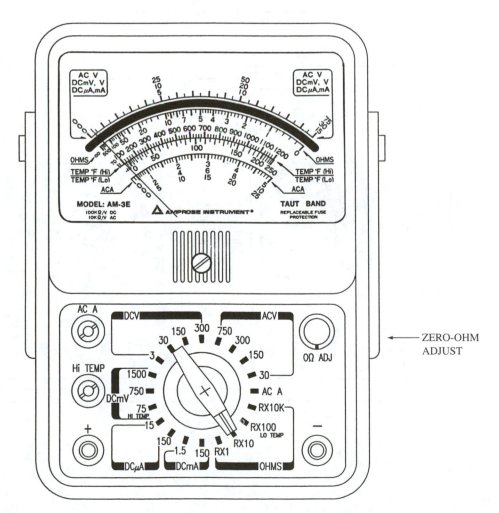

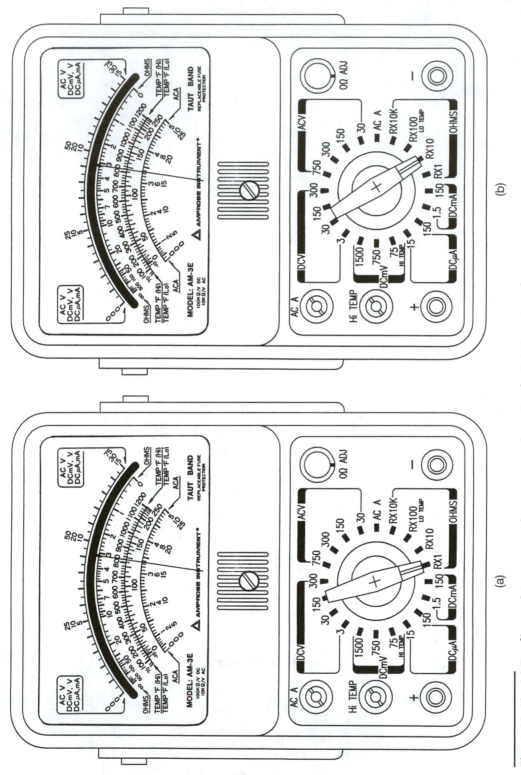

Figure 14–46 (a) Reading resistance on various range settings. On the R × 1 setting, the reading is 3.4. *(Courtesy of Amprobe Promax)* (b) Reading resistance on the R × 10 setting gives 3.4 × 10 = 34. *(Courtesy of Amprobe Promax)*

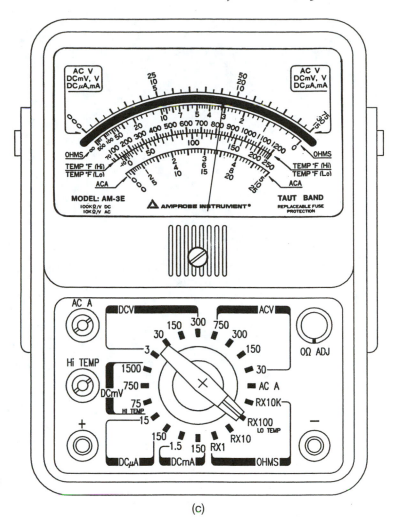

Figure 14-46 (c) Reading resistance on the $R \times 100$ setting gives $3.2 \times 100 = 320$. *(Courtesy of Amprobe Promax)*

(c)

Figure 14-47 Assume that the power is disconnected to the circuit prior to placing the ohmmeter on the component. Components should be isolated or removed from the circuit to reduce reading errors, as shown. The compressor motor windings could be open and the ohmmeter will measure resistance through the outdoor fan motor (°FM) because of the parallel arrangement of the circuit.

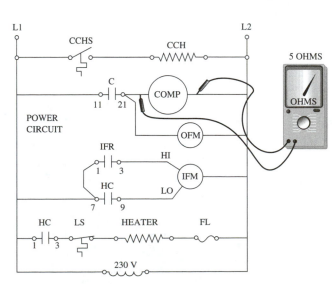

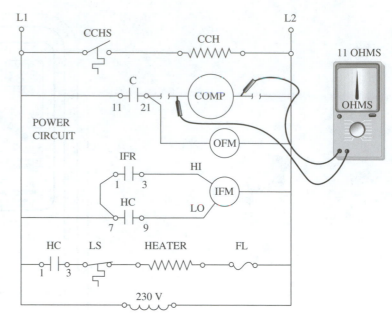

Figure 14-48 Remove the component from the circuit before checking it with an ohmmeter. This prevents reading resistance through other parallel components.

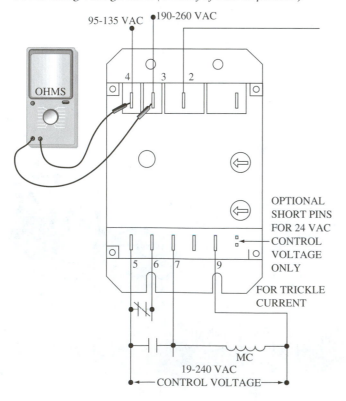

Figure 14-49 An ohmmeter provides DC voltage that may damage circuit boards. Test the operation of a circuit board using voltage tests. *(Courtesy of ICM Corporation)*

SUMMARY

- In this chapter you learned information that is basic to the foundations of troubleshooting. A technician has to be able to identify the differences between series, parallel, and series-parallel circuits. This is fundamental information that requires the identification of a circuit in order to determine how it operates.

- Uses of Ohm's law and the derivative of Ohm's law known as the watts formula were discussed. Knowing this basic information will make you a better technician. Ohm's law and the watts formula tests circuit knowledge and how to make decisions based on what is operating or not operating in a circuit. At this point in your studies you may not realize how important understanding these formulas can be. Processing this information is the beginning of the formation of your troubleshooting philosophy.

- Finally, this chapter discussed the uses of analog and digital multimeters. There are advantages and disadvantages to using each type of meter. It is best to have a good-quality analog and digital meter. As the chapter states, there is a reason to pack each type of instrument when tackling troubleshooting problems. Doctors have more than one instrument to troubleshoot your body. You will need professional instruments to complete your diagnosis. You want to be the best!

PROBLEMS AND QUESTIONS

1. List the basic components of an electrical circuit.
2. What are the advantages and disadvantages of a digital multimeter?
3. What are the advantages and disadvantages of an analog multimeter?
4. Describe the difference between series, parallel, and combination circuits.
5. Name several safety measures that can be used to reduce electrical shock.
6. List several electrical loads found in HVAC/R circuits.
7. How does the current flow in a series circuit?
8. How does the current flow in a parallel circuit?
9. How many volts are supplied to a circuit that has an amperage draw of 12 A and a resistance of 11 Ω?
10. How many volts are supplied to a circuit that has an amperage draw of 20 A and a resistance of 11 Ω?
11. How many watts are drawn by the circuits in questions 10 and 11?
12. How many ohms are found in a series circuit that has three 10-Ω electric heat strips? Show your calculations.
13. How many ohms are found in a parallel circuit that has three 10-Ω electric heat strips? Show your calculations.
14. What is the total resistance of a parallel circuit that has four electric heat strips? The heat strips measure 4, 6, 8, and 10 Ω respectively. Show your calculations.
15. How many watts will a circuit draw if you have the following information: It is a series circuit that has 240 V applied across two heat strips that measure 5 Ω each. Show your calculations.
16. How much resistance will *each* of four heat strips have in the parallel circuit that has a total resistance of 2.5 Ω and an applied voltage of 240 V? Show your calculations.
17. How much power is consumed in Question 16?
18. What is the power consumed in a circuit that has two 15-Ω heaters in parallel with a current draw of 15 A per heat strip? Show your calculations.
19. An air-conditioning technician is troubleshooting an air-conditioning system. The technician uses a digital meter and measures 151 V at the condensing unit power supply. The equipment is not operating. The technician expects a reading in the range of 208 V to 240 V. The technician takes an analog meter and reads 0 V across the same terminals. What would you tell the technician?
20. Describe how to measure low current (less than 1A) with a clamp-on Amprobe.
21. As a professional technician, why is it important to have more than one multimeter?
22. Describe how to use a clamp-on ammeter.
23. Describe how to use an in-line ammeter.
24. What precautions must you take when using an ohmmeter?
25. When checking compressor motor windings to case ground, what is the best resistance setting on the ohmmeter?

CHAPTER FIFTEEN

TROUBLESHOOTING ELECTRICAL COMPONENTS

AFTER STUDYING THIS CHAPTER, THE STUDENT WILL BE ABLE TO:

- Describe the operation of the components discussed in this chapter.
- Discuss how to troubleshoot the various components discussed in this chapter.
- Be able to troubleshoot a transformer.

- Be able to troubleshoot a capacitor.
- Be able to troubleshoot a contactor.
- Be able to troubleshoot a general-purpose relay.
- Be able to troubleshoot a single-phase compressor.
- Be able to troubleshoot a three-phase compressor.

INTRODUCTION

Once you have decided that a specific component is defective, it must be checked out before it is replaced. Many components are replaced because the technician misdiagnosed the problem or simply guessed that a particular component was defective. Component manufacturers complain that a high percentage of their warranty-return items are perfectly good components. Customers complain that unnecessary changes of parts end up on their bills. The problem is that some technicians do not know how the component operates or they do not know how to troubleshoot the component operating in the system or removed from the system. This chapter addresses these issues. In this chapter we review how these basic components operate and how to troubleshoot them isolated from the electrical circuit. Troubleshooting these components "live" is covered in this chapter as well as other troubleshooting chapters in this textbook.

This chapter describes in detail how to troubleshoot many of the common components found in air-conditioning, heating, and refrigeration equipment. The chapter covers compressors, motors, capacitors, transformers, and various types of relays. The following chapter discusses system troubleshooting using schematic or wiring diagrams as the road map to success.

TECHNICIAN TOOLS: DIGITAL VERSUS ANALOG MULTIMETERS

In troubleshooting it is important to know how to use a good-quality multimeter. Quality multimeters start at around $100. A quality multimeter is not valuable unless the owner thoroughly reads and understands the operating instructions. An analog meter or digital meter can be used. Each meter has its advantages and disadvantages. When checking individual components, a digital meter with a capacitor checker is the best (Figure 15–1). When conducting troubleshooting on electrically hot systems, an analog meter is the best bet (Figure 15–2).

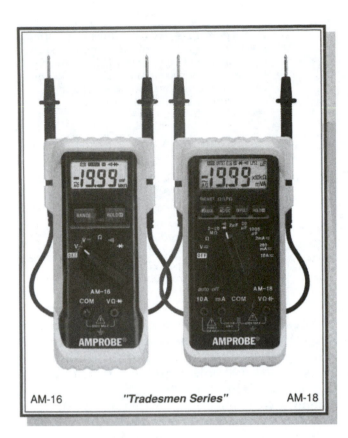

Figure 15–1 Amprobe digital multimeter. *(Courtesy of Amprobe Promax)*

Figure 15–2 Amprobe analog multimeter. *(Courtesy of Amprobe Promax)*

Digital meters are very sensitive devices and have little or no loading effect on the circuit being tested for voltage. Digital meters can measure *ghost voltage*. Ghost voltage is voltage that is in a circuit but does not have enough power or current flow to operate the circuit components. Ghost voltage can also be voltage induced between wires that are close together in the supply voltage or control circuit. Ghost voltage is usually not the normal voltage a technician would expect when measuring a circuit. Most HVAC/R circuits use 24-V, 120-V, 240-V, or 460-V circuits. A technician should expect voltage within ±10% of these common voltages. The detrimental effect of measuring ghost voltage is that it confuses the technician by giving a voltage reading when, in reality, the measured voltage will not drive the components to operate. Another example of a ghost voltage is found in a nearly dead car battery. The dead battery may measure 12 V and operate the car radio. Measuring the voltage while trying to start the car reveals that there is not enough voltage and current flow to move the starter. The measured 12 V will drop significantly when the starter tries to engage.

Digital meters are designed with high-impedance inputs so as not to load down the circuit and change its operating characteristics. Digital meters work well when checking voltages in solid-state electronics circuits such as computers, televisions, or radios. Pure electronic circuits are sensitive to low-impedance loading such as that found in analog meters. When a voltmeter checks voltage, it is in parallel with the load it is checking. The low-impedance circuitry in an analog meter may change the operation characteristics of many electronic circuits; therefore, it is better to use a digital meter when troubleshooting solid-state circuits and an analog meter in conventional air-conditioning circuits.

👉 **SAFETY TIP** When working around HVAC/R equipment, your clothes and shoes should be dry. Dry clothing will not conduct electricity as well as wet or moist clothes or shoes. Carry an extra set of clothing to change into should your clothing become wet. A professional technician carries on extra set of clothes or uniform to change in to, if necessary, to present a groomed appearance for the next service appointment. If you are working on wet ground or concrete, you can enhance your safety by laying a dry piece of plywood over the area in which you are standing or kneeling.

CONTROL TRANSFORMERS

Before troubleshooting any component, a technician must understand the operation of the component. Each of the following sections reviews the basic operation of the component being discussed.

OPERATION

The purpose of a transformer is to do one of the following:

- Step up secondary voltage, which steps down the secondary current.
- step down secondary voltage, which steps up the secondary current.
- Isolate the voltage between the primary and secondary of the transformer, in which case the input voltage and current will nearly equal the output voltage and current.

In a transformer, voltage is induced from the primary to the secondary by magnetism. Transformers are constructed using the induction characteristics of alternating current. When current flows through a coil, a magnetic field is produced. When a second coil is placed in the field of the current-carrying coil (primary), electric current can be transferred to the second coil (secondary), as shown in Figure 15–3. The process is made more efficient by wrapping the coils around a common metal core. The voltage transferred is directly proportional to the ratio of the number of turns on the secondary coil. More than one secondary coil can be used if additional voltages or circuits are required.

The transformer used most often in air-conditioning circuits is the step-down or step-up transformer, shown in Figure 15–4. The amount of voltage induced in the secondary winding depends on the ratio of the number of turns in the primary winding to the number of turns in the secondary winding.

Figure 15–4   A step-down transformer reduces voltage in the secondary side. The ratio between the 120-V primary and 24-V secondary is 5:1. This is the same resistance ratio between the primary and secondary coil.

A step-up transformer increases voltage on the secondary side. The ratio between the 240-V primary and 480-V secondary is 1:2. This is the same resistance ratio between the primary and secondary coil.

A step-down, or low-voltage, transformer (Figure 15–5) is used in air-conditioning-control systems to reduce voltage to operate the control components. Inside a simple step-down transformer are two unconnected coils of insulated wire wound around a common iron core (Figure 15–6). To go from 120-V primary to 24-V secondary, there are five primary turns to one secondary turn. For a 240-V primary, the ratio is 10:1. Thus, the induction ratio is in direct proportion to the voltage supplied by the secondary. Step-up transformers are just the reverse case. You can measure the primary and secondary winding with an ohmmeter and arrive at the same ratio. If the transformer is not labeled, checking for resistance may be the only way of determining the primary from the secondary or the voltage ratios between the two sides.

Energy, or power (watts), is *not* lost when reducing or stepping down voltage on a 120-V to 24-V stepdown transformer. The value of the current on the secondary side is five times higher than the current on the primary side, therefore the power (or wattage) is the same on both sides of the transformer,

Figure 15–3   Typical control transformer with 208/240-V primary and 24-V secondary. Magnetism couples primary and secondary windings.

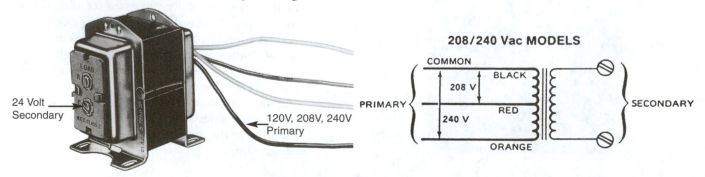

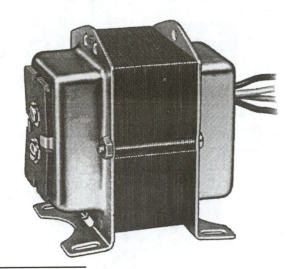

Figure 15–5 AC voltage transformer for low-voltage controls. *(Courtesy of Honeywell)*

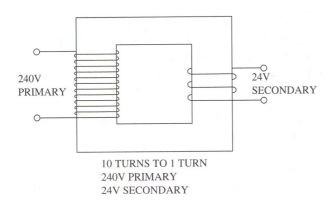

10 TURNS TO 1 TURN
240V PRIMARY
24V SECONDARY

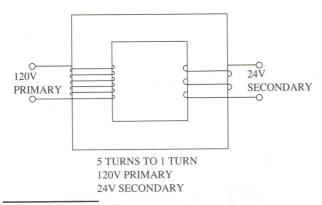

5 TURNS TO 1 TURN
120V PRIMARY
24V SECONDARY

Figure 15–6 Construction of a step-down transformer.

assuming that the transformer is 100% efficient. The ratio can be expressed by the following formula:

$$\text{Volts} \times \text{amperes} = \text{volts} \times \text{amperes}$$
$$\text{(primary)} \qquad \text{(secondary)}$$

or

$$V_p \times A_p = V_s \times A_s$$
$$120 \text{ V} \times 1 \text{ A} = 24 \text{ V} \times 5 \text{ A}$$
$$120 \text{ VA} = 120 \text{ VA}$$

VOLT-AMPERE RATING

In addition to knowing the primary and secondary operating voltages, a technician must know the **VA**, or volt-ampere rating, of the transformer. The VA rating is the maximum power output the secondary of the transformer can handle without overheating and burning out. Common 24-V control transformers have a rating of 40 VA. Control transformers can be rated as 20, 40, 75, or 125 VA or higher. Power transformers are rated in **KVA**, which is 1000 VA.

WORKING EXAMPLE

A common transformer has a primary voltage of 120 V and a secondary of 24 V. Assuming that the transformer is rated at 40 VA, what is the primary current and the secondary current?

SOLUTION

$$\text{Primary (amps)} = \frac{40 \text{ VA}}{120 \text{ V}} = 0.333 \text{ A}$$

$$\text{Secondary (amps)} = \frac{40 \text{ VA}}{24 \text{ V}} = 1.67 \text{ A}$$

Therefore, using the formula,

$$120 \text{ V} \times 0.333 = 24 \text{ V} \times 1.67 \text{ A}$$
$$40 \text{ VA} = 40 \text{ VA}$$

A 40-VA transformer can handle 1.67 A on its secondary side before it overheats and burns out. Notice that the primary-to-secondary amperage ratio is 1:5, as is the primary-to-secondary turns ratio. ∎

Twenty-volt-ampere (20 VA) transformers are usually found only in forced-air heating systems. Furnaces used in combination with air-conditioning systems will have a 40-VA transformer. Heavier VA ratings are used for air conditioning because electrical devices containing a coil and iron, such as solenoid valves and relays, have a power factor of approximately 50%. Thus, for secondary circuits with such inductive controls, the capacity of a transformer must be equal to or greater than twice the total nameplate wattage of the connected loads.

The most common control transformer used in air conditioning and heating is the 24-V, step-down transformer. These transformers are used on the control voltage side of the system. A low-voltage circuit is desired because it is safer. Components and wiring are less expensive because the voltage and current requirements are less, compared to systems that operate on higher control voltages. Low-voltage thermostats provide closer temperature control than do line-voltage thermostats. In most installations, low-voltage wiring is not required to be placed in electrical conduit. The benefits of operating with low control voltage are safety, better system control, and the reduced cost of components and wiring.

Some circuits do use 120 V and 240 V in the control circuit. These high voltage control circuits are common on domestic and commercial refrigeration applications and some commercial and industrial HVAC/R systems.

An electrical diagram of three types of common transformers is shown in Figure 15–7. The single-voltage model transformer (upper diagram, Figure 15–7) has a 120-V input to the primary side of the transformer. As shown other input voltage options are available. The secondary side has a 24-V output (Figure 15–8). The input and output can vary as much as 10% and still be in compliance. Some 24-V relay or contactor coils operate on voltage as low as 15 V, but the performance is not as good as with a 24-V supply. Low voltage may create relay or contactor chatter. Chatter is a term used to indicate a rapid opening and closing of the relay or contactor contacts. The component can be opening and closing more than 60 times a minute. The rapid cycling can damage the equipment and the contacts. The chattering sound is unnerving to the technician trying to find out what is happening. The power should be disconnected to the unit to prevent this damaging short-cycling action.

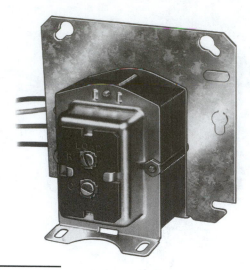

Figure 15–8   Honeywell multitap transformer. Facing side is the secondary. *(Courtesy of Honeywell Corporation)*

PRIMARY SIDE

The 208/240-V model (Figure 15–7, middle diagram, and Figure 15–9) is used when the circuit designer wants the options of either input voltage. The primary of the transformer must be wired to match the voltage supplied. If 208 V are supplied to the transformer but the transformer is wired for 240-V operation, the secondary side of the transformer will provide less than the desired 24-V output. With low control voltage, the controlled components, such as relays or contactors, may operate, operate erratically, or not operate at all. Improper wiring of the primary voltage may shorten transformer life.

Figure 15–7 *(Courtesy of Mars Components)*

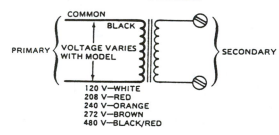

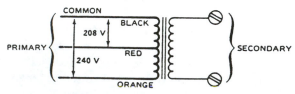

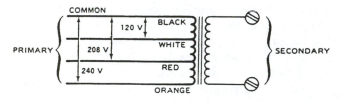

Figure 15–9   Mars multitap transformer. Facing side is the primary. *(Courtesy of Mars Components)*

Notice that the primary of the 208-V transformer (Figure 15–7) has fewer windings and less resistance as compared to the 240-V side. This is one way to determine which is the 208-V and which is the 240-V side of the primary. The 240-V side will have the highest resistance to common transformer lead. The least resistance is between the 208-V and 240-V windings. Simultaneously supplying power to the 208-V and 240-V connections will cause the transformer to burn out quickly.

The voltage tap *not* selected should be taped off to prevent electrical shock or grounding out of the transformer. The unused tap is electrically hot and should be treated accordingly.

The 120/208/240 model shown at the bottom of Figure 15–7 and in Figure 15–9 is a universal transformer that can be used in most control-voltage applications. It is similar to the 208/240-V transformer, except that it has an additional tap between common and the 208-V circuit. If you know that the black lead is common, then the lowest primary resistance is between common and the 120-V tap. This information is useful if the transformer diagram is not available.

The color code on these transformers is common but not universal. Do not take chances by simply using this color code. Measuring the resistance of the primary will determine how to wire the transformer for 120 V, 208 V, or 240 V. The lowest primary resistance will be between the 208-V and 240-V windings. The highest resistance will be between the common and the 240-V winding. The second-highest resistance will be between the common and the 208-V winding. There will not be much difference in resistance between the common and 240-V winding and the common and the 208-V winding. Read the ohmmeter carefully using the R×1 scale. The next-lowest reading will be between the common and the 120-V winding. The key to correct winding identification is to assure that the common has been identified accurately. If a schematic of the transformer is not available, you can determine the voltage-input ratings to this type of multitap transformer by using the ohmmeter.

SECONDARY SIDE

The secondary side of a 24-V transformer has less resistance than the primary side of transformer. The resistance can be less than 1 Ω without being considered shorted. There should be some resistance. The resistance of the secondary side of a step-up transformer is proportionately lower than the resistance of the primary side. The energized and loaded output voltage of the transformer varies between 22 V and 28 V.

TROUBLESHOOTING CONTROL TRANSFORMERS

There are three ways to troubleshoot most components. First, you can replace the component with a like component and see if the problem is solved. There is not much skill involved in the "parts-changer" method of troubleshooting, and truck stock limits this procedure. Customers complain about this type of approach because of a high service cost and lack of a professional approach. Equipment manufacturers know this is happening because good components under warranty are being returned.

The next way to identify a component problem is to troubleshoot the component hot, with voltage applied under its normal operating conditions. The third way to troubleshoot is to test the component with an ohmmeter and the power removed. All these methods have limitations, and in some instances two or all three methods will need to be used to nail down the diagnosis.

TRANSFORMER TROUBLESHOOTING STEPS

Professional transformer troubleshooting can be broken down into the following steps:

- Check primary voltage with the system operating under a load.
- Check secondary voltage with the system operating under a load.
- Remove power and wiring and check the primary for resistance.
- Remove power and wiring and check the secondary for resistance.
- Measure secondary winding current.

These steps are not necessarily listed in sequence or by priority. Let us examine each step in detail.

CHECKING VOLTAGE TO THE PRIMARY AND SECONDARY

The primary needs the correct input voltage to operate with the correct secondary output. Low input voltage or incorrect voltage can be a problem. Low-voltage input will produce low-voltage output at the secondary. It is best to check the input voltage with the transformer hooked to the incoming power supply. In some instances, loading down the voltage causes a significant drop in the power supply to the component being tested. A weak power supply disconnected from a load may read normal voltage. A weak power supply can be defined as a power source with poor connections or in which the wire supplying the component is undersized. The quality of the input voltage should be checked with the transformer in the circuit. Hooking the weak power source to a load such as a transformer primary, relay coil, heat strip, or compressor will cause the voltage to drop to an unacceptable level for operating the component. It is important to check the voltage with all components attached to the circuit.

The input voltage may be correct, but the secondary voltage may be inadequate under a normally operating load. It is important to check the secondary voltage under the operating load with all relays and coils attached and energized.

Checking Primary and Secondary Resistance

First, disconnect the power and wiring to both sides of the transformer. Checking resistance with the power supplied will damage the ohmmeter. The safest way to check any component is with the power off and the wiring removed. Take all safety precautions to assure that the voltage is removed.

Remember that the resistance in a step-down transformer is higher on the primary side and proportionately lower in the secondary. For example, if the transformer is a 240-V to 24-V transformer, the ratio will be approximately 10:1. The resistance on the secondary side of the transformer will be about 10 times less than the resistance on the primary side. If the primary resistance measures 10 Ω, then the secondary resistance will be around 1 Ω. The wiring must be disconnected from both sides of the transformer to be certain that the resistance reading is that of the transformer and not some other component in the circuit. Check the resistance of the primary separate from the secondary. There should be no resistance reading if you check resistance between the primary and secondary windings.

PRECAUTION When using the ohmmeter, *do not* touch both lead tips with your fingers. This will not affect low-resistance readings, but it will affect high-resistance readings because the ohmmeter will measure resistance through your body. You can touch the tip of one lead while holding the wire to the probe. Use an alligator clip to hold the other transformer wire in place for an accurate reading. Zero-adjust the ohmmeter before using it. This step assures that the meter and probe are working and calibrated properly. An analog meter will need to be zeroed each time the resistance scale is changed. Change batteries if the meter does not zero.

To complete the transformer troubleshooting, test the primary and secondary lead to the case ground of the component. Some transformers are painted or covered with a clear protective coating. Paint and coatings need to be scraped off in order to obtain a good connection to measure ground. Select the highest resistance scale and measure from each lead to case ground. A good transformer will have an infinity (∞) or no resistance reading from each wire to ground. Always zero the ohmmeter when switching scales and before using. This behavior will assure that the ohmmeter batteries are adequate and the meter is functioning properly.

Measuring Secondary Winding Current

A transformer may burn out due to several factors:

- The transformer may have incorrect voltage applied to the primary.
- The secondary side may be connected to a shorted load.
- The secondary side may be overloaded.
- The transformer may have finished its useful life.

For the longevity of the transformer, a secondary-side fuse is recommended. The fuse should be sized for 150% to 200% of the maximum current draw in the secondary. This is protection from a short-circuit condition, not an overloaded condition. Sizing a transformer fuse for an overloaded condition is difficult because it requires a smaller fuse that may open on start-up of the system due to high inrush current. A 40-VA transformer will tolerate 1.67 A on the secondary of the transformer before overheating. A 3-A fuse in series with the secondary will give adequate protection from a short-circuit condition. A fuse should be installed on the transformer secondary side. This would save many burned-out transformers. Some technicians temporarily install a fuse and fuse holder in the transformer circuit with alligator clips when they are working on anything in the control-voltage circuit. Various-size fuses can be installed in the fuse holder, depending on the VA rating of the transformer. Fuses that burn out can easily be replaced. Changing a fuse is much easier and less expensive than changing a transformer. A small, 3-A circuit breaker can be used in place of a fuse. This breaker can be purchased at an electronic supply house and can be used anytime you are working on a control circuit.

When replacing a burned-out transformer, a technician needs to explore the reasons for the damaged component. Many technicians simply replace the transformer and hope that it does not burn out. This is *not* the recommended way. The technician should determine if the damage was caused by a short or overloaded transformer secondary. Of course, the primary voltage should also be checked.

Check short circuits in the control circuit after removing the power and transformer from the circuit. Place an ohmmeter in the circuit where the secondary side of the transformer was located. The ohmmeter should be zeroed and set to the R × 1 range. Operate the thermostat through all its functions, including cooling, off, heating, fan on, and fan auto. When checking the heating mode, turn the thermostat to the highest temperature setting. When checking the cooling mode, turn the thermostat to the lowest temperature setting. The ohmmeter should be checked for each thermostat function. The resistance should be greater than 1 Ω. If the resistance is less than 1 Ω, investigate a possible short circuit in wiring, thermostat, and all system loads.

Relay or contactor coils, solenoid coils, and gas-valve coils can have shorted wiring, causing the transformer secondary to draw high amperage and burn up. Control wiring may also be shorted.

Check for a short to the unit's case ground. Some systems are designed by grounding one side of the control circuit and transformer to the case. Figure 15–10 is a diagram showing that the right side of the control design is case grounded. In this design you would check the nongrounded side of the transformer to the case. It is difficult to find grounded-out wires or exposed wiring that is creating a short-circuit condition. You will need to disable one part of the circuit at a time and check to see if the short-circuit condition has been isolated.

An overloaded secondary can cause the transformer to overheat and burn out. Check the secondary amperage with the power on. The amperage should not exceed the amperage rating in all heating, cooling, and fan functions. Check them all. The amperage draw will be very low. You will need to use an Amprobe clamp-on meter and wrap one of the secondary wires several times to achieve a reading on the meter scale. Divide this meter reading by the number of wraps on the Amprobe jaws to find the amperage drawn by the control circuit. The example in Figure 15–11 shows two wraps around the Amprobe clamp-on meter jaws; therefore, the amperage reading is divided by 2.

An in-line ammeter can be used to check secondary amperage. A quality multimeter will have an amperage range high enough to satisfy the amperage output of most control transformers. Be careful to place the in-line ammeter in series with the transformer or the meter will be damaged. Set the meter to the highest amperage setting. A quality multimeter can handle up to 10 A.

Some transformers are mounted on a square metal plate, along with a switching or fan relay (Figure 15–12). These transformers are checked in the same way as a regular transformer. This transformer and relay arrangement is called a fan center. The transformer cannot be easily replaced on the fan center. If needed, a replacement transformer can be mounted somewhere else on the equipment.

SUMMARY

Transformers are found in most air-conditioning systems. The most common transformer is the step-down transformer. The input voltage can be designed to handle 120 V, 208 V, or 240 V. The output or secondary voltage will be near 24 V. Twenty-four volts is the most common control voltage.

Troubleshooting a transformer can be done with the power applied using a voltmeter. Troubleshooting can also be done with the power and wiring disconnected using an ohmmeter. It is best to use both these methods to assure that the transformer is defective. A transformer can check out with an ohmmeter and still

---

ON THE JOB

SPARKING A TRANSFORMER

Sparking a transformer is a practice used by some air-conditioning technicians to determine if the 24-V secondary voltage is present. The practice involves removing the secondary wires from the load and quickly touching them together to create a small spark. The spark is an indication that voltage is present. The technician does not know if it is 18 V or 30 V or something in between. The transformer or transformer fuse can be damaged using this procedure. Many transformers have an internal fuse or external fuse. Sparking the transformer will cause excessive current in the secondary, thus blowing the fuse. An external fuse can be easily changed. An internal fuse will require replacement of the transformer. Sparking a transformer is a poor substitute for a voltmeter.

D3, D4CG/DC, DDUC 036, 048, 060
SINGLE PACKAGE GAS/ELECTRIC AIR CONDITIONERS
208/203-1-60

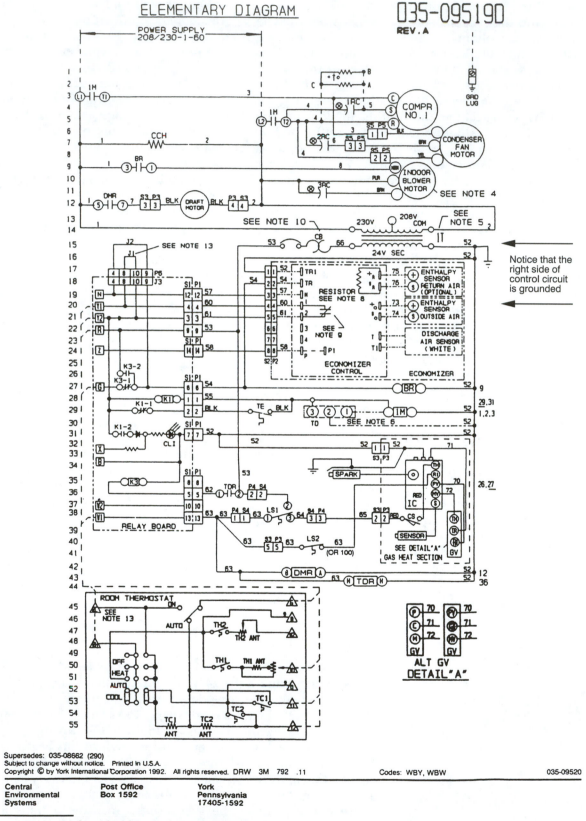

ELEMENTARY DIAGRAM

035-09519D
REV.A

Notice that the right side of control circuit is grounded

Supersedes: 035-08662 (290)
Subject to change without notice.   Printed in U.S.A.
Copyright © by York International Corporation 1992.   All rights reserved.   DRW 3M 792 .11

Central
Environmental
Systems

Post Office
Box 1592

York
Pennsylvania
17405-1592

Codes: WBY, WBW

035-09520

Figure 15–10   Right side of control circuit is grounded. Not all electrical diagrams are designed this way. *(Courtesy of York International Corporation)*

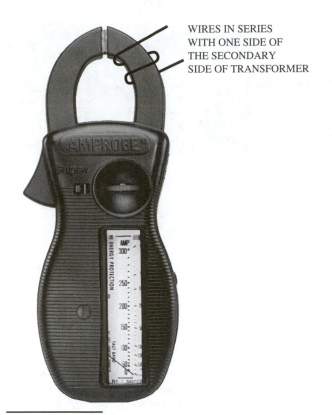

WIRES IN SERIES
WITH ONE SIDE OF
THE SECONDARY
SIDE OF TRANSFORMER

Figure 15–11 Low current can be measured by a clamp-on ammeter. Divide the number of times wire is wrapped around jaws by the number registered on the scale. The Ambrobe will be on the lowest scale. Amperage less than 3 A can be accurately measured by using this procedure. *(Courtesy of Amprobe Promax)*

be defective. Occasionally a transformer will burn out due to unknown causes. A professional will investigate the circuit before replacing the transformer. It is best to install a fuse in the secondary of the transformer. Carry spare fuses and fuse holders for this transformer-saving behavior. It is embarrassing to burn out a transformer in the middle of the troubleshooting process. Temporarily fusing the secondary will eliminate this situation.

When purchasing a transformer, a technician needs to know the primary and secondary voltage and VA rating of the replacement transformer.

TROUBLESHOOTING CAPACITORS

OPERATION

A capacitor is an electrical device that is used to change the phase relationship between the current and the voltage. This effect can be used to increase the starting power (torque) of an electric motor. This type of capacitor is called a starting capacitor (Figure 15–13, right side) and it is in the circuit only a fraction of a second. The starting capacitor is used to create a 90° phase shift between the current and voltage. This phase shift creates the maximum starting torque for the motor. A run capacitor (Figure 15–13, left side) is used to improve the power factor and running efficiency of a motor. The run capacitor also provides some minimal starting torque to the motor. A run capacitor is designed to operate continually in a circuit.

Figure 15–12 Fan-control center with transformer and fan relay. *(Courtesy of Mars Components)*

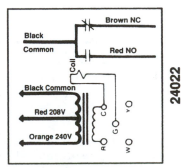

(a) RUN CAPACITORS   (b) START CAPACITOR

Figure 15–13 Most run capacitors have a metal case that is molded into the shape of a cylinder, oval, or rectangle. All start capacitors have a black plastic case to house the plates and dielectric materials. Some capacitor manufacturers use an interruptor design when a capacitor bulges and becomes unsafe. *(Courtesy of Mars Components)*

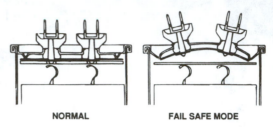

NORMAL   FAIL SAFE MODE

PHYSICAL INTERRUPTER

(c) CAPACITOR SAFETY FEATURES

A capacitor consists of a layer of insulation called a dielectric, which is placed between two plates of highly conductive metal. A capacitor is usually connected in series with the load, as shown in Figure 15–14.

No current can flow through the capacitor because of the dielectric. The current does flow through the series circuit, as shown in Figure 15–14. When the switch is closed, the supply voltage is applied across the capacitor. At that instance the electrons flow rapidly from the source to the right side of the capacitor and from the left side of the capacitor to the source, causing current to flow through the load. The capacitor quickly reaches peak current. It is described as charging during this period.

The symbol for a capacitor is commonly drawn in one of the two configurations shown in Figure 15–15. The top symbol could be mistaken for a set of normally open contacts, which has a similar electrical symbol. The dot on the right-hand symbol indicates this is the side for the power hookup of the run capacitor (Figures 15–15 and 15–16). The power should go to this side of the capacitor in order to prevent damage to the motor if the capacitor shorts to the case. Start capacitors do not have marked terminals.

RUN CAPACITORS

Most run, or oil-filled, capacitors can be identified as having an oval, square, or round shape surrounded by a metal case (Figure 15–17). Today, with the improvement of thermoplastic technology, the technician may find a run capacitor with a plastic case. The capacitor has a microfarad (μF or mfd) rating and

voltage rating. The common run capacitor is usually below 50 μF, although some applications may have higher ratings. Start capacitors have much-higher microfarad ratings. It is important to have the correct microfarad replacement.

Since run capacitors are continually in the circuit, the voltage rating of a run capacitor must be 100 V higher than the applied voltage or the capacitor will overheat and become damaged. Some common voltage ratings of run capacitors are 370 V and 440 V. Using a capacitor with a higher voltage rating is acceptable, but it may cost more than the correctly rated capacitor. Higher voltage rating translates into longer-lasting capacitors, and the capacitor will be larger in physical size.

The voltage across the capacitor is the combined potential of the applied voltage and the voltage generated by the movement of the motor. Measure the voltage across the run capacitor when the motor is operating. The voltage from a 240-V system will be around 300 V or more across the capacitor terminals. The capacitor must be sized to handle the higher voltage or it will be damaged after a period of running time.

Again, the main purpose of the run capacitor is to provide improved running efficiency, which will lower the amperage draw on the motor and reduce the operating cost. Run capacitors are used on single-phase motors. Three-phase motors do not use capacitors. Run capacitors are used on permanent split-capacitor motors (Figure 15–18) and capacitor-start–capacitor-run motor designs (Figure 15–19).

Some run capacitors have bleed resistors across the terminal connections. These capacitors are more

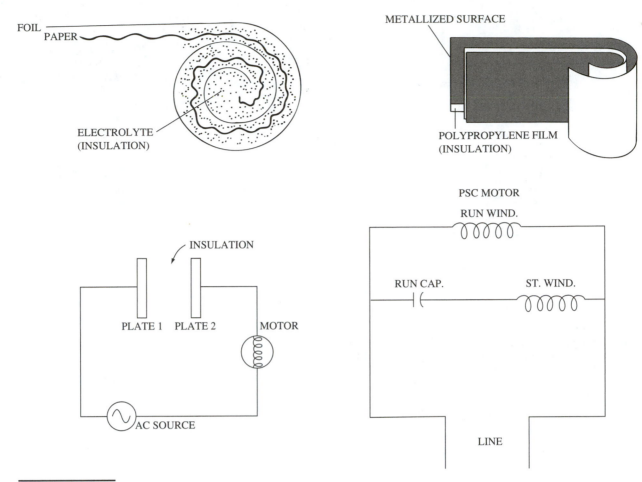

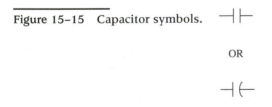

Figure 15–14   A run capacitor creates a phase shift between the starting and running voltage and current. This phase shift improves starting torque and running efficiency. A run capacitor also limits the current flow through the start winding, which prevents this winding from burning out. *(Courtesy of Mars Components)*

Figure 15–15   Capacitor symbols.

⊣⊢

OR

⊣⊢

Figure 15–16   The marked terminal on the run capacitor should be connected to the run terminal of the compressor.

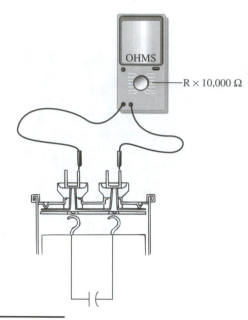

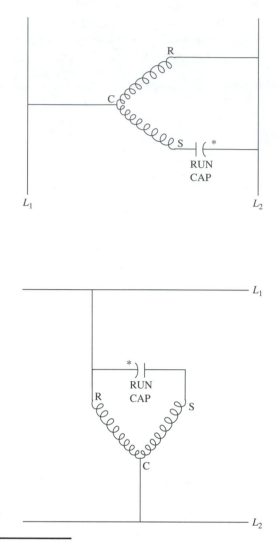

Figure 15–17 When testing a capacitor with an ohmmeter, set the meter to read the R × 10,000 scale. Reverse the leads after the first reading. Reverse the leads several times during this test. The needle or digital readout should be near or below 0 Ω and increase in value. The higher the microfarad rating, the greater the deflection is. Check run capacitor from each terminal to case for grounds or resistance. The reading should be infinity (∞) to case ground. The best meter to use in checking a capacitor is a capacitor checker.

Figure 15–18 A permanent split capacitor (PSC) has a run capacitor in series with the start winding and in parallel with the run winding. The marked side (*) of the capacitor should be installed toward the power supply run winding.

likely to be found in internally fused components. The bleed resistor prevents the capacitor from blowing the internal fuse when it is accidentally shorted or when it is being discharged.

Run capacitors with marked terminals should have the marked side connected to the power supply, as shown in Figure 15–18. The mark is designated with an asterisk (*), plus sign (+), minus sign (−), or a dot of red or white paint.

START CAPACITOR

The start, or electrolytic, capacitor is cylindrically shaped and has a black plastic or Bakelite case (Figure 15–13b). The microfarad rating is much higher than that of a run capacitor of equal size. Microfarad ratings begin as low as 20 μF and can go higher than 1000 μF. Common ratings are between 88 μF and 200 μF. Start capacitors have a range of microfarad ratings, not a specific rating, as found on a run capacitor. For example, a technician will find many 88-μF to 125-μF rated start capacitor at 250 working volts. As long as the capacitor maintains the range of 88 μF to 125 μF, it is considered good.

The voltage rating of a start capacitor needs to be equal to or higher than the supplied voltage. Remember that a start capacitor is in a circuit for less than a second; therefore, the voltage rating and heat-dissipating characteristics of the start capacitor can be less than those of the run capacitor. A higher-voltage-rated start capacitor can be used in the circuit.

Many start capacitors have a bleed resistor soldered across their leads (Figure 15–20). The purpose of the bleed resistor is to discharge the capacitor quickly after it is disconnected from the starting circuit. A fully charged start capacitor, in a short-cycling condition, will impress a higher voltage on the start winding. This could lead to a damaged start-motor winding. Bleed resistors are sized for the application and the capacitor. Bleed resistors are rated in the range of thousands of ohms. They must be disconnected before checking the start capacitor with a capacitor checker or ohmmeter.

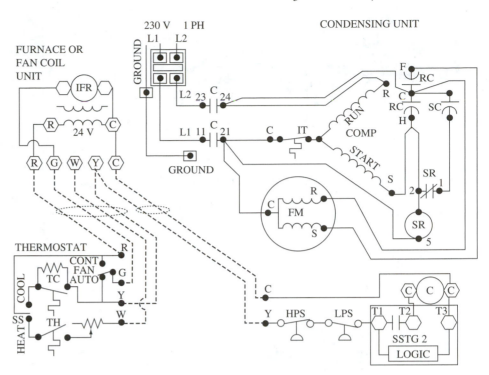

Figure 15–19 A dual-run capacitor (RC) is used to improve the running efficiency of the compressor (comp) and outdoor-fan motor (fan). A start capacitor (SC) is used to increase the starting torque of the compressor.

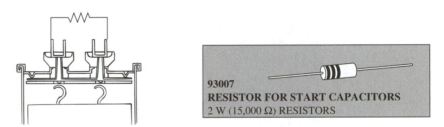

Figure 15–20 A bleed resistor is installed across the terminals of a start capacitor to discharge the capacitor immediately after it is removed from the circuit. The resistor is sized for the application. *(Courtesy of Mars Components)*

93007
RESISTOR FOR START CAPACITORS
2 W (15,000 Ω) RESISTORS

CAPACITOR TROUBLESHOOTING OPTIONS

An air-conditioning technician has the following troubleshooting options:

- Replace the capacitor with a like component.
- Check the capacitor with a capacitor checker.
- Check the capacitor with an ohmmeter.

REPLACING THE CAPACITOR WITH A LIKE COMPONENT

This method is simple; a technician replaces the suspect capacitor with a like capacitor. The problem with this method of analysis lies in the number and variety of capacitors a service vehicle would need to stock. A leaking, bulging, or damaged capacitor does need to be replaced. Some capacitor manufacturers design a physical interrupter to reduce the possibility of a capacitor rupture (Figure 15–21). Yet replacing a capacitor without checking the cause of the damage may be a mistake. In some instances a capacitor may be damaged by a start relay or defective motor or because it was the wrong size or applied incorrectly.

The replacement capacitor may meet the same fate if these conditions are not explored.

If the correct capacitor is not available the technician can wire up capacitors in series or parallel to make the match. Parallel capacitors are additive in microfarad valve. Treat series capacitors like parallel resistors using the formula:

$$C_T = \frac{C_1 \times C_2}{C_1 + C_2}$$

The microfarad rating of series capacitors will also be lower than the smallest capacitor in the set.

CHECKING WITH A CAPACITOR CHECKER

Diagnosing a capacitor with a quality capacitor checker is the best way to determine if the component is good. The capacitor checker can be built into a digital multimeter, as found in the Amprobe AM-18 (Figure 15–22), or it can be a self-contained instrument (Figure 15–23). A quality self-contained capacitor checker costs around $75. A built-in capacitor checker can be purchased as part of a multimeter for little as $20 more. The built-in option is the most

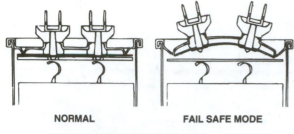

NORMAL   **FAIL SAFE MODE**

PHYSICAL INTERRUPTER

Figure 15–21   Capacitor safety. *(Courtesy of Mars Components)*

Figure 15–22   Amprobe AM-18 digital multimeter with built-in capacitor checker. *(Courtesy of Amprobe Promax)*

economical and is the recommended way to purchase this important instrument.

A capacitor checker has many advantages when it comes to troubleshooting. The capacitor checker indicates the exact microfarad rating. A capacitor should be within ±10% of the rating of a run capacitor and within the ranges listed on a start capacitor. A weak capacitor should be replaced. When checking a dual capacitor, attach one probe to common and the other probes to each of the other terminals. The terminals are generally labeled *common, herm, and fan. Common* is the same voltage connection point for both sides of a dual capacitor. *Herm* is the connection to the start winding of the compressor. *Fan* is the connection point to the start winding of the condenser fan motor. If you check from terminal to terminal, not using the common terminal, the reading will be inaccurate. The symbol for a dual capacitor is shown in Figure 15–24. Figure 15–25 shows a picture (top) of a dual-run capacitor.

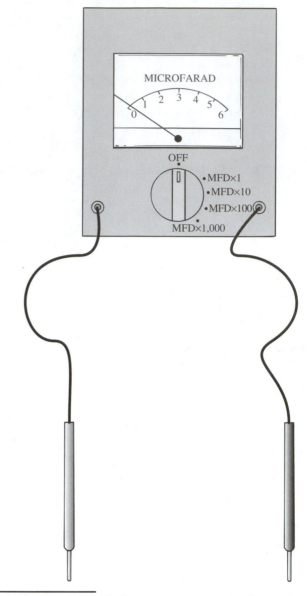

Figure 15–23   Single-function capacitor checker with scale option.

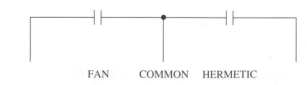

FAN   COMMON   HERMETIC

Figure 15–24   Dual-capacitor hookup.

A run capacitor with a metal case should be checked from each terminal to ground. The ohmmeter function of a multimeter on the R × 10,000 range should be used to check for shorts to the case. No resistance reading to case ground should be indicated if the capacitor is good. The reading should be infinity.

FAN COM HERM

Dual capacitor with labeled terminals used as a run capacitor

(a)

Start capacitor without bleed resistor

(b)

Figure 15–25 Dual capacitor with labeled terminals used as a run capacitor (left); start capacitor without bleed resistor. *(Courtesy of Mars Components)*

CHECKING WITH AN OHMMETER

A capacitor can be checked with an analog or digital ohmmeter. When using either type of ohmmeter, it is important that the range be set to R × 10,000 or the reading will be inaccurate. Always zero-adjust the meter to assure meter operation. What a technician is looking for is a deflection in an analog meter movement (Figure 15–26). The needle should go toward zero and slowly return to the upper or midscale of the meter faceplate. The needle may go below zero and stay there for a few seconds before returning toward the other side of the scale. This procedure should be repeated several times, switching the probes between terminals each time the reading is taken. This is very important! Take a reading; watch for the needle deflection. Reverse the leads on the capacitor and notice the meter movement again. Reverse the probes and do this again. If the needle movement registers a low resistance and stays there, then the capacitor has shorted plates. If the meter reading remains at infinity, the capacitor is open. If you suspect an open capacitor, zero the meter to be sure the meter is working properly. An open meter lead or defective meter will read infinity just like an open capacitor.

It is imperative that patience prevail in this effort. Hurrying this action will lead to misdiagnosing a good capacitor. Take several readings; it is worth your time.

Check the capacitor for shorts to the case. Attach one probe to the case and one probe to each of the terminals. No resistance reading or an infinite reading (∞) is expected on a good component. No needle deflection should be observed.

Using a digital multimeter (without a capacitor checker) is more challenging. Select the R × 10,000 range or higher and zero the meter to assure correct meter operation. Hook the probe to the terminals on the capacitor. The digital readout will rapidly go to

or near 0 Ω and increase to a higher resistance. Some digital meters go into a negative resistance range, rise to 0 Ω, and continue to increase in resistance. This is confusing because it all happens so quickly. Reverse the probes several times and notice the readout each time. If the digital readout drops to a low resistance and stays there, then the capacitor has shorted plates. If the ohmmeter shows no reading, the capacitor is open. Do not be in a hurry when checking capacitors! Take your time for an accurate diagnosis.

Check each capacitor terminal to the case when testing a metal capacitor. Any resistance reading on the R × 10,000 range will indicate a short from the terminal to the case. This is a defective component. Scrape off any paint or coating found on the case before measuring to ground. These coatings prevent a good probe connection.

CAPACITOR REPLACEMENT

Any capacitor should be replaced with a like capacitor. The only exception is the voltage rating. The voltage rating on a replacement capacitor can be higher than the original.

Figure 15–26 Using an ohmmeter to test capacitors.

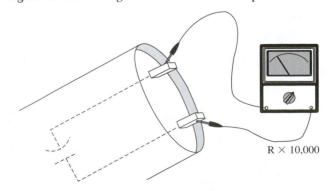

R × 10,000

Capacitors can be arranged in series or parallel to change their microfarad rating. Capacitors wired in parallel increase in capacitance because there is more capacitor plate, as shown in Figure 15–27. Hooking capacitors in parallel is analogous to resistors hooked in series. The microfarad rating of parallel capacitors is directly additive. The following formula can be used to calculate capacitors in parallel:

$$C_T = C_1 + C_2 + C_3 + \cdots$$

Capacitors wired in series decrease in capacitance because there are fewer plates, as shown in Figure 15–28. The plates are stacked and shared, thereby reducing the actual area of the plates and reducing the microfarad rating. The microfarad rating of series capacitors is similar to that of resistors in parallel. Capacitors in series have a microfarad rating smaller than the smallest capacitor. The following formula can be used to determine the total microfarad rating of capacitors in series:

$$C_T = \frac{C_1 \times C_2}{C_1 + C_2}$$

See Figure 15–28 for a sample problem involving capacitors in series. Even if a technician does not have the correct capacitor, wiring more than one capacitor in parallel or series can produce the needed results.

SUMMARY

Contrary to popular belief, capacitors do not give a motor a bump or a boost in power. Start capacitors are designed to create a 90° phase shift between the starting current and voltage. This phase shift gives the motor its maximum starting torque. Having a starting capacitor that is too large creates a larger phase shift and reduces the starting torque. A hard-start kit is a correctly matched potential relay and start capacitor. The hard-start components obtain the 90° phase shift needed for optimal starting torque. Some component manufactures market a so-called hard-start device that has a motor capacity range of 5000 to 100,000 Btuh. These claims are highly unlikely in that the capacitor and starting relay must be sized for a motor in a much narrower capacity band. Do not trust these devices. Obtain information on sizing the correct potential relay and start capacitor designed for the motor that has difficulty starting. Many compressors are condemned as locked up when the correct starting components could solve the starting problem.

A defective start capacitor or potential relay will cause the motor to stall and draw excessive amperage on start-up. The excessive amperage will cause the motor to trip out on its overload protection. Sometimes a fuse or breaker may also open due to this excess amperage condition.

Run capacitors are used to improve the running efficiency of a motor. They do provide a minimal amount of starting torque assistance. A defective run capacitor will cause a motor to stall on start-up. If the motor is successfully started, it will draw higher-than-normal amperage and probably cut off on its overload device. A fan motor that has a defective run capacitor can be started by hand. Spin the fan blade in any direction, and it will start. However, will draw excessive amperage and overheat.

Troubleshooting a capacitor is easy with the proper test instrument. Remove the wiring from the capacitor prior to checking. A quality capacitor checker or a multimeter with a built-in capacitor checker is the best instrument for successful troubleshooting. Inexpensive capacitor checkers are a waste of money. Replace all starting components when replacing a motor or compressor. Inspect the contactor points and coil for damage.

Figure 15–27 Parallel capacitors increase the microfarad rating because the plate area is expanded.

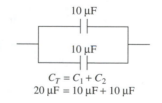

$$C_T = C_1 + C_2$$
$$20\ \mu F = 10\ \mu F + 10\ \mu F$$

Figure 15–28 Series capacitors decreases the microfarad rating because plates are shared, reducing the total plate surface area.

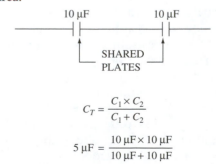

$$C_T = \frac{C_1 \times C_2}{C_1 + C_2}$$

$$5\ \mu F = \frac{10\ \mu F \times 10\ \mu F}{10\ \mu F + 10\ \mu F}$$

TROUBLESHOOTING CONTACTORS AND GENERAL RELAYS

OPERATION

Contactors and relays are electromagnetic load-controlling switches. The purpose of a contactor is to control components that have high current draw—such as a compressor—with low voltage, usually 24 V. The complete symbol for a contactor is shown in Figure 15–29. As with other symbols, these are shown in their deenergized state. A picture of a common contactor is shown in Figure 15–30.

The complete symbol for a relay is shown in Figures 15–31 and 15–32. Relays have normally open or normally closed contacts. They may have several contacts, some normally open and some normally closed. Relay manufacturers make a variety of relay configurations to meet most applications. A picture of a general-purpose relay (sometimes called a fan or control relay) is shown in Figure 15–33.

Relays and contactors are two load-controlling devices commonly used in air-conditioning circuits. Basically, relays and contactors have the same function, except the contactor is designed to handle

larger current loads. Relays are used to handle loads less than 15 A. Some relays are rated for as much as 20 A. Contactor contacts can handle more amperage, are more rugged, and have larger contact surfaces as compared to relays. Another type of load-controlling device is called a motor starter. A motor starter is a contactor designed with motor-overload protection added. The relay, contactor, and starter are combination units. They are classified as combination units because they all comprise

- A load device or coil
- One or more switches that change position when the coil is energized

Figure 15–34 shows how relays and contactors operate. When the coil is energized (supplied with current), a magnetic field is set up that attracts a metallic armature. As the armature moves to the center of the coil, it closes or opens the contacts (switches) attached to it. As long as the current flows through the coil or the contactor is energized, then the switches are in the energized position. Contactors have normally open points, or contacts. Relays can have normally open or normally closed contacts or a combination of both on the same relay.

A contactor can have one set of movable contacts. Many contactors have two, three, or four sets of movable contacts. A set of contacts has two stationary contacts and two movable contacts linked together (Figure 15–34). In contrast, a relay has one stationary contact and one movable contact (Figure 15–32). If the contactor has one set of contacts, it is classified as a single-pole contactor. Two sets of contacts are classified as a two-pole contactor, etc. A single-pole contactor (Figure 15–35) has metal bar that replaces one of the contacts. A two-pole contactor is shown in Figure 15–36 on page 233.

A single-pole contactor is less expensive compared to other multipole contactors. The disadvantage of the single-pole is that the contacts wear out or build up carbon more quickly because the total current load is imposed on closing and opening one set of contacts. Every time a set of contacts open and close under load, an electric arc is created (Figure 15–37, page 233). The arc creates carbon deposits on

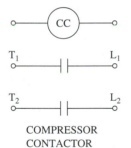

Figure 15–29 Compressor contactor showing contacts and coil.

Figure 15–30 Double-pole contactor. *(Courtesy of Honeywell)*

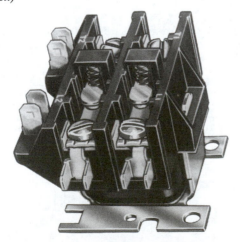

Figure 15–31 Control relay with one normally open and one normally closed contact. They switch positions when the coil is energized.

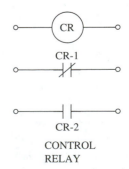

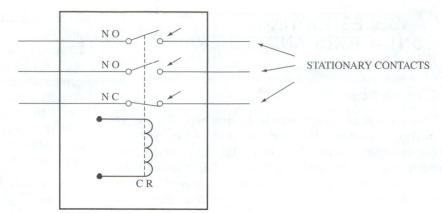

Figure 15–32 Control relay diagram with two normally open (N.O.) contacts and one normally closed (N.C.) contact. Energized CR coil causes contacts to switch positions.

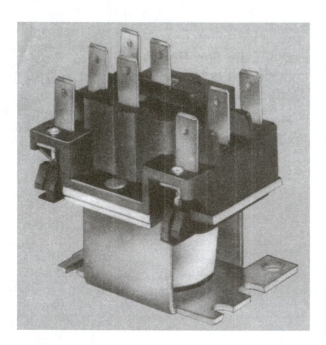

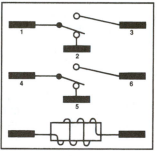

Figure 15–33 General-purpose relay. Also known as a fan relay or control relay. *(Courtesy of Honeywell)*

Figure 15–35 Coil terminals on the side of a single-pole contactor. Some contactors are designed with the coil terminals in each side of the contactor.

Figure 15–34 Energized double-pole contactor. Voltage is supplied to the load (left). Deenergized double-pole contactor. No voltage is supplied to the load.

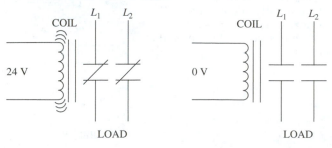

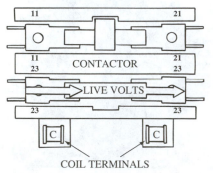

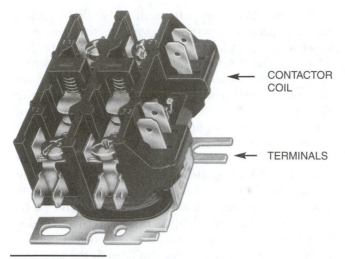

Figure 15–36 Double-pole contactor. The coil terminal is on each side of the contactor.

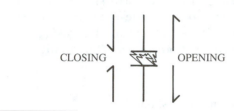

Figure 15–37 Arc between opening and closing contacts.

contact surfaces. Carbon creates resistance on the contactor surface. Having two sets of contacts (double pole) will split the starting and stopping current between the contacts, so the two-pole contactor will have longer life contacts. In some circuit designs, a single-pole contactor is required. The design of the circuit is such that one hot leg is required to keep the compressor start winding energized through a run capacitor or fixed resistor. The purpose of this operation is to use the compressor winding as a crankcase heater. If the single-pole contactor were replaced with a double-pole contactor, the compressor winding crankcase heater would be lost. An external crankcase could be wired in the circuit when converting to a double-pole contactor.

Generally, it is better to install a double-pole contactor in place of a single-pole contactor. Some air-conditioning contractors offer enhancement options on new installations. The double-pole contactor is a good option if the original equipment has a single-pole contactor.

TROUBLESHOOTING SEQUENCE

The contactor and relay can be checked

- By doing a like component change-out
- With voltage applied using a voltmeter

- By disconnecting from the circuit and checking it with an ohmmeter

In some cases the technician will need to check these components both ways to be positive that a component is defective. A like component can be substituted if the technician is certain that the part is defective. The following are indications of a defective component:

- Burned coil
- Contacts that are welded shut
- Contacts that are mechanically bound in the open position

The coil voltage should be checked prior to replacing a contactor with a damaged coil.

CHECKING THE COMPONENTS LIVE

The first thing to check is whether the control voltage is supplied to the coil. Twenty-four volts is the common control voltage, but other voltages can be found; 120 V and 240 V are not uncommon, especially when energizing relay coils. This is a good place to start. Without the proper control voltage, the coil will not energize the contacts. Control voltage that is higher than normal will burn out the coil. Measure the control voltage with the coil attached. Sometimes the control voltage is weak; when it is disconnected from the coil, it measures the desired voltage. Hooking the control voltage to the coil will load down the control voltage, and this may identify the problem as a control-voltage problem and not a problem with the contactor or relay coil. It is best to use an analog meter for this troubleshooting procedure. The voltage should be in the range of 22 to 27 V for low-voltage circuits. Higher control voltages should be within ±10% of their required voltage. Lower control voltage may cause the contactor to chatter. Chatter is a rapid opening and closing of the contacts due to low control voltage or a bad connection. This is damaging to the contactor contacts as well as the controlled component.

CHECKING A CONTACTOR WITH VOLTAGE APPLIED

If the control voltage is within range, next check the line voltage input to the contactor. Sometimes this is call the *line side,* or the L-side of the contactor. The voltage leaving the contactor is called the *load side,* represented with a *T* (Figure 15–38). The line-side voltage, under normal operating load, should be within ±10% of the supply voltage. The voltage on the load side of the contactor should be measured under load and should be practically the same voltage as the input voltage.

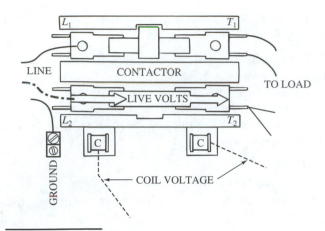

Figure 15–38 Single-pole contactor. One side of the contactor is electrically hot.

At this point, you have checked several voltages on the suspect contactor. The control voltage has been checked. The input voltage has been checked. The output voltage has been checked. There is one final voltage check before going to the resistance check.

Next, check the voltage directly across each of the energized contacts. The voltage drop across a good set of energized contacts should be 0 V (Figure 15–39). A good set of contacts is like a good piece of wire or a fuse—there should be no voltage drop and the voltage will read 0 V. If carbon has built up on the contact surfaces or the contacts are not mechanically making a good bond, then there will be a voltage drop across the closed contacts. Up to 5 V can be tolerated. If the voltage drop across a set of closed contacts is 5 V or greater, then the contactor needs replacement. Some technicians do not tolerate any voltage drop and recommend replacement at the smallest voltage drop. This is not a bad practice. Carbon deposits on the contacts create resistance. This resistance causes higher current flow in the circuit.

Finally, disconnect the control voltage (turn thermostat to the off position) and measure the voltage across each of the contactors. The applied voltage is measured on a single-pole contactor (Figure 15–40). No voltage is measured across contactors that have two or more poles. The circuit on these

Figure 15–39 Checking voltage drop across a closed set of contacts. Voltage drop should be less than 5 V.

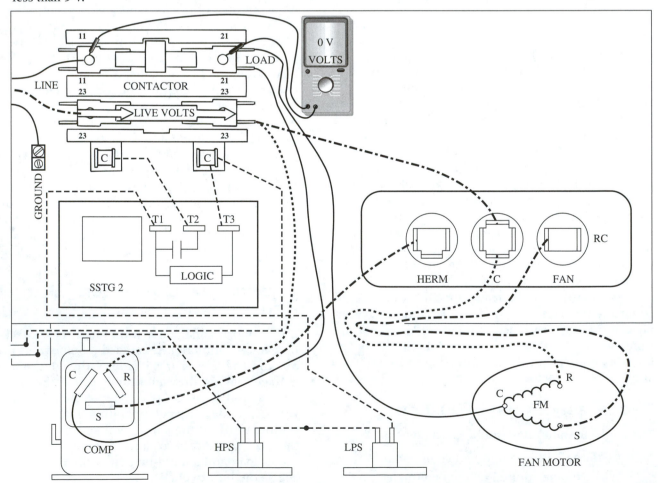

types of contactors is completely broken. If voltage is measured on an open contactor that has two or more poles, there is something wrong with the contactor, or voltage is being fed into the circuit from another voltage source. A contactor may have a set of contacts that are welded (stuck) together or the contactor may not be mechanically releasing to the open position. A visual inspection may help you decide what is happening. If the contactor passes a visual and voltage check, it should be working properly.

CHECKING A RELAY WITH VOLTAGE APPLIED

The same process is used to troubleshoot a relay. The control voltage should be measured with the voltage attached to the coil. A relay has some normally open and some normally closed contacts. When energized, the contacts should switch position. The normally open contacts will close and the normally closed contacts will open. Many relays have extra contacts

Figure 15–40 Checking voltage across an open set of contacts. The applied voltage is measured when using a single-pole contactor that is not energized.

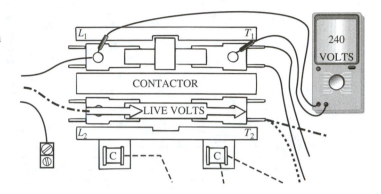

Figure 15–41 Zero-ohm adjusted knob. Select the desired resistance scale and place probe tips together to adjust for zero ohms. (*Courtesy of Amprobe Promax*)

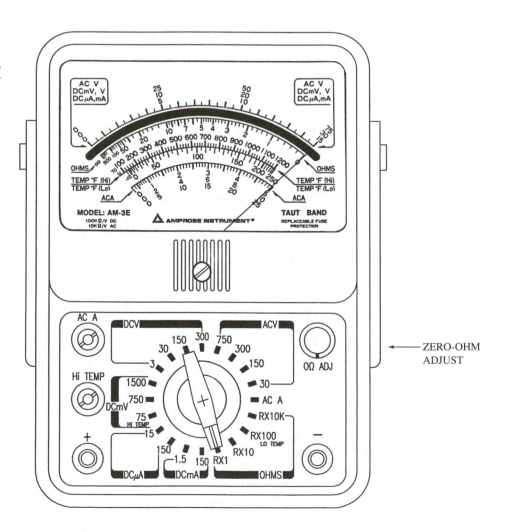

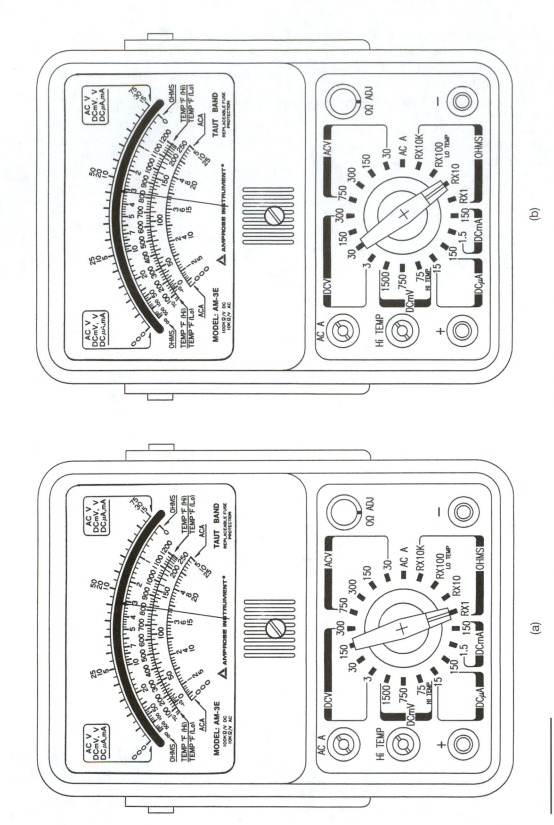

(b)

(a)

Figure 15–42 If there is no resistance reading on the R × 1 scale move to the R × 100 or next higher scales until a resistance is obtained.

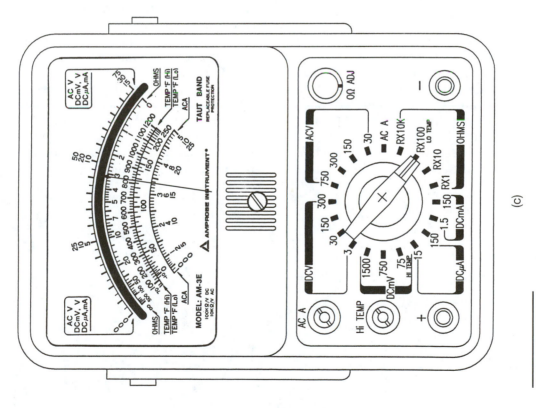

Figure 15-42 *continued.*

(c)

237

that are not used. These do not need to be checked. They can be used in case one of the other sets of contacts becomes defective. The only contacts that need to be checked are the ones that are wired in the circuit. With control power applied to the circuit, the normally open contacts will close and the voltmeter should read 0 V across the closed contacts when the circuit is energized. Up to a 5-V drop is tolerated across a closed set of contacts. With the main power applied but without the control voltage to the coil (thermostat in the off position), the volt meter will read the applied voltage. The applied voltage is read because the contacts are open.

With the control voltage applied to the circuit, the normally closed contacts will open and the voltmeter should read the applied voltage when the circuit is energized. With the main power applied but without the control voltage (thermostat in the off position), the voltmeter will read 0 V because the contacts are closed, providing a path for current flow. If voltage is measured, check to make sure the coil is not energized. If voltage is still being measured, then the contacts are open and considered defective.

CHECKING CONTACTORS AND RELAYS WITH AN OHMMETER

A technician can use an ohmmeter in conjunction with a voltage check to identify contactor and relay problems. For example, if the control voltage is applied to the coil and the contactor or relay does not switch positions, the coil may be shorted or open. Use an ohmmeter to check the coil. Zero-adjust the ohmmeter to assure yourself that the meter is functioning properly (Figure 15–41). A quality digital or analog meter is recommended. Check the resistance

of the coil. The resistance can be a few ohms or be as high as several thousand ohms. First, use the R × 1 range; if there is no reading, use the R × 100 range (Figure 15–42). The resistance of coil will depend on the manufacture and design function. A resistance reading of less than 1 Ω indicates that the coil is shorted (Figures 15–43 and 15–44). A visual inspection may reveal that the coil is burned, thus shorting the coil winding together. Any overheated coil should be replaced.

Next, check the resistance of the contacts. Take the contactor; push in the contacts while measuring the resistance across the each set of contacts (Figure 15–44). The resistance should be less than 1 Ω. The best reading is 0 Ω. Inspect the contacts. The contacts should be clean and free from carbon buildup. Most contacts cannot be cleaned and must be replaced by new units. Larger and more expensive contactors and motor starters have contact-replacement options.

The recommended way to check the resistance of the contactor or relay contacts is to energize the coil with all the power wires disconnected from the contacts. Measuring resistance with power applied to the contacts will damage the ohmmeter. Measuring resistance with wiring hooked up may cause a reading through other components and may give a technician a misdiagnosis of the problem. Disconnecting the power and the wiring to all contacts is important for an accurate measurement.

After energizing the coil, measure the resistance of each of the contacts. The resistance of normally closed relay contacts should be 0 Ω but no more than 1 Ω. Apply power to the coil only. The normally closed contacts will open. The resistance of normally closed relay contacts should be infinity when the coil is energized.

Figure 15–43 Generally contactor and relay coils with less than 1 Ω of resistance are considered shorted.

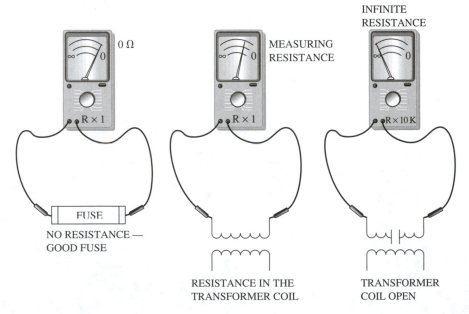

0 Ω

R × 1

FUSE

NO RESISTANCE —
GOOD FUSE

MEASURING
RESISTANCE

R × 1

RESISTANCE IN THE
TRANSFORMER COIL

INFINITE
RESISTANCE

R × 10 K

TRANSFORMER
COIL OPEN

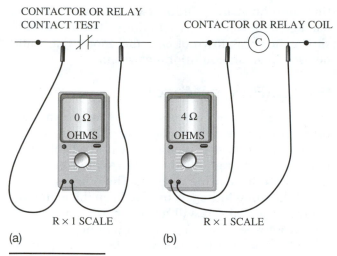

Figure 15–44 (a) A closed set of contacts should measure less than 1 Ω of resistance. Assure that no voltage is applied to the contacts when measuring with an ohmmeter. (b) A coil should measure some minimum resistance. Some coils measure several thousand ohms of resistance. Start with the R × 1 scale and raise the scale until a midscale reading is obtained.

REPLACING DEFECTIVE COMPONENTS

There is some important information you will need to know in order to replace a contactor or relay. It is best to bring the part to the supply house for the exact or substitute replacement. Before removing the component, lock out and disconnect all the power. Remove all the wiring from the component. Tape exposed wiring. Instruct the customer about your action. It is always best to call the supply house to check stock before making the trip for the part.

Next, determine the ratings of the replacement components. The following information is needed:

- Contactor/relay application or load
- Coil voltage
- Number of contacts
- Position of contacts, normally open or normally closed
- The applied voltage
- Rated-load amps (RLA) of the contacts
- Lock-rotor amps (LRA) of the contacts

You will need to know how the component is used. Is it used for inductive loads, such as compressors, motors, or transformers? Is it used to control a resistive load, such as an electric heat strip or a crankcase heater? The contactor amperage rating will depend on whether the load is inductive or resistive. Contacts can handle larger resistive loads, such as electric heat strips.

TABLE 15–1
RELAY MODEL 123, TWO-POLE CONTACTOR, 24-V COIL

	Inductive Load	Resistive Load
Max RLA @ 277 V	30 A	60 A
Max LRA @ 277 V	120 A	120 A
Max RLA @ 480 V	15 A	30 A
Max LRA @ 480 V	60 A	60 A

Reviewing Table 15–1, contactor model 123 will handle different current ratings. Using a contactor that has underrated contacts will cause the points to be overloaded and fail prematurely. Notice that inductive loads (motors) handle lower rated-load amps (RLA) loads as compared to resistive loads (heat strips). Inductive loads have high starting current known as temporary locked rotor amps (LRA). This high starting current will shorten the life of the contacts compared to a resistive load, which reaches its peak current flow at start-up and does not rise further. The component should be selected based on the RLA and LRA operation and whether it is an inductive or resistive load.

The operating voltage is an important factor. The higher the operating voltage, the lower the operating current rating will be. Higher operating voltages create a contact arc sooner on the closing of contacts and the arc last longer when the contacts are separating.

Oversized contactors will last longer but are more costly to replace. Common-size contactors and relays should be stocked on a service vehicle to reduce the number of trips to a supply house. Contactors are components that need replacement more often than general-purpose relays.

SUMMARY

Contactors and relays are very similar in operation. Contactors handle loads above 20 A, whereas relays have loads less than 20 A. Contactors have large contact points to carry large current loads.

These components can be diagnosed in the circuit with power applied, disconnected from the circuit and checked with an ohmmeter, and checked via a visual inspection of points and coils. The visual inspection is least valuable. In some component designs, the coil and contacts are sealed and are not visible for inspection.

Start with a visual inspection followed by live-voltage troubleshooting; then proceed with the ohmmeter check. This sequence may reveal that the component is not defective, but the proper control or line voltage is not reaching the contactor or relay. The ohmmeter diagnosis is a follow-up and should be used to verify what was found in the visual and voltage checks.

TROUBLESHOOTING COMPRESSORS

OPERATION

This section reviews the general operation and specific troubleshooting of single-phase and three-phase compressor motors. The explanation is limited to the vast majority of compressors that have their motor windings sealed in the compressor shell. The operation of external-drive motors is similar to that of motors sealed in the compressor housing, except that the external motors are air cooled, whereas the internal-drive motors are cooled by refrigerant passing around them. Oil acts as a cooling agent, but the returning refrigerant is the major contributor to keeping the motor windings from overheating. The main purpose of the oil is to act as a lubricant, which extends the life of the moving mechanical parts.

The design of a compressor motor is like any other motor. There are a stator and a rotor (Figure 15–45a). The stator or stationary windings are energized and create an electromagnetic field that moves in a circular orbit in this winding. This moving magnetic field causes the rotor winding and shaft to rotate. The motor shaft is connected to the drive mechanism that transfers motion to the pistons, scroll, rotary vanes, or other types of compression device. The rotor moves inside the stator's magnetic field without touching the stator windings. The tolerances are very close. Sleeve or ball bearings keep the rotor the correct distance from the stator. A rotor cut-

ting into the stator windings will short-circuit and damage the stationary windings. Figure 15–45b shows the breakdown of an actual motor that could be used in an air-conditioning system.

SINGLE-PHASE MOTORS

Most single-phase compressor and fan motors are called permanent split capacitor (PSC) motors. Permanent split capacitor motors have two sets of windings. One winding, called the start winding, is used to assist in the starting rotation. The other winding, called the run winding, is used to keep the motor operating. A technician will need to determine the difference between the start and run windings.

The start winding is smaller in diameter and is longer in length compared to the run winding. The smaller and longer wire has a greater resistance compared to the run winding. This aids in determining which is the start winding and which is the run winding.

Compressor windings terminate through a Fusite connector in the shell of the compressor (Figure 15–46). The terminals are labeled *common, start,* and *run,* abbreviated C, S, and R (Figure 15–46). Many times the label is not found on the compressor terminals and a technician must ohm out the windings in order to find the correct common, start, and run terminals. This is not difficult to do and is useful in diagnosing an electrically sound compressor.

Read the resistance across each pair of terminals. The ohmmeter should be zero-adjusted on the R × 1

Figure 15–45 Makeup of a common motor. The stator windings are the stationary winding that create the magnetic field to move the rotor. The rotor rotates within the magnetic field and converts the magnetic energy into mechanical motion.

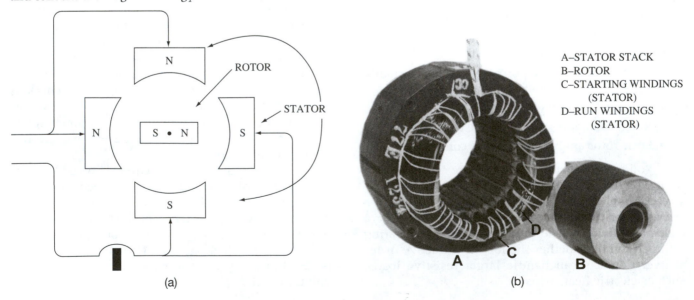

A–STATOR STACK
B–ROTOR
C–STARTING WINDINGS
    (STATOR)
D–RUN WINDINGS
    (STATOR)

(a)

(b)

Figure 15–46 Compressor motor terminals.

scale. For a good set of windings, three resistance readings are needed. Assume that the compressor windings are not identified. Label the windings 1, 2, and 3. The order of numbering is not important. In this example

- Terminals 1 to 2 read 5 Ω.
- Terminals 1 to 3 read 8 Ω.
- Terminals 2 to 3 read 13 Ω.

Note that the terminal 2 to terminal 3 reading is the highest resistance because it is the sum of the other two readings, or the sum of the run and start windings in series. The highest reading, between terminals 2 and 3, automatically lets you know that terminal 1 is the common winding (Figure 15–47). Terminal 1 is common, or shared between the run and the start winding.

Next, we need to determine which winding is run and which is start. This is not difficult now that we know that the common winding is 1.

The lowest resistance is the run winding. The resistance between 1 and 2 is 5 Ω; therefore the run winding is terminal 2. Remember that we established that terminal 1 is the common terminal. The next highest resistance is between common and start. The resistance between 1 and 3 is 8 Ω; therefore, the start winding is terminal 3. A good rule of thumb is

that the resistance of the start winding is three to five times that of the run winding, even though this is not the case in our example. Another rule of thumb is that the common terminal is usually the top terminal or one of the two top terminals (Figure 15–48). If this is not what you determined, then double-check the resistance readings and the calculations.

TROUBLESHOOTING SINGLE-PHASE COMPRESSOR MOTORS

It is important that a technician understand how to check the voltage, amperage, and resistance of a compressor. Many compressors are replaced when there is nothing wrong with them. Customers should get a second opinion when it comes to this major expense. Some air-conditioning contractors do not charge or charge a reduced rate for a second opinion. The advantage to offering this service at no cost or reduced cost is that a number of compressors are misdiagnosed as defective. The second contractor may be able to find something as simple as a bad connection, bad capacitor, or open internal overload. Even if the compressor is defective, this gives the second contractor an opportunity to submit a second bid on the compressor-replacement job. A second opinion can be financially important to the customer because even if a compressor is under warranty, the customer usually has to pay all or a portion of the labor charges. The problem with a second opinion is that it takes time. The customer is hot and does not want to wait another hour or longer for a second opinion. Customers do not realize that spending the night at a cool hotel can save them hundreds of dollars while waiting for the second opinion or bid.

THE FIRST CHECK

If a defective compressor is suspected, the first thing that should be checked is the continuity of the motor windings. A shorted compressor could cause an electrical explosion at the contactors, in the wiring, or at the compressor terminals.

 WARNING! It is extremely dangerous to push in the contactor contactors when a compressor is shorted. Many technicians immediately push in the contacts to determine that the main power is being supplied to the compressor. This practice is dangerous because the compressor winding may be shorted, and manually closing the contacts places the short-circuit condition at the contactor points. The in-rush of short-circuit current can cause the contactor to explode in the face of a technician. Stay to the side of contactors, capacitors, and—especially—compressor terminals when first

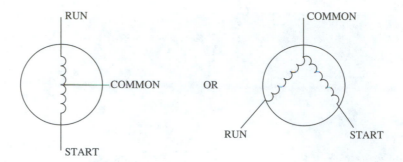

Single–phase compressor configurations

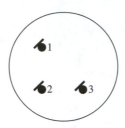

If the common terminal, start, run windings are
not known, label the compressor in any order
1, 2, and 3. Next measure the resistance between
each terminal. Example:

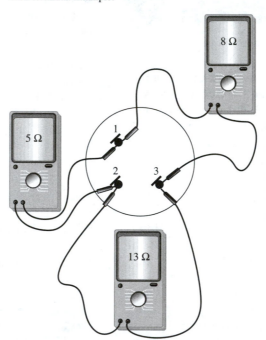

The highest reading is between
terminals 2 and 3; therefore, 1 is
the common terminal.
From the common terminal, measure
resistance to terminals 2 and 3. The
lowest resistance is between 1 and 2;
therefore, 2 is the run winding.
Terminal 3 is the start terminal
leading to the start winding.

Figure 15–47 Determining compressor,
common, start, and run.

starting an unfamiliar compressor. Do not
manually energize the contactor!

Remove the power from the condensing unit
by opening the breaker or disconnect and with a
voltmeter check the power source (Figure 15–49)
and compressor terminals for power. Once it is clear
that no power is supplied, use an insulated wire or
insulated screwdriver to short each compressor ter-

minal to ground. If power is still being supplied to
the compressor, an electrical arc and flash will occur
when you ground the compressor terminals. Investigate the voltage source should this occur. Being assured that no power is supplied to the terminals, remove the wires from the terminal by grasping the
wire terminal with a pair of pliers. Do not pull on
the wire; this will pull the wire away from the wire
connector. Pull the wire connector near the compres-

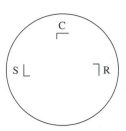

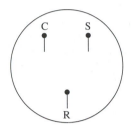

Figure 15–48 Identifying compressor terminals. Notice that the common terminal is near the top on most compressors.

Figure 15–49 Checking the power to the service panel. The next step is to check the voltage to the breaker supplying power to the condenser.

sor terminal. Label the wires *common, start,* and *run.* Hooking the compressor wiring improperly can cause a good compressor to be electrically damaged.

Check each compressor terminal for the correct resistance. Zero the ohmmeter and use the R × 1 scale (Figure 15–50). Record and calculate each of the terminal readings. Check to be certain that the common, start, and run windings are measured and the compressor is wired correctly. Shorted windings, measuring less than 1 Ω on single-phase compressors, are an indication that the compressor is electrically defective. Resistance readings outside the R × 1 range indicate that the windings have higher-than-normal resistance, and so the compressor is probably defective.

Next, check each terminal to ground. The ohmmeter should be zeroed on the R × 10,000 scale for this check. The reading should be infinity from each terminal to ground. The best compressor ground is on unpainted surface or on copper refrigerant tubing (Figure 15–51). Some of the digital meters measure resistance in the millions of ohms. On this high-resistance scale, the digital meter is likely to measure resistance to case ground. From each terminal to ground, resistance may be in the hundreds of thousands of ohms, and yet the compressor may be good. It is best to use no more than the R × 10,000 Ω range. Actually all compressor stator windings have some resistance to ground. The readings may be in the millions of ohms. A special instrument, called a megohmmeter (Figure 15–52), is used to measure high resistance from the motor winding to ground. When using an ohmmeter a reading lower than 1000 Ω per operating volt (1000 Ω/V) indicates that there might be a problem.

WORKING EXAMPLE

A technician is completing a biannual preventative maintenance checkup on an air-conditioning system. The compressor operates on 208 V. The technician measures 250,000 Ω from each winding to case ground of the compressor. Is the compressor in satisfactory condition?

SOLUTION The resistance to ground should be higher than 1000 Ω per applied volt. The compressor is operating on 208 V.

$$208 \text{ V} \times 1000 \text{ } \Omega/\text{V} = 208,000 \text{ } \Omega$$

This is the minimum resistance to be considered good. The technician decides that the compressor is acceptable because the measured resistance is greater than 208,000 Ω. ∎

A truly shorted compressor will measure very low resistance to ground. The resistance may be as low as 0 Ω or as high as a hundred ohms or so. The resistance will be low enough to cause the compres-

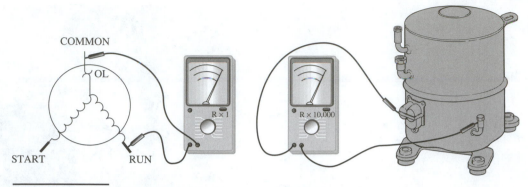

Figure 15–50   Check resistance of windings on the R × 1 scale. No resistance (∞) is an indication that one or more windings are open. No reading between common and run *and* common and start indicates an open internal overload. Allow compressor to cool before condemning a motor. Check for shorts to ground. Measure from terminal to refrigerant line. Resistance to ground should be more than 1000 Ω per operation volt.

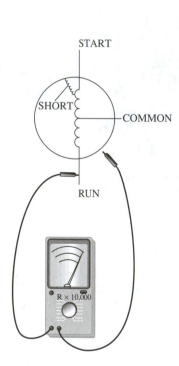

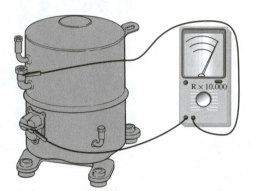

Figure 15–51   Measuring for a grounded motor. The stator can be shorted to ground when the motor bearings wear out due to liquid dilution of the oil. In this case, the rotor grinds into the stator and shorts the windings. Oil and refrigerant contamination can cause a resistance reading to ground. The oil and refrigerant can be changed to improve this situation.

Figure 15–52   Amprobe megohmmeter with a reading of 1. This indicates a reading over 2,000 MΩ at operating voltages up to 500 V. *(Courtesy of Amprobe Promax)*

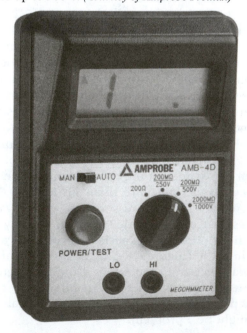

sor overload to open, a circuit breaker to trip, or a fuse to blow.

If a compressor has high resistance to ground but is lower than 1000 Ω/V and the compressor is operating under normal amperage (RLA), then it is time to take measures to increase the resistance reading to ground. Little or nothing can be done for a truly shorted compressor, except for replacement. However, action can be taken to improve resistance-to-ground readings. Something is causing the motor to read an abnormally low resistance. The exact cause of the problem may not be known, but the technician should try to clean the refrigeration and lubrication system. Contamination could be developing between the stator and rotor winding. Contamination may be found in the form of metal particles or "moist" refrigerant. Conduct an acid and moisture test on the refrigerant and lubrication. This test kit is available as a combination test. One sample of refrigerant will determine the condition of the oil and the refrigerant. If the compressor has an oil sight glass, shine a flashlight into the oil to inspect for contamination. Removing a few ounces of oil is the way to visually inspect for oil problems. An oil-analysis kit is available, where the technician mails an oil sample to a test lab. The problem with this method is that it may take a week or two in order to obtain the test results. This can be done as routine PM on larger systems or compressors that share the same lubrication system.

If the refrigerant or oil is contaminated, begin by changing or recycling the refrigerant. Change the compressor oil. Change the liquid-line filter-drier and add a suction-line filter-drier. The suction line drier should be changed after a day or two of operation. Some manufacturers recommend suction-line filter drier replacement after 10 h of operation. Suction-line filters are not made for permanent installation.

This cleanup process involves a lot of work for it not to be successful. However, motor bearing wear can also cause the low resistance- (but not shorted) to-ground readings. Bearing wear reduces stator/rotor clearances and allows the rotor to drag on the stator. This drag slowly damages the stator and the rotor windings until they are shorted and unusable. It is not easy to determine if the lowering resistance is caused by contamination, bearing deterioration, or

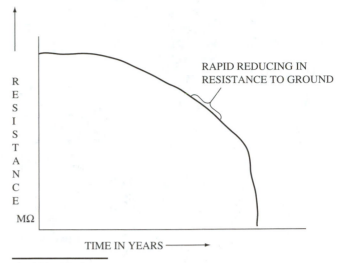

GRAPH 15–1

possibly a combination of both. The customer may not want to take any action until the compressor windings short out. The resistance to ground could be tracked periodically to determine how rapidly the motor windings are failing. A megohmmeter is a good instrument to track this normal motor deterioration. A chart can be developed that tracks the resistance reduction to ground (Graph 15–1). The graph shows that the resistance to ground starts off in the high-megohm range and decreases as the motor ages. A rapid downturn in resistance indicates that the motor is going to fail soon.

Allowing a motor to short out may cause additional clean-up challenges when the failure occurs. A slow burn will take place, spreading contamination throughout the total refrigeration system. A sudden short will more than likely trap the contamination inside the compressor.

The final resistance check should be conducted with all the wiring installed back on the compressor terminals and the power disconnected (Figure 15–53). From the load side of the contactor, check to assure that the resistance is the same as measured on the compressor. A higher-resistance reading indicates that there is a bad connection or defective wire. Occasionally a loose wire connection will burn out because of high-amperage draw experienced through the faulty connection or connector. Loose connections

---

ON THE JOB

When checking common, start, and run windings on a motor, disconnect the wires from the terminal as shown in Figure 15–53b. This action will prevent measuring resistance through other components such as the fan motor (FM). This practice is recommended when checking any component in an air-conditioning circuit.

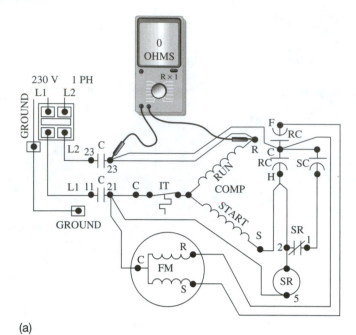

(a)

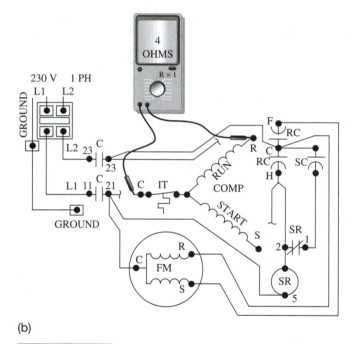

(b)

Figure 15–53   (a)Testing wire resistance between contactor and compressor. (b)Testing compressor windings.

create high-amperage-draw conditions. Generally, the burned connection is replaced with a new connector. The best practice is to replace the wire and the connector. The wire should be replaced because it has also experienced a high-amperage condition through the loose connection. The wire may have been overheated or burned inside the insulation jacket without visually damaging the outer insulation. The wire will continue to cause a high-amperage condition and will

cause the wire or connector to burn open again. Replacing the wire back to its contactor connection point will save having a service callback.

The final warning when checking the resistance of any compressor is to be sure that the motor is not condemned because an external or internal overload protective device is open (Figures 15–54 and 15–55). An overload is sized to protect the motor from a high-current condition or an overheating condition. It is located in series with the common terminal. Overload devices respond quickly to overcurrent conditions. The indication that the overload is open is when the only resistance reading is between the start and run windings. If the overload is defective, it will not reset and close. Some external overload devices have to be manually reset before they close the contacts.

There is nothing a technician can do if an internal overload permanently opens. The technician must give the compressor ample time to cool down and allow the overload to reset. Some compressors get so hot that it takes hours for them to cool down and for the internal overload to reset. Running water over the compressor shell can indirectly cool the compressor windings. Care must be taken to electrically disconnect the compressor during this cooling procedure.

What causes the compressor to trip out on overload protection? There are many possible causes. Some are related to a high-amperage condition and some are related to overheating of the motor windings. Here is a partial list that the technician should check:

- What is the amperage on the common winding in starting and running operations?
- Is the unit charged properly?
- Is the overload opening prematurely?
- Is the supply voltage correct?
- Is the condenser rejecting heat into the air or water?
- Are there loose connections?
- Are the starting/running components defective?

A defective external overload can be replaced with a like component. Sizing is very important. Do not bypass an external overload. If the external overload is defective, a properly sized temporary fuse may be used in place of the overload. Always check the compressor windings for C, R, and S and for shorts to ground prior to installing a fuse or replacing a defective external overload.

The purpose of an overload is to protect the motor from an overamperage or overheating condition. Either of these conditions can keep the motor from operating. Let us troubleshoot overload-related problems.

Internal (Line-Break) Motor Protectors
Internal Line-Break Overloads
Completely Internal
Tamper-Proof
Cannot Be By-Passed

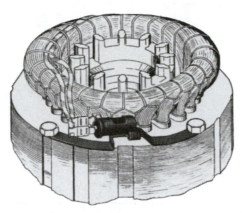 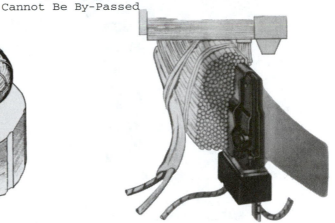

Located adjacent to the motor windings, this device employs the latest technology in detecting excessive heat and/or current draw

Located precisely in the center of "heat sink" portion of motor windings this device also detects excessive motor winding temperature and current draw.

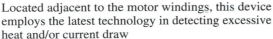

Figure 15–54 Line-duty, internal overload protector. This overload cannot be replaced if it permanently opens. The motor will need to be changed. *(Courtesy of Tecumseh Compressor Co.)*

AMPERAGE CHECK

After the extensive resistance check, it is important to check the compressor amperage. The amperage on the *common* winding should not exceed the rated-load amperage (RLA). The RLA might be exceeded when outdoor ambient temperature is above 100°F. The Amprobe should be attached before starting. The initial current draw on start-up may be an important indication of the compressor operation. A digital Amprobe with a peak hold option can record the peak LRA developed on motor start-up.

Check the amperage on all the terminals of a PSC compressor motor. The common amperage should nearly equal the sum of the amperages of the run and start windings. Figure 15–56 shows a PSC compressor with an amperage reading of 21 A on the clamp on Amprobe X; 3 A on Amprobe Z; and 18 A on Amprobe Y. The RLA of the compressor is 28 A. These reading are generally considered good.

Other factors affect the amperage draw of the compressor. The greater the charge or load on a compressor, the higher the amperage draw is. Shorted windings or contamination also cause high amperage conditions. Contractors that have a heavy carbon buildup will cause high resistance and increase the amperage draw. Low-input voltage will cause high-amp draw.

VOLTAGE CHECK

The correct voltage input to the compressor is important. As stated in the preceding section, a low-voltage condition will create high-amperage demand and the motor will cut out on its overload protection. A higher voltage will reduce current draw. Typically, motors are designed to operate within ±10% of nameplate voltage. Dual-rated motors such as 208 V/240 V may be rated +10%, −5%. Manufacturer's information should be referenced. Table 15–2 shows the range of voltage allowed for commonly applied voltages found in air conditioning.

First check the voltage with the power applied and the operating systems turned off. Next, check the voltage with the system energized. The applied voltage should be within the ±10% range when the system is running. A big difference in voltage between the energized and deenergized state should be investigated. Undersized wiring feeding the panel box or a loose connection in the condensing unit may cause a large voltage drop. A bad connection, such as a loose terminal screw or loose wire nut, will also create an excess voltage drop in the panel box, disconnect, or on the contactor.

Finally, check the voltage drop across each set of closed and operating contacts. Ideally, the voltage drop should be 0 V. The voltage drop should be less than 5 V or the contactor should be replaced.

THREE-PHASE COMPRESSOR MOTORS

This section discusses the operation of a three-phase compressor motor. The motors discussed are inside

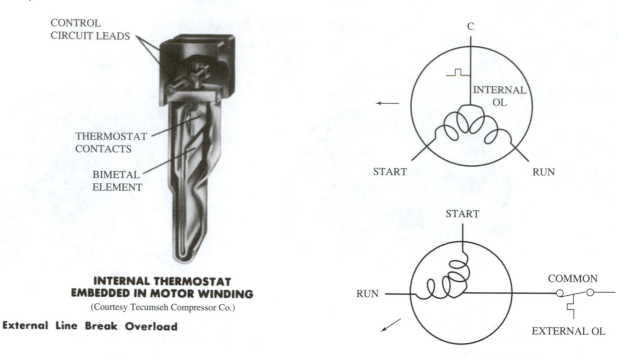

CONTROL CIRCUIT LEADS

THERMOSTAT CONTACTS

BIMETAL ELEMENT

INTERNAL THERMOSTAT EMBEDDED IN MOTOR WINDING
(Courtesy Tecumseh Compressor Co.)

External Line Break Overload

C

INTERNAL OL

START   RUN

START

RUN   COMMON

EXTERNAL OL

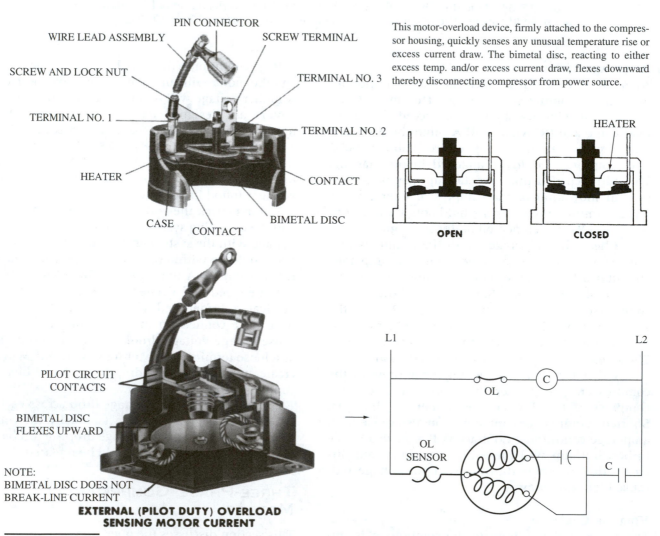

WIRE LEAD ASSEMBLY

PIN CONNECTOR

SCREW TERMINAL

SCREW AND LOCK NUT

TERMINAL NO. 3

TERMINAL NO. 1

TERMINAL NO. 2

HEATER

CONTACT

CASE   CONTACT

BIMETAL DISC

This motor-overload device, firmly attached to the compressor housing, quickly senses any unusual temperature rise or excess current draw. The bimetal disc, reacting to either excess temp. and/or excess current draw, flexes downward thereby disconnecting compressor from power source.

HEATER

OPEN   **CLOSED**

PILOT CIRCUIT CONTACTS

BIMETAL DISC FLEXES UPWARD

NOTE: BIMETAL DISC DOES NOT BREAK-LINE CURRENT

EXTERNAL (PILOT DUTY) OVERLOAD SENSING MOTOR CURRENT

L1   L2

OL   C

OL SENSOR   C

Figure 15-55 Internal and external overloads. *(Courtesy of Tecumseh Compressor Co.)*

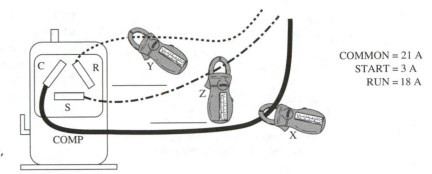

COMMON = 21 A
START = 3 A
RUN = 18 A

Figure 15–56 Identifying the common, start, and run windings by Amperage draw.

the compressor shell and cooled by returning suction gas. A three-phase motor that is used to drive external or open compressors, fans, or pumps can be checked in the same way.

Three-phase motors are simpler compared to single-phase motors. There are many advantages of three-phase motors over single-phase motors. Three-phase motors do not require starting and running components like capacitors or potential relays. Three-phase motors are generally more efficient and have excellent starting torque compared to their single-phase counterparts.

A three-phase motor operates much the same as a single-phase motor. The stator has three sets of motor windings that rotate a magnetic field and cause the rotor to move and drive the compressor mechanism. The three motor windings are supplied with three hot power leads, thus the motor windings are 120° out of phase with each other for optimal starting torque and running efficiency.

The three-phase motor is wired in one of two configurations. These wiring configurations are called wye and delta (Figure 15–57). To the technician, it is not important whether the motor is wye or delta; the operation and troubleshooting techniques are the same. Both wye and delta configurations are represented by a symbol with a circle around it.

Like a single-phase compressor, a three-phase motor also has three terminals. The differences is that a three-phase motor has three power connections from $L1$, $L2$, and $L3$. The letter L means line voltage. The letter T means motor terminal. The resistance between each of the windings should be nearly the same. The resistance is lower than with single-phase motors. Because three-phase motors

draw higher current and create a stronger magnetic field the motor windings are larger in diameter. Large-diameter windings have low resistance. Many three-phase compressor windings measure less than 1 Ω between each of the terminals. Figure 15–58 shows a resistance of 0.5 Ω between T_1 and T_2, T_1 and T_3, and T_2 and T_3. Electrically, this part of the windings is good. A ground check also must be conducted on all terminals.

It is important to know whether you are checking a single-phase compressor or a three-phase compressor. Getting them confused in your mind during the diagnosis process will cause the compressor to be

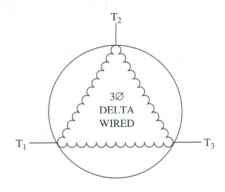

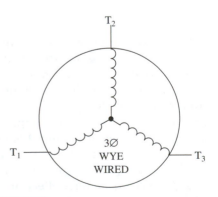

Figure 15–57 Three-phase (3ϕ) motor configurations. The resistance between each wiring is the same. With the use of large motor-winding coils, the resistance may be less than 1 Ω between terminals.

TABLE 15–2
Voltage Handled by Common Motors

Nameplate	Upper Limit	Lower Limit
208	228	188
230	253	207
460	506	414
575	633	518

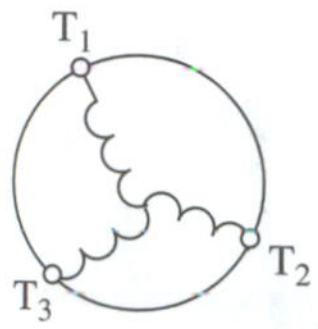

T_1 TO $T_2 = 0.5\ \Omega$
T_1 TO $T_3 = 0.5\ \Omega$
T_2 TO $T_3 = 0.5\ \Omega$

THREE-PHASE
COMPRESSOR

Figure 15–58 Three-phase motor windings have nearly the same resistance when measured from terminal to terminal.

condemned. A single-phase motor with the same resistance reading between terminals means that the motor is electrically defective. If it is a 3ϕ motor the readings would be acceptable. If a three-phase motor has readings that measure more like common, start, and run, the motor is defective. Of course, if it is a single-phase motor, it is good. There are small and large single-phase motors and there are small and large three-phase motors. The size does not always determine the phase operation. Review the power supply, nameplate rating, and schematic diagram to aid in this decision.

TROUBLESHOOTING THREE-PHASE COMPRESSORS

Troubleshooting three-phase compressor motors is similar to troubleshooting single-phase motors. The technician should *not* push in the contactor to energize the compressor. The technician should stand to the side of the contactor or compressor terminal contacts when the system is first energized. This is a safety practice; a shorted compressor can blow out the compressor terminals creating a blast of fire, oil, and refrigerant that could seriously injury the technician.

If a defective compressor is suspected, the technician should conduct a continuity check before applying power.

CONTINUITY CHECK
Lock and tag the power supply. Remove the power and wiring from the compressor. Use all safety precautions. Tape exposed wiring. Measure and record the resistance between each of the terminals after zeroing the ohmmeter on the R × 1 range.

Check each of the terminals to compressor case ground. The best place to check ground is on a bare refrigerant line to each of the terminals. Zero the ohmmeter on the R × 10,000 range and take careful measurements. Expect no meter deflection or reading. The best reading is infinity (∞). A shorted compressor will have a reading of 0 Ω or very low resistance. The resistance to case ground should be higher than 1000 Ω per applied volt. A megohm-

meter should be used to track the resistance to ground during the preventive maintenance checkup.

Three-phase motors also have external or internal motor protection. Some are designed with both external and internal protection. Figures 15–59 and 15–60 shown both types of overload protection. These figures also show the difference between pilot-duty and line-duty overload devices. Pilot duty (Figure 15–59) uses overload heaters in series with the motor windings. The heaters will cause the overload switch (OL) to open the control circuit in the event of high-current draw or an overheating condition. The control circuit is usually low voltage. Low-voltage circuits allow for smaller contactors, thereby reducing the cost of the component.

Line-duty overload protection places the device in series with the total current load. The contacts are designed to momentarily carry locked-rotor amps without opening but open when the rated-load amps are exceeded by 115% to 125% of the normal operation current. An over-temperature condition will cause the overload to open. It is important to understand the operation of these devices when completing a continuity check and measuring amperage draw. A defective, open internal overload can be repaired only by replacing the compressor. It is important not to rush to conclusions when diagnosing an open internal overload. Allow the compressor ample time to cool before making the final condemnation. Larger compressors take many hours before heat is dissipated, thus allowing the overload to reset to take a resistance reading.

The only indication that an internal overload is open is when resistance is measured from two terminals and not from the third terminal.

WORKING EXAMPLE

The right-hand bimetal internal overload in L_3 (Figure 15–59b) is open. What resistance readings should a technician expect?

SOLUTION Using the zeroed R × 1 range, the technician makes the following readings:

L_1 to $L_2 = 0.6\ \Omega$
L_1 to $L_3 = $ infinity
L_2 to $L_3 = $ infinity

Based on these readings the technician is certain that the overload protecting L_3 is open or the L_3 windings are open. The exact determination is not known. The compressor will not operate and is, therefore, replaced. ∎

AMPERAGE CHECK
As in any air-conditioning troubleshooting sequence, before the equipment is operated, the refrig-

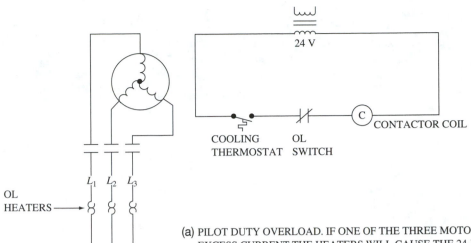

(a) PILOT DUTY OVERLOAD. IF ONE OF THE THREE MOTOR WINDINGS (LEFT) DRAWS EXCESS CURRENT THE HEATERS WILL CAUSE THE 24-V PILOT-DUTY OVERLOAD TO OPEN AND BREAK THE 24 V TO THE CONTACTOR COIL, WHICH OPENS THE THREE-POLE CONTACTOR, STOPPING THE MOTOR.

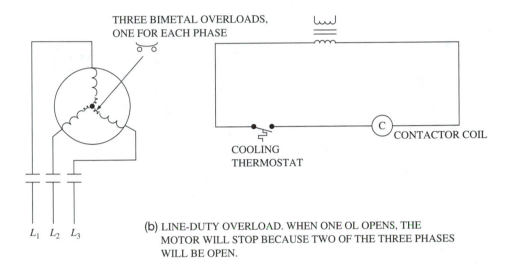

(b) LINE-DUTY OVERLOAD. WHEN ONE OL OPENS, THE MOTOR WILL STOP BECAUSE TWO OF THE THREE PHASES WILL BE OPEN.

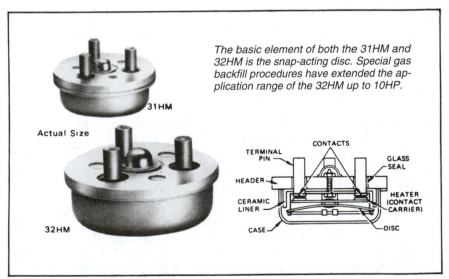

The basic element of both the 31HM and 32HM is the snap-acting disc. Special gas backfill procedures have extended the application range of the 32HM up to 10HP.

(c) THE 31HM AND 32HM ON-WINDING MOTOR PROTECTORS ARE THREE-PHASE LINE BREAK, AUTOMATIC RESET DEVICES WIRED IN SERIES WITH EACH PHASE AT THE NEUTRAL POINT AND MOUNTED ON THE WINDINGS.

Figure 15–59 Pilot- and line-duty overloads

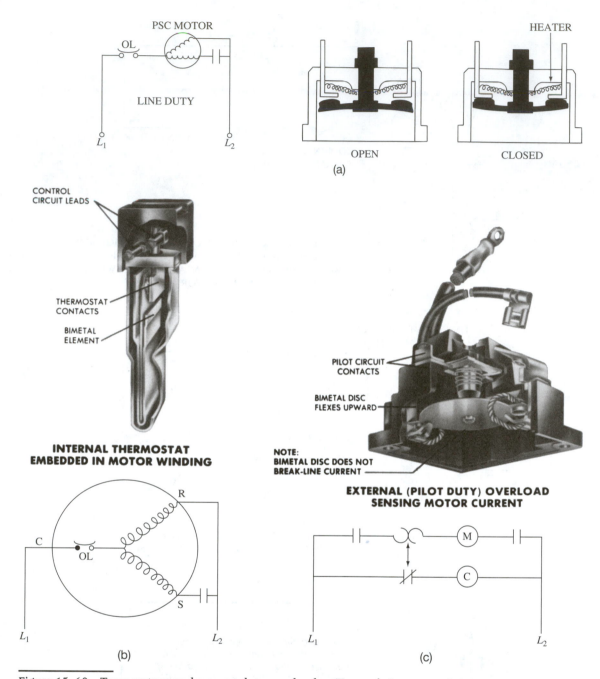

Figure 15–60 Temperature and current duty overloads. *(Tecumseh Compressor Co.)*

eration gauge set and Amprobe must be properly installed. If the equipment is not operating, it should be controlled and not allowed to operate until the technician is ready for pressure and amperage readings. The technician will need to see the valuable information provided by these instruments when the compressor initially starts. The compressor may operate for a short time and cut off. Without the instruments connected, the technician may have missed out on a time-saving diagnostic opportunity. The gauges and ammeter will be hooked up before the technician leaves the job site, so why not just hook them up at the initiation of the visit. Valuable information can be missed in this short period of operation.

First, the refrigerant pressure may be low enough to cause the low-pressure switch to open the control circuit. It may take anywhere from 1 to 5 min before the system pressure is high enough to operate again. Larger and better-quality equipment will have timing circuits that prevent the unit from operating for time periods up to 5 min or longer. The timer is used to allow the system pressures to equalize and power fluctuations to stabilize.

Another potentially lost opportunity occurs when the overload takes the equipment off-line. A high-amperage or overheated condition could open the overload. The clamp-on Amprobe would be an immediate indicator of a high-current condition on start-up. The compressor may have been cycling on overload prior to the technician's arrival at the site. The compressor may be hot. The one opportunity the technician has to gain data is wasted because the ammeter is not on the line. The motor may take minutes or hours to cool before the overload is reset. An external overload can be removed and allowed to cool and reset. Internal overloads do not reset so quickly. Again, time is wasted because a professional setup sequence was not used at the beginning of the troubleshooting process.

Under normal operating conditions each winding of a three-phase motor should have nearly equal amperage (Figures 15–61 and 15–62). This is different when compared to single-phase motors, where the amperage readings are not the same. In Figure 15–61, the current reading in each leg is 34.64 A. In Figure 15–62, the current in each leg is 20 A. The im-

portant point is the amp draw should be nearly the same in each leg.

Unbalanced current may be caused by voltage imbalance between phases, a shorted motor winding, or a high-resistance connection. Current imbalance for three-phase motors should not exceed 10%. The formula for calculating current imbalance is

$$\begin{array}{l}\text{Percent} \\ \text{current} \\ \text{imbalance}\end{array} = \dfrac{\begin{array}{c}\text{Maximum deviation} \\ \text{from the average voltage}\end{array}}{\text{Average voltage}} \times 100$$

WORKING EXAMPLE

For currents of 25, 30 and 25 A the average current would be 26.7 A. The maximum deviation would be 3.3 A and the current imbalance would be 12.6%, which is unacceptable. ■

If the measured current is imbalanced, the reason for the imbalance must be determined before damage is done to the compressor. Current imbal-

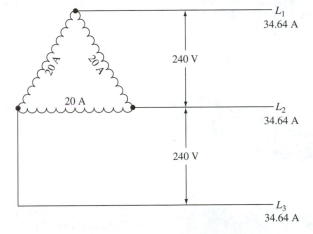

Figure 15–61 Delta configuration, three-phase motor.

LINE VOLTAGE EQUALS PHASE VOLTAGE.

LINE AMPERAGE IS 1.732 TIMES PHASE AMPERAGE.

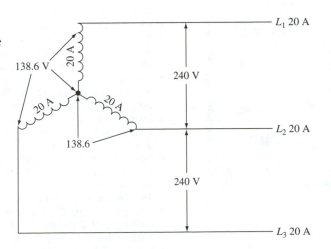

Figure 15–62 Wye-connected three-phase motor. 138.6 V are measured across each phase.

LINE VOLTAGE EQUALS 1.732 TIMES PHASE VOLTAGE.

LINE AMPERAGE EQUALS PHASE AMPERAGE.

ances can be caused by voltage imbalance. Check the input voltage when the motor is operating. A short in the motor winding or a short to case ground can cause a current imbalance. This can be detected with an ohmmeter.

Voltage Check

The voltages between L_1 and L_2, L_1 and L_3, and L_2 and L_3 should be nearly the same. The voltage should be measured with the equipment operating at full capacity. Some causes of voltage imbalance are a partial short in the motor stator or to ground.

Voltage imbalance can be caused by load imbalance. The three phases should be split equally across the load at the site. Some sites take all their single-phase lighting from two phases. This loads down or reduces the voltage available to these two phases, thereby producing a voltage imbalance. Reduced-phase voltage is also created by undersized power feeders or loose connections in the panel box, in the junction box, or at the contactor. Check for loose connections in these places.

Voltage imbalance for three-phase motors should not exceed 2%. To calculate voltage imbalance use the following formula:

$$\text{Percent voltage imbalance} = \frac{\text{Maximum deviation from the average voltage}}{\text{Average voltage}} \times 100$$

Figure 15–63 Compressor test kits. *(Courtesy of Thermal Engineering)*

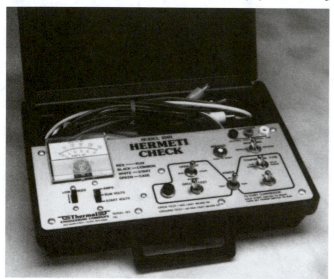

(a)

- Locked-up compressors may be freed by reversing motor rotations.
- Tests windings for opens, continuity and grounds.
- Tests start and run capacitors and system relays by substitution.
- Measures running amps, 0–10, 50A range.
- Measures start and run volts, 0–150, 600V AC ranges.
- Compressors are tested in the manner in which they were designed to operate: Split-phase, capacitor start or PSC.
- Has provisions for connecting external run capacitor.
- 30A breaker protected to allow high starting current with adequate overload protection.
- Complete with 220V adapter cord and capacitor leads.

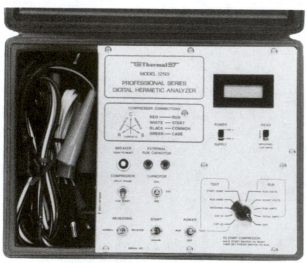

(b)

- Easy to read digital LCD display.
- Tests and runs 110, 220 or 277 volt 50/60 Hz single phase compressors up to 5 H.P.
- Locked-up compressors may be freed by reversing motor rotation.
- Tests windings for opens, continuity and grounds.
- Checks capacitors for leaks, opens and shorts.
- Measures capacitor values, 0–200, 0–2000 MFD ranges.
- Automatic capacitor discharge.
- Measures compressor amps, 0–30A.
- Measures start and run volts, 0–600VAC.
- Measures winding resistance, 0–2000 ohm.
- 500V megger for testing compressor windings for grounds.
- Compressors are tested in the manner in which they were designed to operate:
- Split-phase, capacitor start or PSC. Multiple internal starting capacitors: 100, 200 and 400 MFD values.
- Has provisions for connecting external run capacitor.
- 30A breaker protected.
- Complete with 220V adapter leads and capacitor leads.
- Power: 110, 220 or 277 VAC 50/60 Hz
- Dimensions: 13 1/2″ × 10″ × 4″
- Weight: 5.5 lbs.

WORKING EXAMPLE

Voltages of 449, 470, 462 are measured and the average voltage is 460 V. The maximum deviation from the average would be 11 V. The percent imbalance would be

$$\frac{11}{460} \times = 2.39\%$$

This voltage deviation is unacceptable. ∎

SUMMARY

This section discussed the electrical aspects of troubleshooting single-phase and three-phase compressors. Mechanical troubleshooting was discussed in another section.

The important point to remember is that if the compressor is not operating, and all the troubleshooting leads to a compressor problem, disconnect the power, ohm out the windings, and the check the compressor to ground. Compressor terminals, capacitors, contactors can explode and harm the technician.

The amperage and voltage checks are also important diagnostic tools. Diagnose the problem accurately, because the customer may request a second opinion. Your reputation is on the line. Every technician makes mistakes. Good techs will double-check and even triple-check readings to assure themselves that they are correct. Some companies dispatch a second technician to confirm that the compressor needs replacement. This action will benefit the technician and the company, providing outstanding service.

There are compressor analyzers (Figure 15–63) that go a step beyond the common multimeter. Some of the testers have the ability to free stuck single-phase compressors by temporarily reversing motor rotation. Other features include the ability to

- Test multivoltage motors
- Test-start and run capacitors and relays by substitution
- Test for opens, continuity, and grounded winding
- Check capacitor for leaks, opens, and shorts
- Measure capacitance
- Use a built-in megohmmeter
- Measure running amperage

Compressor analyzers are used to make the final determination if a single-phase compressor has failed. They can also be used to "revive" a compressor by trying different start and run capacitor combinations that might create the 90° phase shift that will start the motor. A compressor analyzer will not resurrect all compressors, but it is an important instrument for the professional technician.

SUMMARY

- This high-impact chapter has provided material that will help you improve troubleshooting action. Each individual develops personal troubleshooting behavior and techniques. There is no wrong way or right way to troubleshoot a problem. Some technicians believe their way is the only way. Do not get trapped in this mentality—it is very closed and can be counterproductive. Your troubleshooting skills will develop throughout your career as an air-conditioning professional. The older technician, author included, learns faster and better troubleshooting techniques all the time. The goal of troubleshooting is to identify and correct a problem as quickly as possible. New techs take longer than experienced techs—this is only to be expected.

- Read the material again; you will be surprised how much information you missed the first time through. This book and this chapter will be a handy troubleshooting reference. Now you know how to troubleshoot single- and three-phase compressor motors. Fan motors and pump motors can be diagnosed the same way. Contactor and relay troubleshooting was an important part of this chapter. This information can be transferred when you troubleshoot other types of relays. Obtain a basic understanding of the component before attempting troubleshooting. However, a complete understanding of a component is not totally necessary to conduct fundamental diagnostics.

- The next chapter builds on the information provided in the previous two troubleshooting chapters. It discusses wiring diagrams and how to use them as a troubleshooting tool. Schematics or wiring diagrams are road maps to use in finding a problem. You will feel a sense of accomplishment when tracking down the problem with a schematic while following an organized sequence of events leading up to that discovery. Practice is the only way a technician can improve. This book provides you with the fundamentals. Practice what you have learned and you will be a successful-air conditioning technician.

PROBLEMS AND QUESTIONS

1. Digital multimeters can measure ghost voltage. What is the disadvantage of measuring ghost voltage?

2. Describe the basic operation of three types of transformers.

3. A technician is troubleshooting a step-down control transformer. It is not clear which is the primary side and which is the secondary side. Describe how the technician can use an ohmmeter to determine both the primary and secondary side of the transformer.

4. What is the maximum secondary amperage a 75-VA transformer can handle? Show your calculations.

5. What are the reasons for using a 24-V control system as compared to a higher-voltage system?

6. What is contactor chatter? What causes chatter?

7. A technician is going to use a multitap step-down transformer. A 120-V primary is required. There is no diagram to determine the voltage taps. Common is the black wire. Which wires will be used for a 120-V primary if the technician measures the following resistances: black to white, 13 Ω; black to orange, 11 Ω; black to red, 6 Ω; and white to orange, 2 Ω?

8. Name *all* the steps for troubleshooting a transformer.

9. Why is it important to check the transformer output under a normal load or attached to a coil?

10. What is the best resistance range when checking for transformer winding continuity? What is the best range for shorts to ground?

11. What size fuse is best installed in the secondary side of a 40-VA transformer?

12. A technician finds a burned-out control transformer and suspects a short circuit in the low-voltage controls. Describe the troubleshooting sequence to locate the short circuit.

13. How can an Amprobe be used to check current draw less than 1 A?

14. How is an in-line ammeter hooked up to measure amperage in a circuit?

15. What is the purpose of a start capacitor?

16. What is the purpose of a run capacitor?

17. Why do run capacitors require a much higher voltage rating than start capacitors?

18. What is the best way to troubleshoot a capacitor?

19. Describe in detail how to check a capacitor with an ohmmeter.

20. What resistance range is used to check for case shorts in a run capacitor?

21. What is the capacitance of three 10-μF capacitors hooked in parallel?

22. What is the capacitance of two 10-μF capacitors hooked in parallel?

23. What is the capacitance of two 10-μF capacitors hooked in series?

24. What is the advantage of a single-pole contactor as compared with the double-pole contactor? What is the disadvantage?

25. Describe how to troubleshoot a contactor that is in a live (hot) circuit.

26. Describe how to troubleshoot a relay with an ohmmeter.

27. What six pieces of information will the technician need in order to replace a contactor?

28. What is the difference between a contactor and a relay? How are they the same?

29. How are the motor windings of a sealed compressor cooled?

30. What resistance range will be needed in order to check compressor motor windings?

31. A compressor has unmarked terminals. An ohmmeter reveals the following readings:

    T_1 to $T_2 = 15\ \Omega$

    T_1 to $T_3 = 13\ \Omega$

    T_2 to $T_3 = 2\ \Omega$

    Which terminal is common, start and run?

32. Why is it *not* a good practice to walk up to a nonworking condensing unit and push in the contactor?

33. What resistance range is used to check for shorts to ground in compressors?

34. Some compressors measure resistance from the windings to the case ground. What is the minimum acceptable resistance to ground to be considered a good compressor?

35. Other than shorted windings, what will cause a low-resistance reading to case ground?

36. What happens to the resistance from the motor winding to ground as the motor is used over a period of years?

37. How does a technician determine that the internal overload is open in a PSC motor?

38. List four specific things that could cause an overload to open.

39. Which terminal on a single-phase motor will have the highest amperage reading?

40. What percentage of voltage deviation is allowed on a motor?

41. What will cause a voltage drop in a circuit that supplies an air-conditioning system?

42. What is the general resistance measurement between terminals on a three-phase motor?

43. A technician suspects that the internal overload of a compressor is open. What does the technician need to do before condemning the motor?

44. Why is it important to hook up the manifold gauges and Amprobe prior to starting the troubleshooting sequence?

45. What is the normal current draw on each of the windings of a three-phase motor?

46. What will cause a three-phase current imbalance?

47. What will cause a three-phase voltage imbalance?

48. What are the advantages of using a compressor analyzer over a multimeter when troubleshooting a compressor?

49. What is the goal of troubleshooting?

50. In this chapter, what information did you learn that was important to you?

CHAPTER SIXTEEN

UNDERSTANDING ELECTRICAL DIAGRAMS

AFTER STUDYING THIS CHAPTER, THE STUDENT WILL BE ABLE TO:

- Identify the common air-conditioning symbols.
- Describe the different types of circuits.
- Draw the different types of circuits.
- Identify the different types of electrical diagrams.
- Describe the different types of electrical diagrams.
- Draw an electrical diagram.
- Discuss the rules for reading wiring diagrams.

INTRODUCTION

Electrical diagrams are a technician's road map into the operation of an air-conditioning circuit. An electrical diagram, like unfamiliar terrain, is foreign to a new technician. Electrical diagrams have symbols that must be recognized in order for a technician to navigate through the maze of wires and components. Electrical diagrams help technicians make sense of what is happening in an air-conditioning system. Even an experienced technician has to be able to identify new and different diagrams. Sometimes it is a matter of using the process of elimination. Experienced techs know many different type of wiring configurations, yet there is always a new one with additional circuitry. Knowing how part of a circuit operates will make it easier to understand the possibly unclear operation of the remaining system. This chapter covers the basic types of circuits, electrical symbols, the different types of wiring diagrams, and how to read these diagrams. This information is important to understand prior to reading the next chapter on troubleshooting electrical diagrams. You cannot troubleshoot something that you do not understand.

When learning to use a common map, a person must know how to read road-map symbols. The next section discusses electrical symbols, which will facilitate reading electrical diagrams, the road maps to successful troubleshooting.

 SAFETY TIPS You should watch out for flying debris, moving parts, and sharp edges when working around HVAC/R equipment. Here again, gloves and safety glasses will reduce potential injury from any of these hazards.

Remember that some of the piping around the compressor will be hot. When in doubt of a refrigerant-line temperature, test the temperature of a line with a quick touch. Temperatures can reach as high as 250°F and can cause severe skin damage. A temperature in excess of 120°F will feel hot.

ELECTRICAL SYMBOLS

Electrical symbols, wiring diagrams, and schematics are the technicians' guide to the operation of heating, air conditioning, ventilation, and refrigeration circuits. Symbols (Figure 16–1) are used in place of a picture of the actual component. Component symbols are shown in their deenergized state as if they were not in the circuit. These symbols are universally accepted, but there are variations. Unfortunately, there is always an exception in the electrical and the refrigeration sides of the HVAC/R industry. Symbols vary somewhat between manufacturers and between countries. Imported equipment has increased in the past decade, and technicians are seeing unfamiliar symbols and wiring diagrams on this foreign equipment. There are always new challenges for the air-conditioning professional.

Wire is used as the pathway or highway for operating electrical circuitry. Even though there is an unofficially accepted color code for a thermostat hookup, it is not always followed; this is true even in wiring diagrams. For example, in the middle of a manufacturing run of condensing units, the blue wire could be depleted, so it is not readily available on the assembly line. The production process is not halted. Another wire color is substituted without a change in the wiring diagram supplied with the unit.

Technicians do not carry a wide assortment of wire colors. When replacing a defective or burned wire, they use what is available on their service vehicles. All we can expect is that the wire is replaced with a wire of the same or larger gauge. Matching the color of the wire is a plus. Be cautious when tracing electrical diagrams because the wire colors may not be what you expect. Finally, in some installations the dye used in the wiring insulation may not be colorfast. Wiring exposed to sunlight may fade. Red wire can fade to an orange color. Black wire can fade to a gray color. White wire insulation can become dirty and change to a tan color. These changes can be confusing to the technician.

Just as a road map has a legend or index, so does an electrical diagram. A legend (Figure 16–2) is used to identify what the abbreviations mean on an electrical diagram. There is not enough room on a diagram to write out the complete name of a component. A legend is usually provided with the electrical diagram. The legend provided here is common to many electrical diagrams. As stated before, there is no such thing as *always* in HVAC/R, and this statement holds true for diagram legends.

Figure 16–3 is a sample of a heating diagram with the legend and appropriate symbols. Many diagrams do not include the symbol translation on the working-circuit diagram. This is an excellent example of how a legend can identify and aid in the understanding of a circuit operation. Notice the various ways the manufacturer depicts wiring types under the heading of legend. This is a common designation (Figure 16–3).

DEFINING CIRCUIT TYPES

It is important for a technician to have a handle on circuit terminology. This material is basic information, but it worth repeating. It is important to know difference between a

- Complete circuit
- Short circuit
- Open circuit

COMPLETE CIRCUIT

A complete or closed circuit must have three basic components:

1. Source of power
2. Load
3. Path for current flow (wires)

An optional control or thermostat may be included, but it is not necessary for the operation of the circuit (Figure 16–4).

An adequate power source is required that will push (voltage) electrons (current flow) through the circuit. A load is required to consume and dissipate some of the electron flow in the circuit. The load should do some useful work, such as a compressor motor pumping refrigerant to cool a building or a fan moving air. The path for electron flow, or wires, interconnects the load and power source for a complete circuit.

Complete or closed circuits can be simple series circuits, parallel circuits, or series-parallel circuits. Series-parallel circuits are also known as combination circuits. Combination circuits are the most common circuits found in the HVAC/R industry.

SERIES CIRCUITS

A series circuit is shown Figure 16–4. The circuit has the same current flow at every point in the circuit. If there are two loads in a series circuit, the current flow will be same everywhere in the circuit, but the voltage will divide across the load. If there are two equal loads, then the voltage drop will be the same across each load. Figure 16–5 shows two electric heating elements of the same resistance in series with the power supply. The amperage will be the same throughout the circuit, and the voltage will be split evenly across the load. This is just an example.

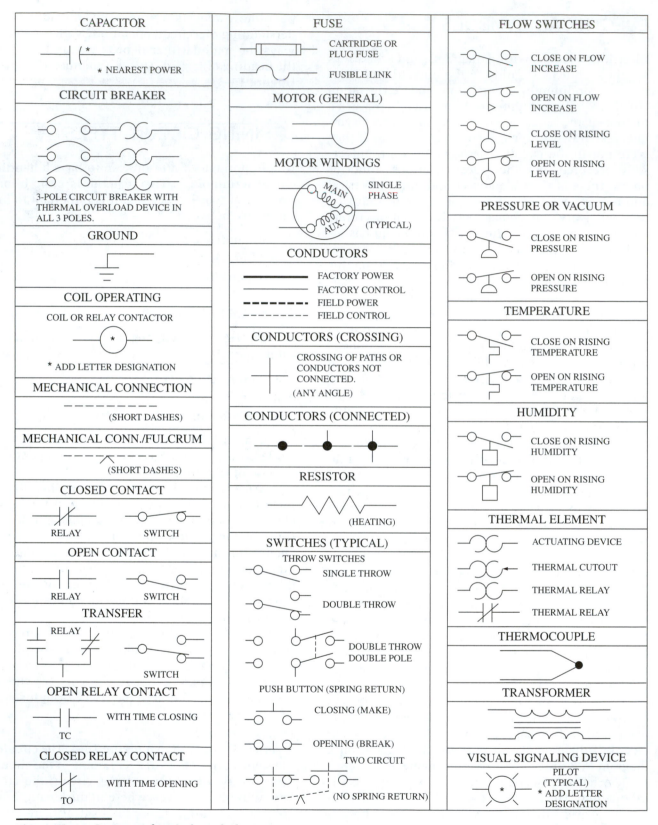

Figure 16–1 Common electrical symbols.

Relays		Switches		Miscellaneous	
R	Relay, General	DI	Defrost Initiation	C-HTR	Crankcase Heater
CR	Cooling Relay	DT	Defrost Termination	RES	Resistor
DR	Defrost Relay	DIT	Defrost Initiation–	HTR	Heater
FR	Fan Relay		Defrost Termination	PC	Program Control
IFR	Indoor Fan Relay		(dual function device	OL	Overload
OFR	Outdoor Fan Relay	GP	Gas Pressure	L	Indicating Lamp
GR	Guardistor Relay	HP	High Pressure	$\oplus$	Manual Reset Device
HR	Heating Relay	LP	Low Pressure	+	Automatic Reset Device
LR	Locking Relay	HLP	Combination High-Low	CAP	Capacitor
	(Lock-in or Lockout)		Pressure		
PR	Protection Relay	OP	Oil Pressure	Compressors	
	(Relay in series with	RM	Reset, Manual		
	protective devices)	FS	Fan Switch	C	Common
VR	Voltage Relay	SS	System Switch	S	Start
TD	Time Delay Device	HS	Humidity Switch	R	Run
THR	Thermal Relay (type)		(Humidistat)		
M	Contactor	TA	Thermostat, Ambient		
MA	Auxiliary Contact	TC	Thermostat, Cooling		
Solenoids		TH	Thermostat, Heating		
		TMA	Thermostat, Mixed Air		
S	Solenoid, General	CT	Thermostat, Compressor		
CS	Capacity Solenoid		Motor		
GS	Gas Solenoid	HT	High Temperature		
RS	Reversing Solenoid	LT	Low Temperature		
		RT	Refrigerant Temperature		
		WT	Water Temperature		

Figure 16–2 Legend for the wiring diagrams.

Most electric heat strips are wired in parallel for maximum heat output.

PARALLEL CIRCUITS

Parallel circuits are arranged so that the power source is across all the loads. In Figure 16–6, the power source is in parallel with three electric heating elements. The voltage supplied across each element is the same. The current flow through each element is the same because the resistance of each heater is the same. The total current flow equals the sum of the amperage draw of each branch or heater. This is represented by the formula $I_T = I_1 + I_2 + I_3$. If the resistance of the heating elements is different, then the current flow through each branch is different. Using Ohm's law, you can calculate that the heating element with lower resistance will develop higher current flow because there is less opposition to current flow. Heating elements with higher resistance develop lower current flow because they oppose the flow of electrons.

SERIES-PARALLEL CIRCUITS

Combination circuits have series components placed in a parallel arrangement. This is the most common type of air-conditioning diagram. Figure 16–7 shows the contactor in series with the series load of the compressor and run capacitor. This line is in parallel with a switching relay that is in series with an indoor fan relay and indoor fan. These fans are in parallel with each other. The primary of the transformer is in parallel with the components above it. The fuse gas valve, limit switch, and thermostat are in series. Some components are in series and some are in parallel, giving the name combination, combining series and parallel circuits together in the same operation.

In combination circuits, the voltage is the same across each parallel branch. The current flow depends on the amount of resistance, or impedance, in each branch. The total current flow is the sum of the current flow in each of the branches. Impedance is the resistance to current flow in inductive components, such as a motor, transformer, or relay coil. Simply measuring the resistance of an inductive component and using Ohm's law does not give a technician the current-flow requirements of a circuit. A special impedance formula is needed to calculate this information. Knowing the exact impedance is not important to the air-conditioning technician.

A complete circuit can be found in many different configurations. It is important to identify the type of complete circuit in order to understand the

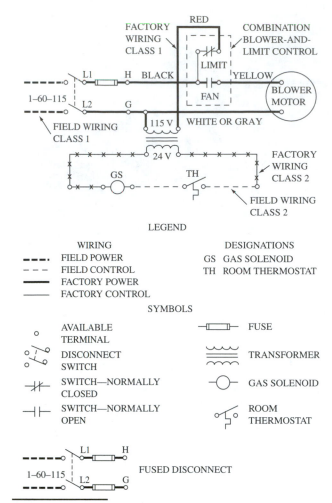

Figure 16–3 Heating circuit diagram for gas-fired up-flow furnace.

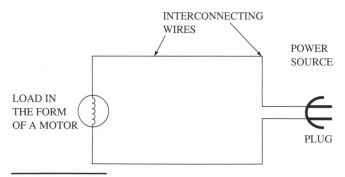

Figure 16–4 A complete circuit includes a power source, load, and interconnecting wires. Switches, thermostats, and pressure-control devices are placed in the circuit as optional devices for convenience and safety in operating the system.

voltage and current flow and requirements. Knowing what to expect in the circuit makes the troubleshooting process easier to follow. The next circuit we discuss is the short circuit.

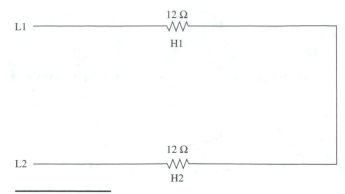

Figure 16–5 Series circuit.

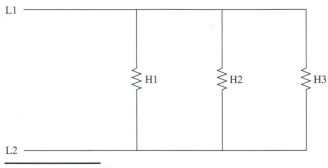

Figure 16–6 Parallel circuit.

SHORT CIRCUIT

The short circuit is the simplest of all circuits. A short circuit is not a circuit by design. A short circuit is usually a wiring mistake or a defect that causes the short circuit. A short circuit bypasses the load(s) in the circuit and causes maximum current to flow, because there is little or no resistance to impede the electron flow. The fuse, wire, or power source is damaged in a shorted condition. A short circuit is similar to placing a wire across the voltage source (Figure 16–8).

A short circuit is generally thought to have 0 Ω of resistance. A circuit can be considered as shorted even if it has a resistance reading. For example, consider what will happen if the resistance or impedance in a parallel branch of a parallel circuit is 5 Ω and a relay coil in that branch shorts out and reduces the resistance to 1 Ω. This low resistance will create a high current flow, which may cause the transformer to overheat or a fuse to open. This becomes difficult to troubleshoot because there is resistance in the circuit. Each component needs to be inspected for abnormal appearance, such as a burned coil. If this does not reap results, each of the parallel branches needs to be checked individually for resistance. Also remember that a wire or terminal can be touching the case ground and cause a shorted condition. This should not be overlooked as a possible cause of a short.

Do not test transformer voltage using the spark method. The spark method involves quickly touch-

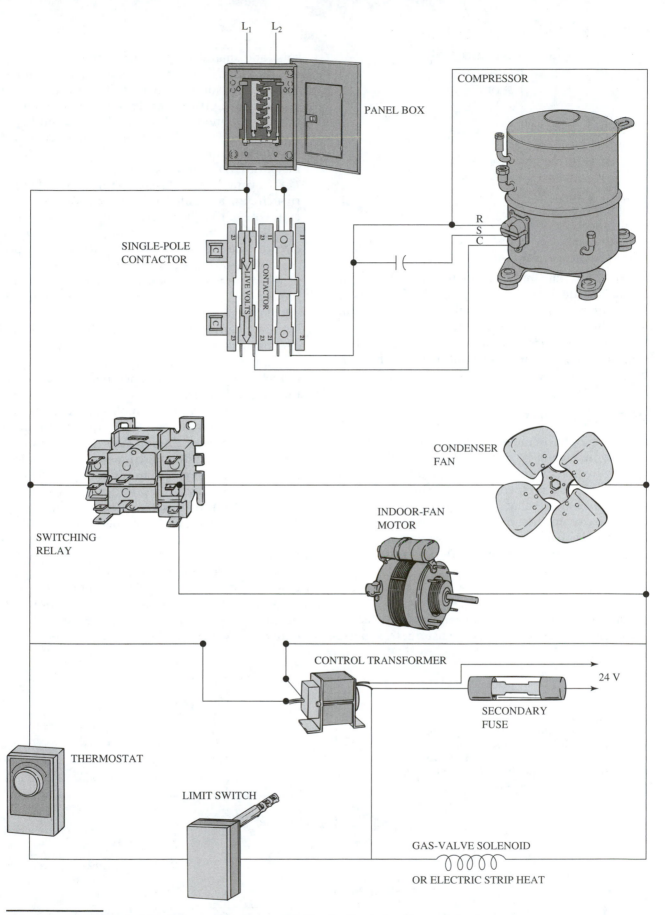

L₁ L₂

PANEL BOX

COMPRESSOR

SINGLE-POLE
CONTACTOR

LIVE VOLTS

CONTACTOR

R
S
C

CONDENSER
FAN

SWITCHING
RELAY

INDOOR-FAN
MOTOR

CONTROL TRANSFORMER

24 V

SECONDARY
FUSE

THERMOSTAT

LIMIT SWITCH

GAS-VALVE SOLENOID

OR ELECTRIC STRIP HEAT

Figure 16–7 Simple ladder diagram showing parallel loads. Control voltage wiring has
been omitted. This is a series-parallel circuit.

263

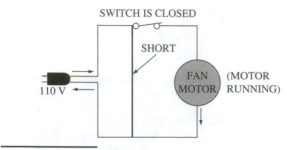

Figure 16–8 Short circuit.

ing the 24-V transformer leads to each other. Some technicians do this to see if a spark is created. A spark indicates that there is some voltage at the output of the transformer. Some transformers have internal or external fuse protection that will be damaged by this procedure. Use a voltmeter to check the secondary voltage.

OPEN CIRCUIT

Another simple circuit is the open circuit. There is a break in the circuit that prevents the current from flowing. Figure 16–9 shows an open circuit because the magnetic switch is open. In some circuits it is the switch or contacts that are open. In other open circuits it is a wire or open load that creates the open circuit. The applied voltage is measured across an open switch, open contacts, open fuse, open wire, or open load.

SUMMARY

It is important to know circuit terminology. Technicians need to communicate with customers, supervi-

sors, and manufacturer representatives. Mixing up the meaning of the terms open, closed, and short will confuse someone who is giving you advice on a problem. The customer will doubt your reputation if told a component is shorted when a second opinion reveals it is open. The component may be defective, but the diagnosis must not be misstated. There is a major difference in these terms.

A complete or closed circuit can be a simple series circuit, a parallel circuit, or series-parallel circuit. One of the characteristics of the series circuit is that the amperage is the same at each point in the circuit. The voltage is applied across the load. If more than one load is present, the voltage is split across the loads. Loads with higher resistance have a greater voltage drop. The sum of the voltage drops equals the applied voltage.

A complete parallel circuit has the same voltage applied across each of the branches. The total amperage is the sum of the current draw in each branch. The lower the branch resistance, the higher the current flow is.

A series-parallel or combination circuit has series components in parallel with other components. This is the most common type of air-conditioning circuit. Series-parallel circuits are more challenging to understand compared to series or parallel circuits.

An open circuit is a circuit that has a broken path for electron flow. The circuit is not complete. Electrons cannot flow in an open circuit.

A short circuit is a circuit that has little or no resistance. The lack of a normal load or resistance causes great current flow that will open a fuse or circuit breaker or burn a wire or component (Figure 16–10). A 24-V transformer that has a shorted component across its leads causes the transformer secondary to overheat and open (Figure 16–11).

Figure 16–9 The contactor is a magnetic switch that opens and closes a contact or contacts to operate the compressor or some other load. A thermostat is used to control the 24 V to the contactor coil.

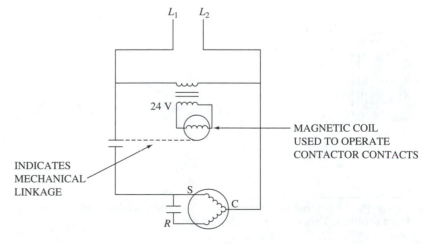

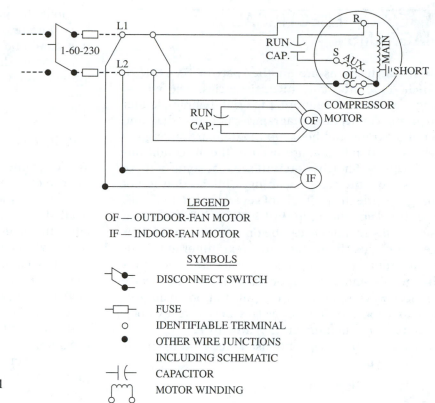

LEGEND

OF — OUTDOOR-FAN MOTOR

IF — INDOOR-FAN MOTOR

SYMBOLS

DISCONNECT SWITCH

FUSE

o   IDENTIFIABLE TERMINAL

●   OTHER WIRE JUNCTIONS
     INCLUDING SCHEMATIC

CAPACITOR

MOTOR WINDING

Figure 16–10   Shorted main winding will cause a fuse or overload to open.

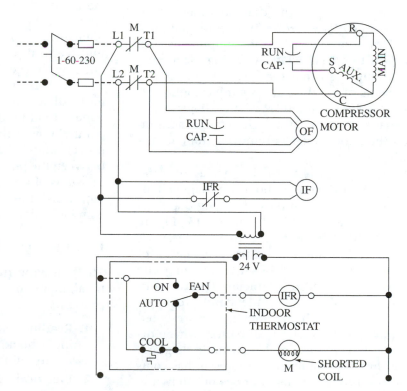

Figure 16–11   Example of shorted coil.

TYPES OF ELECTRICAL DIAGRAMS

Electrical diagrams are another part of HVAC/R in which there are many inconsistencies. The wiring diagram, external diagram, pictorial diagram, ladder diagram, connection diagram, schematic diagram, panel diagram, and hook-up diagram are among the names of circuits that are used in the air-conditioning industry. Shortly, you will see that some of these names have the same meaning. This section is not going to settle the problem of several diagram names with the same meaning, but it does define some of these terms so that we can be on common ground for the rest of this chapter and the next important chapter on troubleshooting. These terms are used interchangeably and it is important that they be clearly defined. Reviewing industry publications, you will find that the terminology varies between textbook authors and in manufacturers' literature.

This section has condensed electrical diagrams into two types:

1. Wiring diagrams
2. Schematic diagrams

WIRING DIAGRAMS

As the name implies, wiring diagrams have to do with how a system is wired. A wiring diagram can be defined as showing how the components and wiring appear to a technician. A wiring diagram is generally arranged so that the physical location of the components in the diagram is similar to the layout of the actual equipment. A wiring diagram is a pictorial snapshot of what a technician expects to find when the panel cover is removed from an air-conditioning unit.

Wiring diagrams are also known as pictorial diagrams, external diagrams, hook-up diagrams, or connection diagrams. Examples of wiring diagrams can be seen in Figures 16–12 and 16–13. Figure 16–12 is a simplified version of a wiring diagram. It is helpful in understanding the basic operation of the circuit but needs more detail if it is going to be valuable to an experienced technician. Figure 16–13 provides the detail needed to understand circuit operation. Notice how each component is represented by a line drawing of the electrical part. A wiring diagram is not as complete as a schematic diagram. Notice that the drawing of the compressor has common, start, and run terminals but does not show the internal winding connecting C, R, and S. Figure 16–14 shows what compressor symbol would be shown on a schematic wiring diagram. A control transformer is shown in the middle of the diagram

without the primary and secondary winding symbols. Figure 16–15 shows the transformer symbol that would be shown on a schematic diagram.

Wiring diagrams can be used to troubleshoot a system. As we will see, schematic diagrams are more useful for performing troubleshooting diagnostics.

SCHEMATIC DIAGRAMS

To determine how a unit is controlled or the sequence of operation of the various components, you can use a *schematic diagram*. If troubleshooting is needed, a schematic diagram is practically a necessity. This diagram divides the system up into a series of individual electrical circuits. Each circuit has a load (a component that requires power), a path for the power to travel, and one or more optional switches. Figure 16–16 is an example of a schematic diagram. The schematic diagram is more detailed and is easier to understand compared to the wiring diagram.

A schematic may be drawn as a ladder diagram. These are clear and easy to follow. Each circuit appears as a ladder rung between two power sources, L_1 and L_2, as shown in Figure 16–17. This is a very simply form of a ladder diagram. Figures 16–18, 16–19, and 16–20 show increasingly complex ladder diagrams. The more complex an electrical diagram is, the greater is the value of the ladder diagram. Although a ladder drawing is usually made with vertical power lines, it can be drawn horizontally, with the ladder lying on its side (Figure 16–18). This arrangement may be more convenient for the diagram designer or may fit better on the unit panel. A ladder diagram can be described as having parallel power lines. These can be considered as the power rungs. The controlling switches and loads are between the power lines. The switches and loads are the steps of operating sequence. In the next chapter you learn how to do hopscotch troubleshooting using a ladder diagram as guidance.

RULES FOR READING DIAGRAMS

The following rules will help you understand electrical diagrams and assist in the troubleshooting process.

1. Reading a schematic diagram is similar to reading a book. Diagrams are generally read from left to right and top to bottom.
2. Electrical symbols are shown in their deenergized positions.
3. Relay contact symbols are shown with same numbers or letters used to designate the relay coil. All contact symbols that have the same

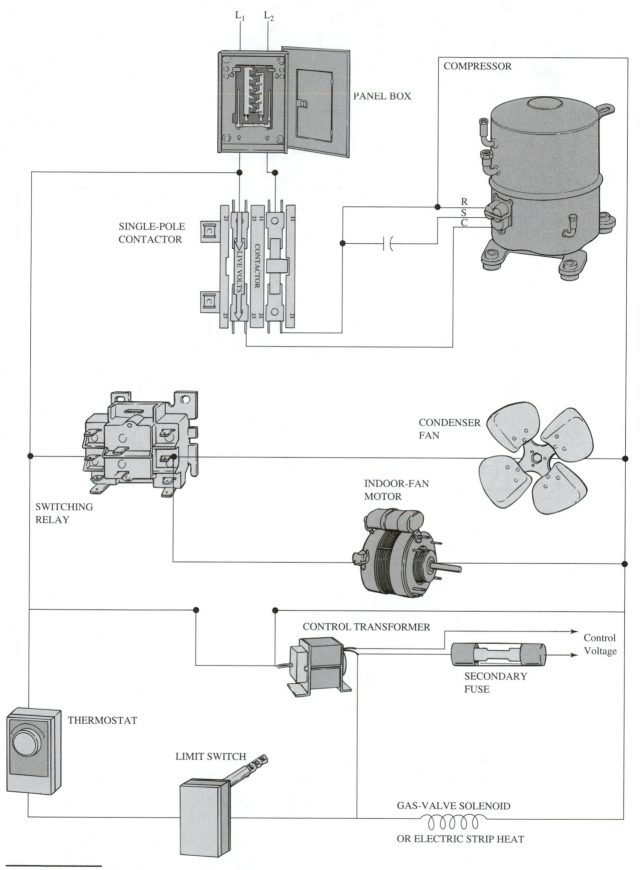

L₁ L₂

PANEL BOX

COMPRESSOR

SINGLE-POLE
CONTACTOR

LIVE VOLTS

CONTACTOR

R
S
C

CONDENSER
FAN

INDOOR-FAN
MOTOR

SWITCHING
RELAY

CONTROL TRANSFORMER

Control
Voltage

SECONDARY
FUSE

THERMOSTAT

LIMIT SWITCH

GAS-VALVE SOLENOID

OR ELECTRIC STRIP HEAT

Figure 16–12 Electrical loads required for a packaged air-conditioning/heating system. Gas or electric heat can be used on package units with heat.

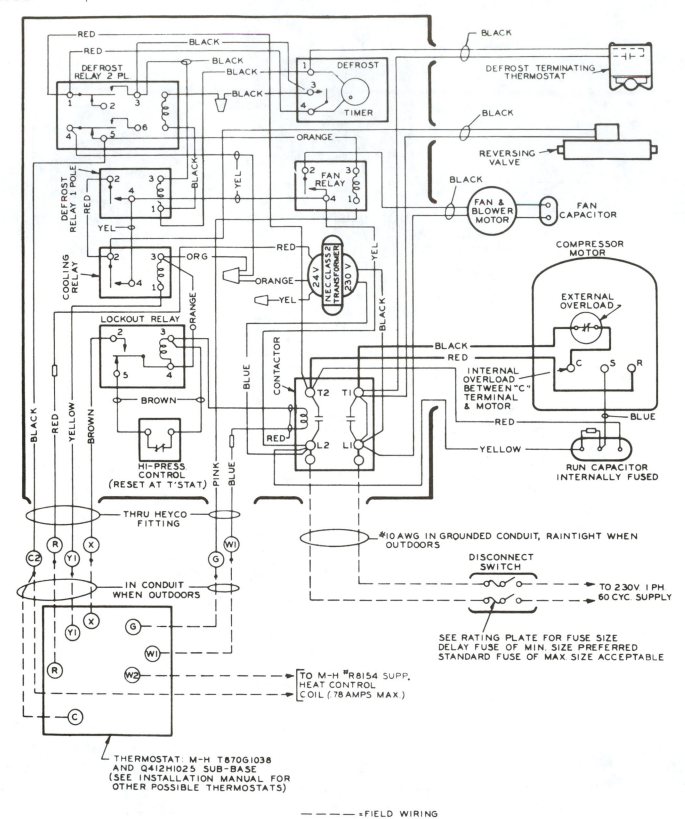

Figure 16–13   Component arrangement enlarged from label diagram.

Figure 16–14 Single-phase motor with start and run windings.

SINGLE-PHASE COMPRESSOR

Figure 16–15 Step-down 24-V control transformer.

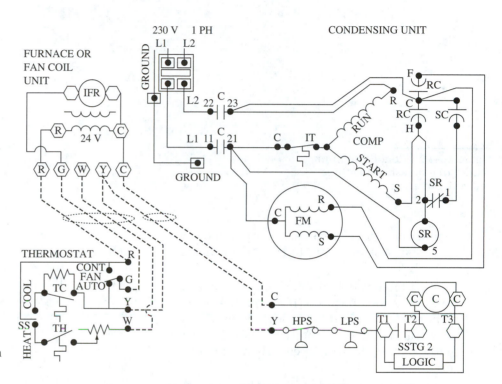

Figure 16–16 Schematic diagram for an air-conditioning system.

Figure 16–17 Ladder diagram showing a power supply and three loads.

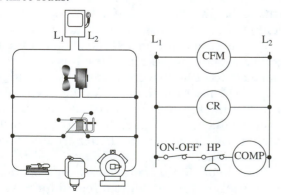

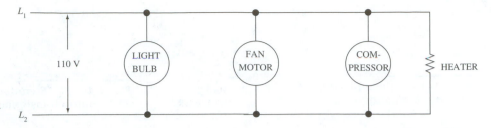

Figure 16–18 Parallel circuit with four different loads.

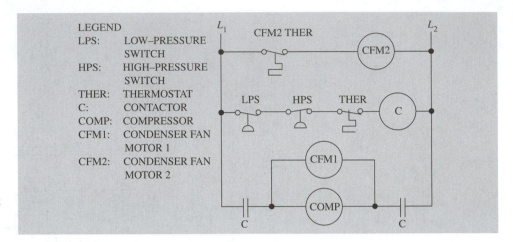

Figure 16–19 Condensing-unit wiring diagram showing three separate units.

number or letter as a coil are controlled by that coil regardless of where in the circuit they are located.

4. When a relay is energized, or turned on, all its contacts change position. If a contact is shown as normally open, it will close when the coil is energized. If the contact is shown as normally closed, it will open when the coil is turned on.

5. There must be a complete circuit before current can flow through a component.

6. Components used to provide a stop function are generally wired normally closed and connected in series with the load it controls (Figure 16–21).

7. Components used to provide a start function are generally wired normally open and connected in parallel (Figure 16–21). If either switch is closed, a current path will be provided to the motor and it will turn it on.

8. Switches, like thermostats or contacts, are in series with the loads they control.

SUMMARY

There are many names for the electrical diagrams used in the HVAC/R industry. This chapter consolidates and simplifies all the terms into two categories: wiring diagrams and schematic diagrams.

Wiring diagrams are used to show the electrical components and interconnecting wiring installed on the equipment. A wiring diagram is a picture of the component layout a technician would expect when opening a piece of equipment. As defined, wiring diagrams can also be called pictorial diagrams, hookup diagrams, and connection diagrams. A wiring diagram is useful in that a technician can see the layout of the components. Knowing the layout out of the components will aid in rapid identification of components in the troubleshooting sequence.

However, a wiring diagram is not always correct. Last-minute manufacturing changes may not be reflected in the unit's wiring diagram. Optional components not on the unit may be shown on the electrical diagram. Previous equipment maintenance may find that the component arrangement has been altered. Never assume that a wiring diagram or schematic diagram is correct. In additional to factory or field changes, the diagram may have a mistake. An experienced technician will be able analyze the diagram and spot the discrepancies. It is easier to locate an electrical diagram error using a schematic diagram. A schematic diagram shows the electrical sequence. Tracing through the electrical sequence will uncover the diagram mistake. The more you practice with electrical diagrams, the better troubleshooter you will become.

The schematic diagram shows components in their electrical sequence without regard for physical

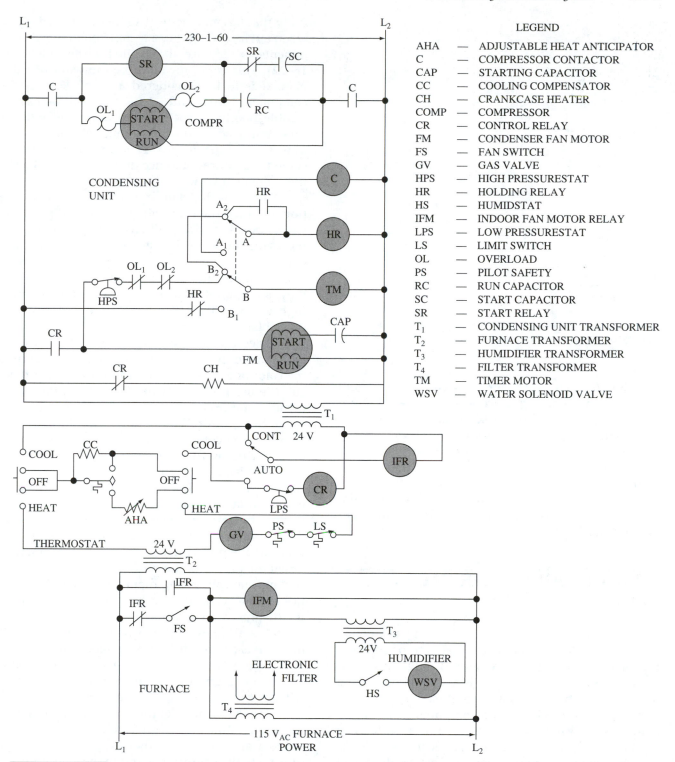

LEGEND

AHA	—	ADJUSTABLE HEAT ANTICIPATOR
C	—	COMPRESSOR CONTACTOR
CAP	—	STARTING CAPACITOR
CC	—	COOLING COMPENSATOR
CH	—	CRANKCASE HEATER
COMP	—	COMPRESSOR
CR	—	CONTROL RELAY
FM	—	CONDENSER FAN MOTOR
FS	—	FAN SWITCH
GV	—	GAS VALVE
HPS	—	HIGH PRESSURESTAT
HR	—	HOLDING RELAY
HS	—	HUMIDSTAT
IFM	—	INDOOR FAN MOTOR RELAY
LPS	—	LOW PRESSURESTAT
LS	—	LIMIT SWITCH
OL	—	OVERLOAD
PS	—	PILOT SAFETY
RC	—	RUN CAPACITOR
SC	—	START CAPACITOR
SR	—	START RELAY
T_1	—	CONDENSING UNIT TRANSFORMER
T_2	—	FURNACE TRANSFORMER
T_3	—	HUMIDIFIER TRANSFORMER
T_4	—	FILTER TRANSFORMER
TM	—	TIMER MOTOR
WSV	—	WATER SOLENOID VALVE

Figure 16–20 Schematic wiring diagram showing the various electrical circuits.

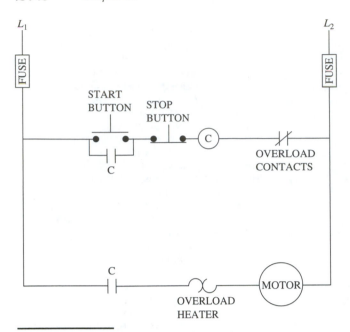

Figure 16–21 Start-stop control circuit.

location. A schematic diagram is an important tool in the troubleshooting process. The ladder diagram is the most useful type of schematic diagram. The next chapter explains the way to troubleshoot using a ladder diagram. Once you learn this troubleshooting method, you will always request a ladder diagram as a diagnostic aid.

Remember the rules for reading wiring diagrams. These rules are the basic foundation for accurate troubleshooting.

DRAWING AN ELECTRICAL DIAGRAM

On some occasions an air-conditioning technician will need to draw an electrical diagram for an existing piece of equipment. A helpful diagram can be drawn if no diagram exists or is available. Without a diagram, a technician may be confused by the problem, the mass of wiring, and components in the circuit. Some equipment circuits have been so heavily modified that the existing diagram is almost useless. In this instance, a new diagram must be drawn in order to proceed with the troubleshooting process.

Developing an electrical diagram can be done in a variety of ways. First, a technician can develop a wiring diagram and convert it to a schematic diagram. The following example describes the basic procedure for developing an electrical diagram of a packaged unit with electric heat. The same example can be used as an example of how to develop a diagram for a furnace, refrigeration equipment, or any-

thing that has an electrical sequence. In order to develop a wiring diagram, a technician draws the components as they are shown inside the unit (Figure 16–22). This is not meant to be the work of a draftsman. It is to be considered a rough sketch of the component layout and wiring that will be useful in troubleshooting process. The drawing will take time, but it will reduce the troubleshooting process now and in the future. A rough draft should be copied for future reference and one copy left inside the condensing unit.

First, a technician draws the high-voltage section and then the low-voltage section. Figure 16–22 shows boxes, circles, and electrical symbols used to represent the high-voltage components in the circuit. Some components use high and low voltage. These components are drawn in the circuit because they are important for the understanding of the high-voltage circuit. The low-voltage section is discussed later.

Once the components are sketched, a technician must draw interconnecting wiring. Identifying the color of each wire will aid in tracing out the circuit (Figure 16–23). This will give a technician a reference point from which to work and help keep him or her in the desired place in the circuit. The components can be considered as landmarks and the colored wiring as roadways between the landmarks. Once a wiring diagram is developed, a technician can transform the information into a schematic diagram. A ladder diagram can be made from the wiring diagram in a matter of minutes. The wiring diagram may be all the technician needs in understanding the problem, but a ladder diagram will nail down the problem more quickly. It is extremely important that the wiring diagram be double-checked prior to proceeding with troubleshooting. Errors in the diagram will lead to confusion, frustration, and failure. A little extra time spent will assure that the diagram is correct. It is easy to make a mistake.

Figure 16–23 shows the technician's progress. The major high-voltage components are sketched. All the connections to L_2 are completed. L_2 is attached to the primary of the control transformer with a red wire. Blue wires run from L_2 to one side of the outdoor and indoor fan motors. A white wire goes directly to the compressor and to the compressor-run capacitor. A red wire connects the run capacitor to the compressor. The wire colors are identified and abbreviated to make tracing easier. L_1 is hooked to the left side of the normally open contacts of the contactor. A blue wire attaches the right side of the contacts to the compressor.

Figure 16–24, shows the remaining high-voltage wiring connections, with the color of the wire labeled for identification. This wiring diagram has

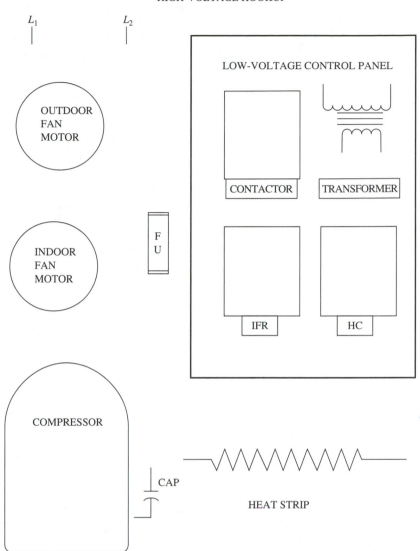

HIGH-VOLTAGE HOOKUP

Figure 16–22 High-voltage and low-voltage components as seen inside a package unit.

been transformed in the schematic diagram in Figure 16–27. Compare the wiring diagram to the ladder diagram. Which diagram is easier to follow if you want to understand and troubleshoot a circuit?

After the high-voltage section is completed, a technician needs to complete the low-voltage, or control-voltage, section of the package unit. It is recommended that they be kept separate at this point in developing a useful diagram. The two diagrams can be combined later. Diagrams should be completed a section at a time in order to reduce mistakes. Try to understand what is being developed while drawing the wiring diagram.

Figure 16–25 illustrates what a technician will draw before connecting wiring and assigning wire colors. These are the low-voltage-control components. The only component not inside the package unit is the thermostat. The thermostat should be

mounted inside the building in a central location, probably near the return-air stream. These components have high- and low-voltage connections. Most of them have high voltage running through part of the component. The only components without high voltage are the thermostat and the fuse that protects the secondary of the transformer. The high-voltage and low-voltage sections are never connected together. If this were to occur, the low-voltage section would be heavily damaged.

Figure 16–26 shows the completed low-voltage wiring diagram with components and wiring hookup. Even though this low-voltage diagram has fewer components and connections compared to the high-voltage diagram, it is more complicated than the ladder diagram shown in Figure 16–27. The ladder diagram has the control section located in the lower third of the page, starting at the secondary of the transformer.

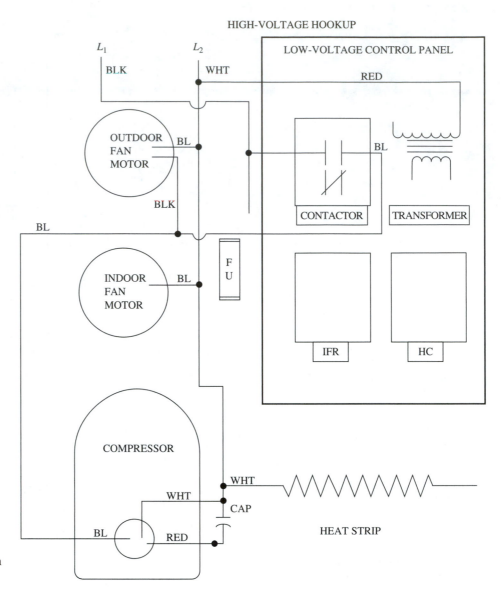

HIGH-VOLTAGE HOOKUP

Figure 16–23   Connecting the components when developing a high-voltage diagram.

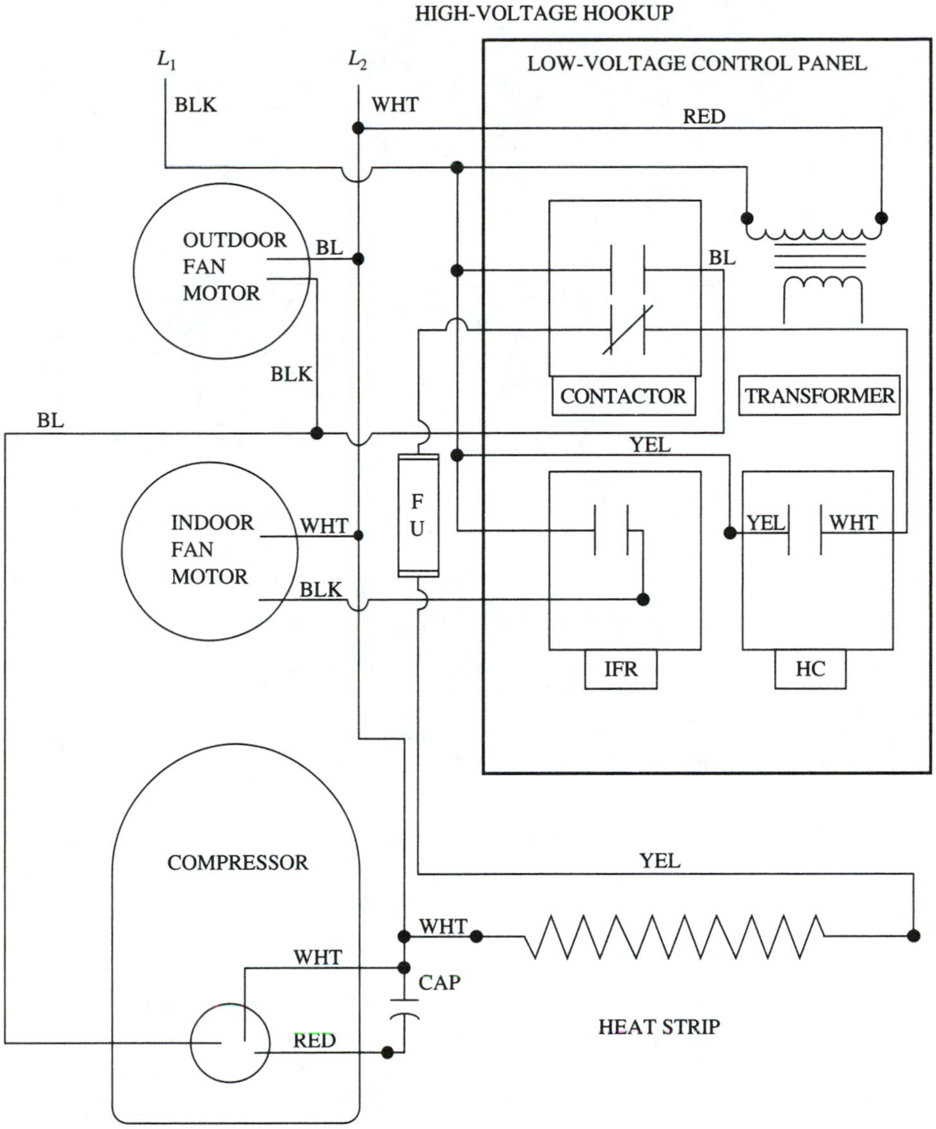

HIGH-VOLTAGE HOOKUP

Figure 16–24 High-voltage circuit completed.

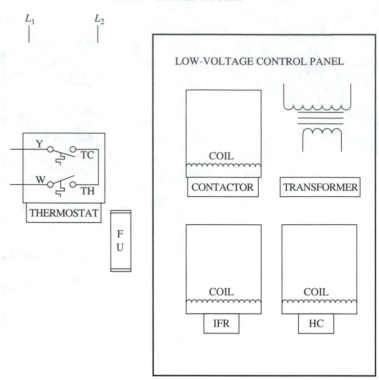

Figure 16–25 Low-voltage units as seen inside a package unit. Thermostat located outside the unit.

Figure 16–26 Connecting the components when developing the low-voltage diagram.

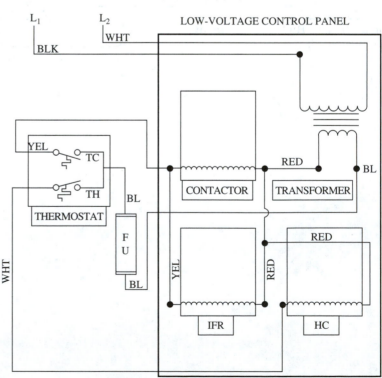

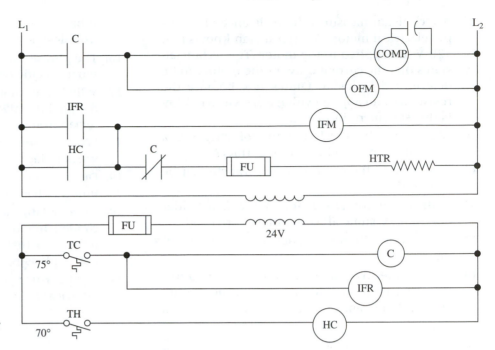

Figure 16–27 Ladder diagram. The high-voltage section is in the upper part of the diagram. The low-voltage section is in the lower part of the diagram.

SUMMARY

- This chapter contains essential information needed before tackling the next chapter on troubleshooting wiring diagrams. You have learned the terminology used when air-conditioning technicians discuss the electrical operation of equipment. Unfortunately, the terminology is inconsistent between technicians, manufacturers, educators, and textbook publishers. This chapter has established the definition of what will be used in the troubleshooting chapter. There is agreement on terms such as a closed circuit, open circuit, short circuit, and series, parallel, and combination circuits. The differences in definitions surface when discussing the types of electrical diagrams.

- Finally, learning to draft wiring and schematic diagrams provides a valuable troubleshooting tool for your future. There are many times when the wiring has been butchered so badly it is a wonder the equipment ever operated correctly. Safeties are bypassed and disconnected. Undocumented alterations make the wiring a complete rat's nest. You open the unit panel door and a bundle of wires fall out. You can spend hours trying to figure out what is going on in the unit. A good solution is to obtain a wiring diagram, remove all the wiring, and start from scratch. Sometimes this is the best way to approach this problem. Drawing the wiring diagram is a time-saving solution if diagrams are

not available and the wiring is difficult to trace or follow. Keep a file of wiring diagrams. These diagrams may come in handy even if the equipment is from a different manufacturer.

- Practice drawing wiring diagrams and converting them to ladder diagrams. Locate a unit with a diagram. Without reviewing the diagram, draft your diagram and compare yours to the existing one. You will learn by completing this exercise on air-conditioning equipment, furnaces, refrigeration circuits, and other HVAC/R related circuits.

PROBLEMS AND QUESTIONS

1. What is the purpose of electrical symbols?
2. Draw, identify, and label 20 electrical symbols.
3. What is a legend as it relates to electrical diagrams?
4. What is the difference between an open circuit, a short circuit, and a complete circuit?
5. What are the basic components that make up a closed circuit?
6. What happens to the voltage and current in a series circuit?
7. What happens to the voltage and current in a parallel circuit?
8. What is the most common type of air-conditioning circuit found in air-conditioning equipment?

9. A technician measures the resistance of a simple electrical motor. The technician knows that 120 V are applied to the motor. The technician states that the current draw of the motor can be determined by using Ohm's law because the resistance and applied voltage are known. Why is this statement incorrect?

10. What will the spark method of checking a transformer do to a fused transformer?

11. How should the secondary of a transformer be checked?

12. Redraw Figure 16–11 and make it a ladder diagram. Assume all components are good.

13. List seven different names for an electrical diagram.

14. This chapter recommended the use of two different terms used to describe air-conditioning electrical diagrams. Define these terms in detail.

15. What type of diagram is most useful when troubleshooting electrical problems?

16. The phrase *power rung* is used when describing what specific type of diagram?

17. Why is the color of the wire according electrical diagram not always what is found when a technician troubleshoots a piece of equipment?

18. When would a technician need to draw a wiring diagram? List several conditions.

19. Describe how to draw a wiring diagram on a unit that has no information available.

20. In your lab, locate and draw the diagram of a condensing unit that has no diagram. Show only the actual components that are located on the condensing unit.

21. List the rules you should follow when using an electrical diagram.

CHAPTER SEVENTEEN

TROUBLESHOOTING ELECTRICAL DIAGRAMS

AFTER STUDYING THIS CHAPTER, THE STUDENT WILL BE ABLE TO:

- Describe how to isolate a problem.
- Define troubleshooting.
- Use a sequence to quickly determine the type of problem.
- Discuss four preparation steps for troubleshooting.
- Describe electrical troubleshooting procedures.
- Discuss the advantages and disadvantages of alternative troubleshooting methods.

- Be able to use electrical diagrams to troubleshoot problems.
- Describe the hopscotch method of troubleshooting.
- Show how to use the hopscotch method of troubleshooting.

INTRODUCTION TO USING ELECTRICAL DIAGRAMS AS TROUBLESHOOTING TOOLS

It is difficult to teach any type of mechanical or electrical troubleshooting. The reason it is so difficult to teach troubleshooting is because there are many problems that are not clear-cut. Problems can blend together. There can be more than one problem. Mechanical and electrical problems can occur simultaneously. A mechanical failure can manifest itself as an electrical problem. Additionally, individuals learn differently. One person will learn and put into practice what is presented in this chapter. Another person will read this material and totally ignore its value and logical sequence of troubleshooting events. Everyone wants hands-on practice. This chapter is designed to give you the tools, logic, and thought processes that can be used in isolating the problem, and if it is an electrical problem, determining the exact defective component.

The first segment defines troubleshooting and discusses how to narrow down the problem quickly using the **ACT** method of troubleshooting as introduced in the chapter "Troubleshooting Air-Conditioning Systems." The next segment introduces basic electrical troubleshooting, followed by specific troubleshooting using the ladder diagram as the primary tool of diagnosis.

WHAT IS TROUBLESHOOTING?

Any operating mechanical or electrical equipment will at some time require service. The repair work necessary to place the equipment back on-line is one of the most important functions of the HVAC/R technician. The basic term that describes this type of work is *troubleshooting*. Troubleshooting is the process of determining the cause of an equipment malfunction and performing corrective measures to bring the equipment back to normal operating parameters. Depending on the problem, this may require a high degree of knowledge, experience, and skill.

279

Basically, there are two types of problems, (1) electrical and (2) mechanical, although there is much overlapping. Whatever the nature of the problem, it is good practice to follow a logical, structured, systematic approach. In this manner the correct solution is usually found in the shortest possible time.

QUICK CHECKS

A seasoned technician will use a set of quick checks to determine what is happening in an air-conditioning system. One way to organize your troubleshooting thought patterns is to develop a plan of *act*ion. Use the word *ACT* as a reminder about what to check on an air-conditioning system. It is important to make these checks so that you do not end up on a wild-goose chase and embarrassing yourself and your company.

The word **ACT** is an abbreviation:

A is for airflow.

C is for compressor.

T is for thermostat.

Let us explore how to use **ACT** as a quick way to check the vital signs of an air conditioning system.

A IS FOR AIRFLOW

Check the outdoor unit airflow. Check the indoor unit airflow.

C IS FOR COMPRESSOR

Is the compressor operating? What is the amperage draw of the common winding? Is there a pressure difference?

T IS FOR THERMOSTAT

Is the thermostat set to the cooling mode and the temperature set 10°F below the indoor temperature?

The ACT method or another method you devise will eliminate parts of the system that are working and help you diagnose the problem quickly. Each job is a little different, and developing a systematic way to start troubleshooting is important. Using the ACT method provides a fast and logical sequence for checking the system and, at the same time, familiarizes a technician with the system under scrutiny.

Because most malfunctioning problems are electrical, it is a common practice to perform electrical troubleshooting first. If the problem is mechanical, the electrical analysis will usually point the technician in that direction. Because this chapter is based on electrical troubleshooting, we now discuss that specific topic.

ELECTRICAL TROUBLESHOOTING

In preparing to analyze electrical problems, four items are important to know or recognize:

1. The operation sequence of the unit
2. The parts of the equipment that are working and those that are not working
3. The electrical test instruments that are needed to analyze the problem
4. The power circuit driving the system

A manufacturer, through the installation instructions, usually supplies the operating sequence. A technician can also determine the operating sequence by studying the schematic wiring diagram. The function of the operating and nonoperating equipment is determined by examination and testing. Necessary test instruments include the voltmeter, the clamp-on ammeter, the ohmmeter, the capacitor checker, and a manifold gauge set. A technician must be proficient in the use of these instruments.

The power circuit is the first to be examined, because power must be available to operate the loads. For example, on a refrigeration system with an air-cooled condenser, the two principal loads that must be energized are the compressor motor and two fan motors. Before proceeding with anything else, a technician must be certain that the proper power can be supplied to the loads.

Each electrical circuit has one or more switches that start or stop the operation of a load. This switching operation is called the control function. In troubleshooting, when a load is not working, a technician must determine whether the problem is in the load itself or in the switches that control the load.

To assist in analyzing an operational problem of a unit, a manufacturer furnishes one or more of the following pieces of information:

1. Wiring diagrams
2. Installation and service instructions
3. Troubleshooting tables
4. Fault-isolation diagrams
5. Diagnostic tests

Wiring diagrams usually consist of connection diagrams and schematic diagrams. A *connection diagram* shows the wires to the various electrical component terminals in their approximate location on the unit. This is the diagram that the technician must use to locate the test points. A *schematic diagram* separates each circuit to clearly indicate the function of switches that control each load. This is the diagram that the technician uses to determine the sequence of operation for the system.

Installation and service instructions supply a wide variety of information that the manufacturer believes is necessary to properly install and service a unit. This bulletin includes the wiring diagram, the sequence of operation, and any notes or cautions that need to be observed in using them. A troubleshooting table, as shown in Figure 17–1, is helpful as a guide to corrective action. By the process of elimination, these tables offer a quick way to solve a service problem. The process of elimination permits a technician to examine each suggested remedy and disregard ones that do not apply or are impractical, leaving only the solutions that fit the problem. These tables offer the solution to the most common malfunctions and are a good starting point for any troubleshooting sequence.

A *fault-isolation diagram* (Figure 17–2) starts with a failure symptom and goes through a logical decision-action process to isolate the failure.

Diagnostic tests (Figure 17–3) can be conducted on electronic circuit boards, at points indicated by the manufacturer, to check voltages or other essential information critical to the operation of the unit.

Some electronically controlled systems have automatic testing features, which indicate by code number a malfunction in the operation of the equipment (Figure 17–4). Some equipment uses a flashing LED to indicate a malfunction. The number of flashes indicates a specific problem. Further tests are usually required to determine the action that is required.

ELECTRICAL TROUBLESHOOTING PROCEDURES

The procedure for electrical troubleshooting is to

1. Complete the ACT or other generic sequence of checks to see what is operating and what is not operating.

Figure 17–1   Typical troubleshooting table.

SYMPTOM	PROBABLE CAUSE	REMEDY
1. IFM NOT RUNNING	1. LINE FUSE(S) BAD; IFC BAD; IFR NOT PICKING UP	1. REPLACE FUSE(S); CHECK IFC; CHECK -A/C SWITCH -THERMOSTAT
2. ETC.	2. ETC.	2. ETC.

2. Take control of the system if it is not operating. Turn off the circuit breaker, remove the fuse, or place the thermostat in the off position. The equipment may operate briefly and cut off without giving you the opportunity to determine the cause of the shutdown. The technician needs to control the unit at the point of the troubleshooting sequence.

If the technician begins at the condensing unit, the thermostat wire can be temporarily removed from the contactor coil to prevent the unit from operating. Use a set of alligator clips to bridge the wire between the disconnected thermostat connection on the contactor coil. This procedure will assure a quick, positive contact when reconnecting the thermostat wire. Trying to start a piece of equipment by connecting a thermostat wire to a terminal board or twisting wires under a wire nut may cause shortcycling because the wire connection is not always positive on the first try. Short-cycling can cause other problems not associated with the original condition. For example, short-cycling can cause the overload to open when the original problem was overheating of the compressor because of lack of charge. Taking positive control of the section of the equipment being investigated is a very important step.

When removing the control-voltage wiring, do not allow the wire to touch the equipment case. This short-circuit condition may cause the control transformer to draw excess current and burn out.

Some technicians use a damaging practice for checking 24 V to the contactor. In this method, the technician quickly touches together the two wires that go to the contactor coil. The idea is that a small spark will occur if 24 V is present at the wires. The problem with this procedure is that many transformers have internal or external fuses that will open at the instant the two wires are shorted together. This swiping action is not a recommended procedure. Use a voltmeter to check for the correct applied control voltage. The thermostat wiring should be hooked to the contactor coil when checking for voltage.

3. Connect the manifold gauge set to the high- and low-pressure sides. This will determine if the system has a refrigerant charge. If the equipment abruptly begins to operate with low charge or a high-pressure condition, the technician will be able to determine that the unit is responding to a pressure switch that will cut the unit off in a matter of seconds.

4. Connect the Amprobe to the common of a single-phase compressor or any phase of a three-phase compressor. If the initial amperage draw is sustained beyond the normal locked-rotor period, the overload will open the power to the compressor. The compressor overload may reset in a short time or it may take

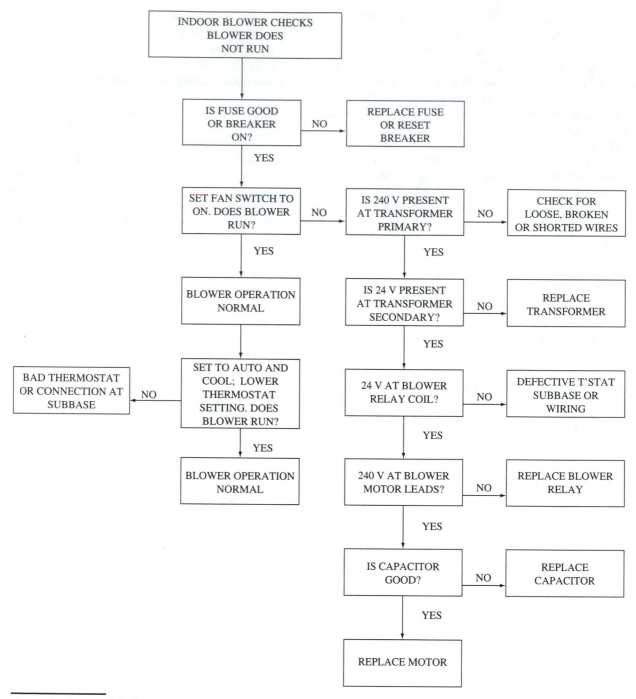

Figure 17–2 Troubleshooting chart. *(Courtesy of International Comfort Products)*

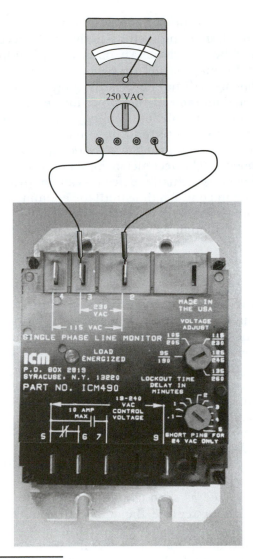

Figure 17–3 Testing a circuit board or module. The input voltage should be measured first (top), and an expected action should occur, such as opening or closing a set of contacts (bottom). *(Courtesy of ICM Corporation)*

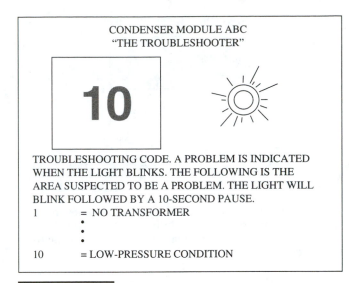

Figure 17–4 Some newer condensers and air handlers have technology that assists a technician in troubleshooting. A number, which refers to a problem, appears on a module. Some troubleshooting modules use flashing lights to indicate a specific problem. The number of flashes indicates the problem. The light will blink followed by a ten-second pause.

several hours of cooling before it closes. It is imperative to have the amperage information. The manifold gauge reading and the Amprobe reading should be installed where both can be read simultaneously.

5. If the system does not operate, check the power to the air handler, control circuit, and condensing unit. The air handler may contain the input power for the control circuit. The proper voltage input is required for all these systems to operate. Many times it is a simple lack of power to these system components that prevents anything from operating. Do not overlook the simple things as the first step of troubleshooting.

On a split system, the building's power supply provides voltage to the indoor-blower section and the outdoor condensing unit. Figure 17–5 shows a common indoor-blower-section voltage of 120 V. A 240-V supply may also be used for the indoor-blower section. The indoor section normally includes the control-voltage components, which operate through a 24-V transformer. Notice that the building power supply is separate for the indoor-blower section and the outdoor condensing unit section. This is common for a split system. Check both supply voltage sources. Each supply voltage source should have its own disconnect and overload protection. Package units have one power source that may divide once inside the unit.

The low-voltage control (Figure 17–6) derives its power from the indoor section. The thermostat should be considered as a switch that automatically controls the on-and-off cycle of the condensing unit

Figure 17–5 The high-voltage power supply provides electricity to the indoor-blower section and the outdoor condensing unit.

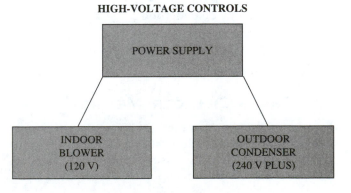

HIGH-VOLTAGE CONTROLS

LOW-VOLTAGE CONTROLS

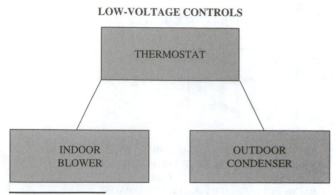

Figure 17–6 The control voltage normally comes from the control transformer in the indoor section. The thermostat is the switch that supplies control voltage to the condensing unit.

and the indoor blower. This is usually a 24-V control. Higher control voltage may be used in larger commercial and industrial applications, especially in refrigeration systems that use high control voltage exclusively.

6. If the power is not available to any of these system components, use an ohmmeter to check fuses, breakers, and transformers. Remove these components from the circuit prior to checking. These are common components that will prevent power from being applied to the system. Remember that a closed breaker can be tripped. Open the breaker and reset it. Stand to the side of a breaker when opening and resetting it. A breaker can explode, so it is best to stand to the side when operating it.

OTHER TROUBLESHOOTING METHODS

There are other troubleshooting methods that should be discussed, because some of them have value and some are used due to a lack of troubleshooting skills. The following troubleshooting methods are discussed:

- Parts-changer method
- Equipment-changer method
- Call someone else method
- Voltage-to-ground method
- Draw a schematic diagram method
- Hopscotch method

THE PARTS-CHANGER METHOD

A person who uses this method is not classified as a technician and has the lowest level of troubleshooting skills. The person guesses what part or parts to change and continues this operation until the system begins to work properly. The unfortunate aspect of this troubleshooting method is that a changed part may not be defective. An open power supply or open wire may be the problem. Replacing all the components will not necessarily solve the problem. Another limitation of this procedure is the lack of truck stock available for changing out parts. Technically speaking, all technicians could be classified as parts changers. A problem may be isolated to two components. The technician may not know which component is actually defective and replaces the suspect parts one at a time until the equipment functions properly. There is a little parts changer in each of us, even though we are not proud to admit it.

EQUIPMENT-CHANGER METHOD

The equipment changer is probably classified lower than the parts changer. In this scenario, troubleshooting the equipment leads to no solutions. The equipment is evaluated and condemned. Equipment replacement is suggested. Some customers are vulnerable to the suggestion of a new system or the replacement of a major component such as a condensing unit or furnace. All customers understand that their equipment has a limited life expectancy. If the unit is old and has been giving the customer increasing problems, then the customer is more likely to fall prey to this solution. The customer should obtain a second opinion. Many customers want the problem fixed as soon as possible and go with the suggestion to replace the unit. They are tired of being hot or cold. One thing they do not realize is that a second opinion may reveal a minor problem that was overlooked by the original technician. It is possible that the equipment could be repaired in a short period of time.

A customer may also change out a unit prematurely because of the technician's persuasiveness. Some technicians are so biased in favor of specific equipment they try to convince a customer that the

customer would be better off with a change in brand or capacity. A case can be made for replacing older, inefficient equipment with new, high-efficiency systems with a 12 SEER or higher. However, in such a case the existing system should be more than 10 years old and have a SEER rating 8 or less. The customer can recuperate the investment by lower utility bills and fewer repair costs. Many companies offer extended repair warranties that include labor and parts for 5 years or in some instances, 10 years. Improved warranties may be another reason to replace a system that has had repeated failures in the past couple of years.

There are some technicians that recommend equipment replacement because they do not understand the equipment operation. This is the case with a heat-pump system. A heat-pump system is a more technical heating and cooling product. There are cases in which the customer thinks that the heat-pump product is not what should be installed. A technician convinces the owner that a conventional heating and cooling system should replace the heat pump. Many technicians do not like heat pumps because they do not understand their operation. If you hear a technician or a company degrading heat pumps, you may find that they do not have the knowledge and experience base to service heat pumps. Additionally, they may have a difficult time trying to repair heat pumps, and it is easy to criticize something that is not understood. A heat pump is a technical challenge. A technician who is familiar with a piece of equipment or a technology will not debase the equipment because it is more difficult to understand and troubleshoot.

CALL SOMEONE ELSE METHOD

Everyone that does troubleshooting will eventually need to call someone for advice. A technician may call the office and speak with a more experienced technician about a problem. The call may be to a manufacturer's technical hotline that offers assistance as it relates to the products.

The entry-level technician starting in the field is more likely to call for advice than an experienced technician. This is expected, but too much time on the telephone indicates that the technician needs more training. Traveling with an experienced technician that is willing to train is one possible answer. Attending community college courses or manufacturer-sponsored courses is another option for gaining knowledge about troubleshooting. A technician can be assigned less challenging jobs until his or her skill level is developed sufficiently. Conducting preventative maintenance (PM) checks is one way to develop these skills in a repetitive and nonthreatening

environment. A technician will find problems when conducting PMs and have the opportunity to solve the problem or reassign it for another service call.

Many technicians call someone else, but the professional technician will limit this practice. Any calls should be made in the service vehicle away from earshot of the customer. The exciting part of the HVAC/R trade is that a technician is always learning something new about equipment and ways to complete a job quickly and accurately.

VOLTAGE-TO-GROUND METHOD

Another method used to troubleshoot air-conditioning circuits is to check components or wiring to ground. This is an *unacceptable* way of troubleshooting a circuit. As a matter of fact, it will give the technician a totally erroneous idea about what is happening in the system.

Review Figure 17–7. A technician is using this schematic (used for troubleshooting purposes) diagram to diagnosis a malfunctioning package unit. The problem is that one of the two condenser fan motors is not operating. The technician notices that condenser motor no. 2 is obviously not turning properly. It is turning slowly backward due to the airflow pattern of the condenser motor that is operating. The technician checks the voltage on the run winding (right side of motor) to ground and it measures 120 V. This is what is expected. Next, the voltage across the common winding (left side of motor) is measured to ground as 120 V. The technician believes that there is correct voltage to the motor because whenever you measure a 230-V power supply to ground you will get a 120-V reading, and 120 V is measured across both sides of the motor to ground. There is a problem with this logic.

Let us assume that the contacts 1R-1 (left side of diagram, line 13) are defective and do not close when energized. The condenser fan motor uses a single-pole relay to control its operation. The power is broken on the left side of the motor. Contacts 1R-1 are located to the left of condenser motor no. 2, on line 13 of Figure 17–7. A 120-V reading to ground will be measured every time the power supply is connected whether this condenser fan is operating or not operating. A 120-V reading to ground is read on the common winding of the motor even when the 1R-1 contacts are open, because the voltmeter reads potential difference. There is a potential difference measured from the right side of wiring diagram through the run motor winding to ground. This is why measuring voltage to ground will only confuse the troubleshooting process. What will the technician deduce when obtaining these voltage readings to ground? Will the motor be condemned without an accurate and

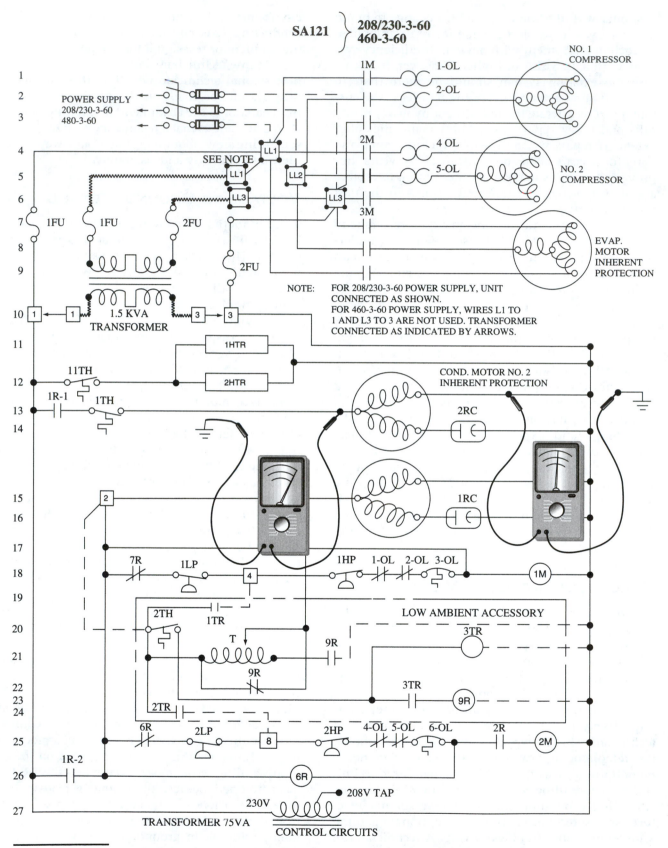

Figure 17–7 Checking voltage to ground will lead to misdiagnosis of an electrical problem.

complete checkout? Examples such as this can occur throughout the circuit and lead to misdiagnosis.

The main reason technicians would check the circuit to ground is to assure themselves that no voltage is present in the circuit before removing a component. If a voltage reading occurs to ground in a circuit, then the equipment should be treated with caution. This circuit is electrically hot. Entry-level air-conditioning technicians with previous background as electricians tend to use the voltage measurement to ground technique as a way of troubleshooting. As you can see, there is definitely an inherent fault with this procedure.

DRAW A SCHEMATIC DIAGRAM METHOD

This method was discussed in detail in the chapter on understanding electrical wiring diagrams. On some service calls a technician does not have a road map or wiring diagram to assist in troubleshooting. A technician might be successful if he or she contacts the equipment manufacturer for a faxed copy of the schematic diagram. If a technician cannot obtain a diagram directly and rapidly, the technician will need to develop their own schematic.

Sometimes the wiring on a unit has been so heavily modified that the actual schematic diagram is of limited value. The recommended action is to remove all the wiring and rewire if the wiring diagram is available. If it is not available, draft a diagram based on the components that are available in the unit. Use a diagram from a different unit if it is close enough in component type and operation sequence. Every technician should keep a file of installation and wiring instructions on the service vehicle to use in situations such as this. This collection can begin with new installation instructions as they are completed.

HOPSCOTCH METHOD

The hopscotch method of troubleshooting requires a ladder schematic diagram and voltmeter. A ladder diagram has two parallel power lines that have the loads and switches applied across the power source. The power source is usually 24, 120, 240, or 480 V. Simple ladder diagrams are shown in Figures 17–8b and 17–9. The diagram in Figure 17–8 shows the circuit without power applied, or in the deenergized state. The diagram in Figure 17–9 shows the circuit energized. Normally diagrams are shown in their normally deenergized state or without voltage applied to the circuit.

In the hopscotch method of troubleshooting, one of the voltmeter probes is attached to the right side of the ladder diagram, and the second probe is used to begin troubleshooting from the far-left side of

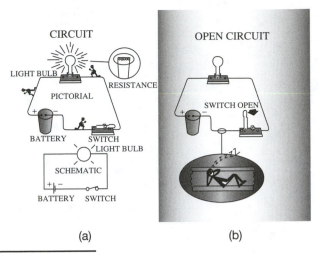

Figure 17–8 When the manual switch is closed, the fan motor runs.

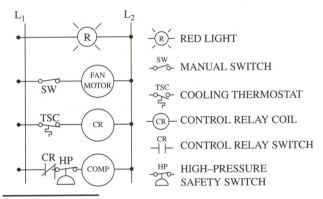

Figure 17–9 Schematic diagram showing thermostat calling for cooling.

the ladder diagram, jumping toward the load. Most diagrams are drawn with the load located on the right side of the diagram and the switches on the left side of the diagram. If the technician follows the procedure as described, successful diagnosis will be the outcome.

An example of how to conduct the hopscotch troubleshooting sequence is shown in Figure 17–10. The voltmeter probe is attached to the right side of the ladder diagram. In this diagram the probe is alligator-clipped or attached to the right side of the contactor coil C2 or L2. Next, check for voltage at L1. All the items between L1 and the C1 are switches or safety devices. CR is a control relay contact that is energized when the system calls for cooling. The CR coil is not shown in this diagram. HP and LP are the high- and low-pressure switches, which are closed if the pressure is not excessive (HP) and if the pressure does not fall below a predetermined setpoint (LP). IT is an internal motor protector that opens if the compressor motor temperature is excessive. OL1 and OL2 are overload devices that will open when the compressor motor draws too much current for an unsafe period of time.

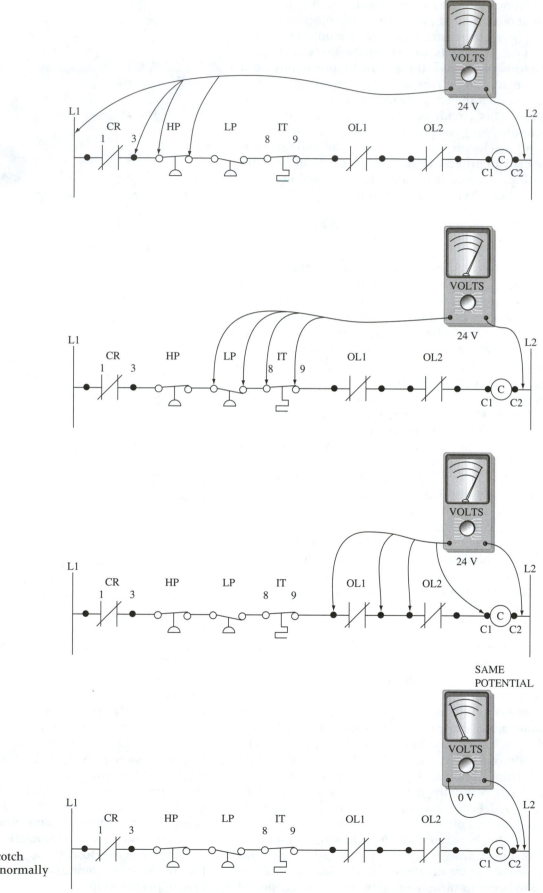

Using the hopscotch method of troubleshooting under normal operating conditions, applied voltage is read from L1 and every point up to the left side of C or to C1. Once the voltmeter probe jumps over to the right of C or terminal C2, the voltage will no longer register. The voltage is still in the circuit. Again, remember that the voltmeter measures potential difference. If both probes are on the same side of the power supply, the voltmeter will no longer register voltage because a potential difference does not exist. The probes are at the same potential in this instance.

There must be a difference in voltage for the meter to operate. If two probes are placed on a bare, 500-V wire, no voltage will be registered on the voltmeter. If one probe is placed in the earth as a ground and the other probe touches the bare wire, voltage will be measured because the bare wire has a different potential to earth ground. Earth ground is 0 V potential, and the bare wire is 500 V. The difference in potential is 500 V, and this can be measured by the voltmeter.

In Figure 17–10 the hopscotch, or jumping, sequence is shown in steps. The top diagram shows the hopscotch procedure with one probe attached to the right side of the coil C2. It is best to attach this probe with an alligator clip or wire it into the circuit. This allows a technician a free hand with which to conduct the jumping procedures. It is also safer to have one hand in a circuit rather than both hands. It is easier for an electrical circuit to be completed through two hands. Use one hand to measure voltage when troubleshooting.

Next, with the free probe, jump from L1 across each switch and safety device until reaching the left side of C or point C1. A 24-V reading should be found across each component until it crosses the contactor coil, C. As shown in the bottom diagram of Figure 17–10, the voltage is lost because the two probes are at the same potential.

THE HOPSCOTCH METHOD AS A TROUBLESHOOTING TOOL

Figure 17–11 shows a diagram with open overload 1, designated as OL1. Because OL1 is open, the contactor coil is not receiving the necessary 24 V to close the compressor contactor. How can a technician use the hopscotch method of troubleshooting to pinpoint this problem?

First, attach a probe to the right side of the ladder diagram (Figure 17–11a). Begin the hopscotch procedure at the left side of the circuit, locating L1 in the unit. The 24-V control circuit will be measured all the way to the left side of OL1. Once OL1 is jumped over, the voltage is lost (Figure 17–11b) be-

cause there is no potential difference between the two test probes. OL1 is open, preventing a complete circuit from forming. The final step in the hopscotch procedure is to verify that the correct defective component has been selected. This is confirmed by placing the probes across the open contacts, OL1. If the overload is open and there is power supplied to both sides of the OL contacts, 24 V will be measured (Figure 17–11c). At this point, disconnect the power to the circuit, remove the connections to the overload and check the OL contacts with an ohmmeter. The resistance will be infinity (Figure 17–11d). Sometimes a very high resistance will be measured. A high resistance creates a voltage drop, which robs the rest of the circuit of the necessary voltage to operate. Generally, a resistance greater than 1 Ω across a closed set of contacts is too high, so that component should be replaced. The hopscotch procedure identified the defective component and made it easy to troubleshoot. This procedure can be used with the most complex ladder diagram schematics. Without the hopscotch method of troubleshooting it would take a technician many hours of hunt-and-check, unprofessional troubleshooting.

WHERE TO BEGIN HOPSCOTCH TROUBLESHOOTING

Before a technician begins to hopscotch troubleshoot he or she must have an idea the problem is electrical or an electrically manifested problem caused by a refrigerant malfunction. Using the ACT procedure or any other logical troubleshooting pattern will reduce the number of components that could be defective and identify the general problem area. A technician should not waste time using the hopscotch procedure if the equipment is in an under- or overcharged condition.

Once the technician decides that the problem is an electrical one that can be diagnosed by the hopscotch method, then the technician must decide whether to start troubleshooting in the high-voltage section or low- (control-) voltage section. In most cases it is important just to get started, whether starting in the low-voltage or high-voltage section. If nothing is operating, check the voltage to the primary and secondary of the transformer, as shown in Figure 17–12. The advantage of doing this quick check is that a technician can determine if high and low voltage exist in the unit. If high voltage is *not* present, the problem is in the high-voltage section. Check the input voltage first. If high voltage is present at the primary and low voltage is *not* present, the problem is the transformer. If low voltage *is* present, start troubleshooting the 24-V section. This simple transformer check will guide the technician to decide in which section of the equipment to begin the troubleshooting process.

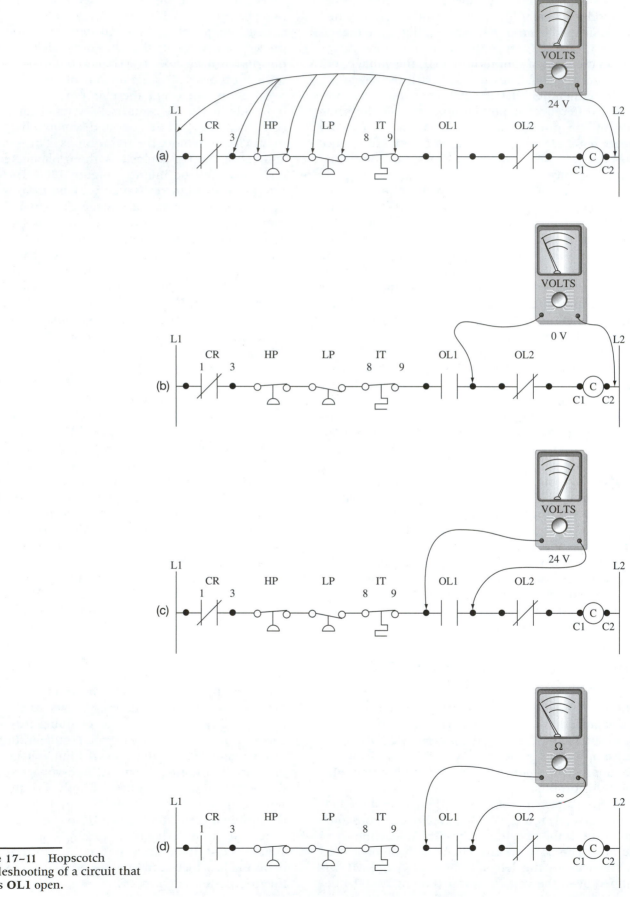

Figure 17–11 Hopscotch troubleshooting of a circuit that shows **OL1** open.

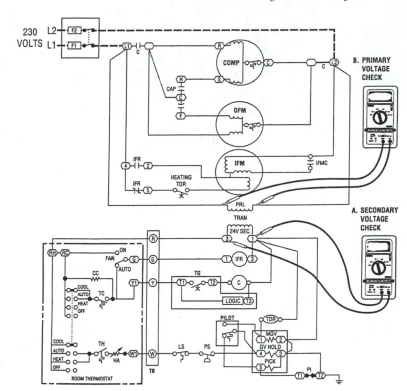

Figure 17–12 Initiate troubleshooting by checking transformer voltage.

It is imperative that a technician understand the circuit operation prior to using the hopscotch method of troubleshooting. Figure 17–13 (see pages 292–93) shows a complex air-conditioning circuit that is energized and operating properly. The voltmeters are labeled 1 through 13 for the high-voltage sections. Voltmeters are labeled A through D in the low-voltage section.

The Challenge Practice your knowledge of schematic reading and record all the voltage readings before proceeding with this section. Study Figure 17–13 and determine the correct voltage readings of all meters. The circuit is energized.

The Answer The diagram in Figure 17–13 is divided into three voltage sections. The upper portion of the diagram receives 480 V from L1, L2, and L3. This is a three-phase circuit. The middle part of the diagram operates on single-phase 230 V. The 230 V are obtained through the 480-V-to-230-V step-down transformer 1. The control voltage in the lower section comes from the 230-V-to-24-V step-down transformer 2.

Voltmeter 1 is located in the upper left-hand portion of the diagram. This voltmeter reads 480 V across L1 and L2. Voltmeters 2, 3, 5, and 6 also read 480 V under normal operating conditions.

Voltmeters 7, 8, 9, 10, 11, 12, and 13 read 230 V. Voltmeters A, B, C, and D all read 24 V. There are voltage readings on all the meters because the measurements are taken across a load. A technician may

need to change the voltage range as voltage readings are taken in the circuit. A very low and inaccurate reading will be obtained if the voltmeter is on the 600-V range when the technician tries to measure control voltage. On analog voltmeters the midscale voltage readings are most accurate.

Now that you understand Figure 17–13, use it for troubleshooting purposes.

Problem A system is not cooling. Using the ACT method to survey the situation, a technician notices that the thermostat is in the *cooling* mode, the fan is set to *auto*, and thermostat is set to 60°F. The building is warm. The indoor fan (IFM) is operating. The condensing unit is silent. The manifold gauges have good pressure readings, but they are equalized. The technician must determine where to start troubleshooting.

Solution Before initiating the tried-and-true hopscotch method of troubleshooting, the technician finds 480-V on voltmeters 1, 2 and 3. This indicates that all phase voltages are present. All three readings are important to establish that the correct three-phase power supply is present. All voltages have a ±10% swing tolerance. A 480-V supply can vary 48 V above or below the 480-V rating and be acceptable.

Next, 230 V is measured on voltmeter 10. This shows the technician that transformer 1 is good. Voltmeter A reads 24 V, indicating that transformer 2 is good and 230 V is getting to the lower part of the

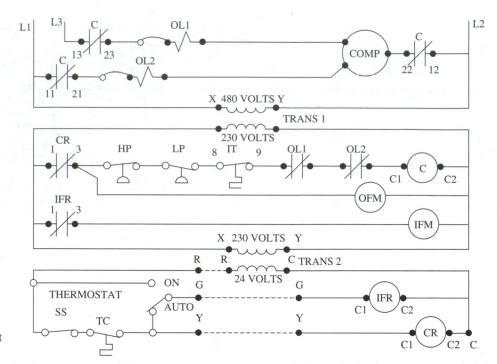

Figure 17–13 Checking circuit voltage.

circuit. From this information the technician must decide on which voltage source to begin the hopscotch troubleshooting sequence.

The technician decides to start troubleshooting the 230-V circuit because it has more switches and safety devices to become defective compared to the other two circuits. Voltmeter 10 gives the first reading, 230 V. Starting from left to right, voltmeter 12 is the next reading in the hopscotch sequence of troubleshooting. Voltmeter 12 reads 0 V. A reading of 230 V on the left side of the contacts and 0 V on the right side of the contacts indicates that the contacts are open. The circuit is broken, and therefore no voltage is read by the voltmeter as it passes to the right side of the open contacts. This must be verified. If the CR contacts are open, they will read 230 V when measured directly across the contacts. A voltmeter reading directly across contacts 1 and 3 reads 230 V (meter not shown). This is important, because 0 V should be the reading across a set of closed contacts, even with voltage passing through the circuit. A 230-V (or the applied voltage) reading across a set of contacts tells the technician that the contacts are open. The open contacts are similar to a load with infinite resistance. It is important to verify this information before going on to troubleshoot the control-voltage section. The voltage to the indoor fan-motor circuit does not need to be checked, because the fan motor is operating correctly.

The next step is to hopscotch the 24-V section of the diagram. Voltmeter A indicates 24 V. Continuing the logical sequence from left to right, voltmeters B and C also indicate 24 V. The problem has been narrowed down to a defective control relay, CR coil, or CR contacts that will not close, so the technician wants to replace the control relay. The exact defect can be determined by component troubleshooting. The coil resistance can be checked with an ohmmeter. Remove the voltage and the relay from the circuit. Resistance should be measured across the coil The resistance could be a few ohms and range as high as a few thousand ohms. Start the test by using the R×1 scale; if there is no reading, switch to the higher R×100 scale. An open coil will show no reading on a digital meter or show no deflection on an analog meter. On some relays the coil is visible for inspection. See if the coil is burned or shows signs of overheating. The coils should have some resistance, or they would be considered shorted and probably burn out the control transformer.

A technician would expect no resistance reading on good relay contacts that are closed. A very high resistance reading would be found on defective contacts. Contacts that are burned or pitted will have high resistance readings. The resistance readings are so high that they drop enough voltage to split it with the load, which is the contactor coil, C, which is in series with the contacts. The contacts need to be checked with an ohmmeter, and the coil must be energized and the contacts should be disconnected from the voltage source. The resistance of the closed contacts should be less than 1 Ω. The contacts can also be checked with voltage applied by using a voltmeter. A reading of 5 V or greater across a set of closed contacts is an indication that the contacts are burned, pitted, or not mechanically bonding together when energized. The relay needs to be hooked up to the rest of the circuit for this measurement. Most contacts are

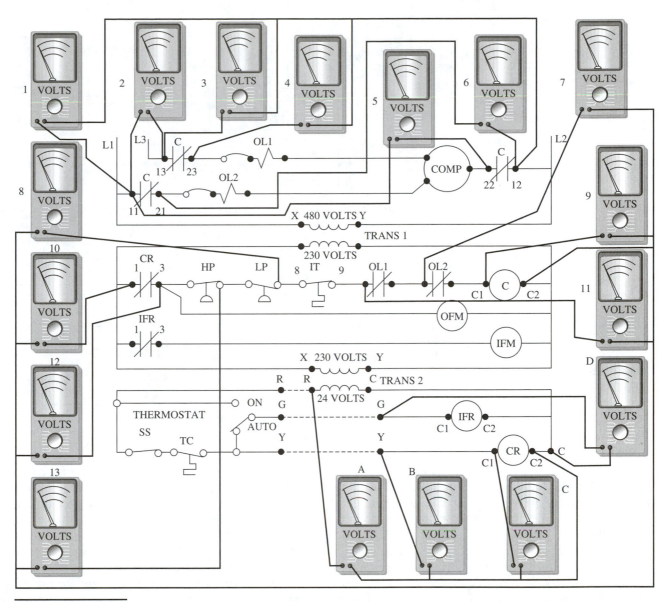

Figure 17–13, *continued*   Checking circuit voltage.

seals and cannot be inspected. If contacts are visible, replace relays and contactors that have damaged contact points.

Learning to use the hopscotch method of troubleshooting takes practice. Review the sample problem again to obtain a better understanding of how to use this valuable troubleshooting tool.

TROUBLESHOOTING A WIRING DIAGRAM

It is best to have a ladder diagram when troubleshooting an electrical problem. A ladder diagram is not always available. In many situations no diagram is available, so even a wiring diagram would be welcomed. Most wiring diagrams can be converted into ladder diagrams, but this exercise takes valuable time not usually available when a customer is hot.

This section uses the wiring diagram in Figure 17–14 to show ways to troubleshoot a wiring diagram. The most important part of troubleshooting any electrical diagram is to know its operation. This explanation discusses the expected voltage and amperage draw found in a normal, operating air-conditioning circuit. The diagram shows the outdoor condensing unit and control-voltage section. The indoor fan-relay coil (IFR) is identified in the diagram, but the indoor fan motor is deleted due to space constraints. The outdoor fan motor (FM) is a permanent split capacitor motor. The compressor motor (comp) is a capacitor-start–capacitor-run (CSCR) motor. This motor has a start relay (SR) and start capacitor in addition to the run capacitor. The compressor and fan motor share a dual run capacitor (RC). The logic

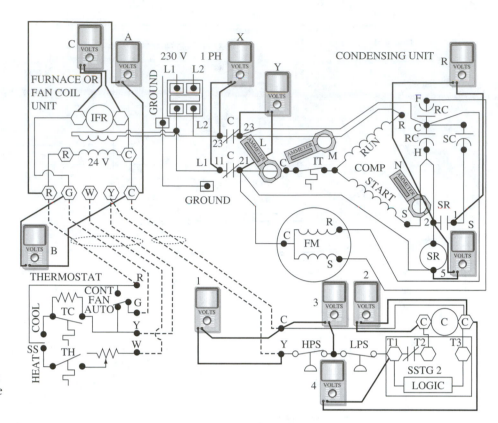

Figure 17–14 Checking circuit voltage and amperages with the circuit energized.

circuit located in the lower right-hand portion of the diagram is a time-delay device. Control voltage must be applied to the logic board terminals T1 and T3 to close the contacts between T1 and T2 and complete the circuit allowing the contactor coil to be energized. The logic circuit prevents short-cycling. The logic time delay between cycles is 3 to 6 min.

Under normal operating conditions the indoor fan, outdoor fan, and compressor are operating. The manifold gauge set should be connected to determine an undercharged or overcharged condition. Voltmeters A, B, C, 1, 2, 3, and 4 measure the control voltage (24 V). Voltmeters R, X, and Y measure the 230-V circuit. Voltmeter S measures the back EMF generated by the compressor motor. This voltage will be above 300 V. The 300 V keep the SR coil energized and SR contacts open. Ammeters L, M, and N measure amperage in the compressor-motor circuit. What reading would be expected if everything were operating normally? Make your decision before continuing.

The thermostat is in the *cooling* mode and the fan switch is set to *auto*. The temperature setting is set low enough to assure that the unit runs continuously. Voltmeters A, B, C, 1, 2, 3, and 4 all read 24 V.

Voltmeters R, X, and Y read the applied voltage, which is 230 V. Voltmeter S reads a voltage of about 300 V or higher. Voltmeter S is reading the voltage across the potential relay coil, SR. The

potential relay consists of coil, (SR), and normally closed contacts SR (⅃⅂). The purpose of the potential relay or starting relay is to remove the start capacitor from the circuit a split second after start-up. The potential relay coil operates on the supplied voltage in combination with voltage generated by the motor. The voltage across the coil is always greater than the supply voltage. The SR coil causes the normally closed SR contacts to open at a predetermined voltage range. For example, the relay may be designed to open the contacts at a range of 260 to 280 V. As long as the voltage supplied to the coil is greater than this range, the contacts will remain open, keeping the start capacitor out of the circuit. Leaving the start capacitor in the circuit too long will overheat and damage the capacitor. It is important that the potential relay and start capacitor be sized for the compressor motor. This is a true hard-start kit.

Voltmeter R reads 230 V because the SR contacts open when the SR coil is energized. The open SR contacts have 230 V applied across the open contacts.

The clamp-on ammeters are measuring current to motor windings *common, start,* and *run*. The ammeter L, measuring the amperage through the *common* winding, will have a reading closely equal to the amperages of the *run* and *start* winding combined. Check the nameplate on the unit to determine that the common amperage reading is not exceeding the rated-load amperage (RLA).

A wiring diagram can be used as a troubleshooting tool. As shown in this section, it is not easy to evaluate the correct voltage, therefore making it more difficult to use the wiring diagram as a troubleshooting tool. Occasionally, a technician will need to draw a wiring diagram because one is not available. A diagram makes the circuit less difficult to troubleshoot. The diagram can be converted into a ladder diagram for a more useful tool. Practice converting wiring diagrams to ladder diagrams when you have the time and opportunity, without pressure, and you will be able to use this knowledge quickly in the field when the need arises. The only way a technician becomes better at reading, understanding, and troubleshooting electrical diagrams is by practice and experience.

SUMMARY

- This chapter is filled with electrical troubleshooting tools that will help a technician in the field. If you do not understand any section of this chapter, it will be beneficial to review it. It is important that a technician understand and use the hopscotch method for troubleshooting ladder diagrams. Most technicians can find an electrical problem if given enough time. Using logical patterns eliminates many dead-end leads. The technician who finds and repairs a problem quickly is more valuable to a company and a customer. Many unnecessary components are replaced each year due to misdiagnosis. Be one of the professionals with zero tolerance for mistakes when it comes to troubleshooting. You will be valued as a technician who can be called upon to solve a problem that someone else cannot. You can be that resource person that every company would like to maintain on its payroll.

PROBLEMS AND QUESTIONS

1. Define troubleshooting.
2. What is ACT? How is ACT used in troubleshooting? Be specific.
3. What is the value of using ACT or some similar method of initiating troubleshooting?
4. What are four important steps to take when starting electrical troubleshooting?
5. List and describe five ways manufacturers provide assistance for troubleshooting.
6. Why is it important that a technician take control of the electrical operation of the equipment when arriving on the job?
7. What might happen if a technician uses the spark method to check a 24-V control circuit?
8. Why is it important to install gauges when beginning to troubleshoot a system?
9. Why is it important to hook up an Amprobe to the compressor when beginning to troubleshoot a system?
10. If a system is completely dead when a technician arrives on the site, what immediate steps should the technician take to check out the problem once the Amprobe and manifold gauge set have been installed?
11. What safety precautions should a technician follow when testing and resetting a breaker?
12. What is the parts-changer method of troubleshooting? What are the advantage and disadvantage of this method?
13. What is the equipment-changer method of troubleshooting? What are the advantage and disadvantage of this method?
14. What is the call someone else method of troubleshooting? What are the advantage and disadvantage of this method?
15. What is the voltage-to-ground method of troubleshooting? What are the advantage and disadvantage of this method?
16. What is the draw a schematic diagram method of troubleshooting? What are the advantage and disadvantage of this method?
17. What is the hopscotch method of troubleshooting? What are the advantage and disadvantage of this method?
18. Describe in detail how to use the hopscotch troubleshooting method on Figure 17–15, assuming that the right contactor contacts (C) are open and everything else is operating normally. Start the hopscotch procedure in the high-voltage section of the ladder diagram.
19. Describe in detail how to use the hopscotch troubleshooting method on Figure 17–15, assuming that the control relay (CR) coil is open and everything else is operating normally. Start the hopscotch procedure in the low-voltage section of the ladder diagram.
20. What voltage reading will a technician obtain if one voltmeter probe is clipped on L2 and the other probe is placed on the right side of the primary of the transformer in Figure 17–15?
21. In what way is hopscotch troubleshooting safer that two-hand troubleshooting?

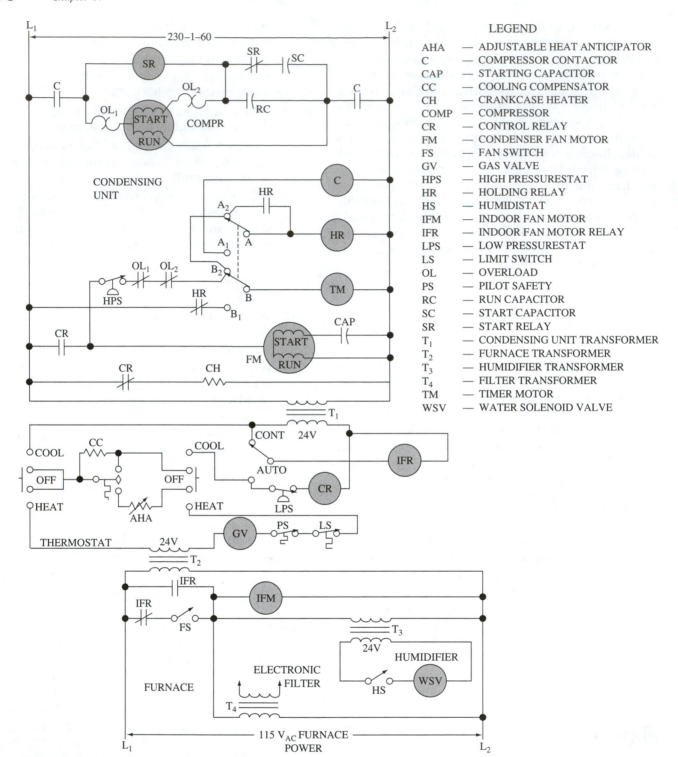

Figure 17–15 Schematic wiring diagram showing the various electrical circuits.

22. Why will checking the voltage to the transformer determine where a technician should begin the hopscotch troubleshooting?

23. A technician has isolated a defective component by use of the voltage hopscotch troubleshooting method. What is the next step in the troubleshooting process?

24. How much resistance is expected across a closed set of contacts? How much voltage drop will occur across a good set of closed contacts that has 24 V passing through them?

25. In Figure 17–14, what will voltmeters A, B, C, 1, 2, 3, and 4 read when the thermostat is *not* calling for cooling?

26. How many volts are read by voltmeter R in Figure 17–14?

27. When measuring the current on a single-phase compressor, what amperage is expected on the common winding when the start winding measures 4 A and the run winding measures 21 A?

28. In Figure 17–12, what will the transformer voltage at points 1 and 3 read when the thermostat is in the off position? When the thermostat is in the on position?

29. In Figure 17–13, what voltage will be read across the CR coil if the coil is burned open?

30. In Figure 17–14, what approximate voltage will be read across the SR coil when the compressor is operating?

CHAPTER EIGHTEEN

PREVENTIVE MAINTENANCE

AFTER STUDYING THIS CHAPTER, THE STUDENT WILL BE ABLE TO:

•Discuss the differences between preventive maintenance (PM) and emergency service.

•Describe the importance of preventive maintenance.

•Outline the important aspects of a maintenance contract.

•Describe the three types of maintenance contracts.

•Discuss how a technician can offer benefits to the customer.

•List the various instruments needed to conduct a preventive maintenance job.

•List the various supplies needed to conduct a preventive maintenance job

•Describe the preventive maintenance routine.

•Select the 10 most important parts of an ultimate tune-up.

•State the benefits of a preventive maintenance report.

•List items that should be checked on a commercial preventive maintenance that are not checked on a residential preventive maintenance.

PREVENTIVE MAINTENANCE VERSUS EMERGENCY SERVICE

A customer should decide if they want preventive maintenance or emergency repairs. What is the difference? What are the advantages and disadvantages of each?

PREVENTIVE MAINTENANCE

Preventive maintenance (PM) means that equipment is scheduled to be cleaned, checked out, and serviced, thus reducing operating time and operating cost. If necessary, repairs are planned or completed at the time of the PM. Maintained equipment lasts longer than nonmaintained equipment because it operates in the designed ranges and does not run as long. Doing preventive maintenance does not guarantee that the system will operate without a

breakdown. A good PM inspection will head off many unwanted problems and allow the customer the opportunity to schedule equipment downtime when it is convenient for the customer's operation.

Preventive maintenance jobs should be scheduled in slow periods to fill in the gaps during fall and spring lulls. Preventive maintenance appointments should be made with the idea that they may be rescheduled with a 24-h notice should a heavy workload occur during the time period in which the PM is scheduled.

EMERGENCY SERVICE

Emergency service is required when equipment needs an unscheduled repair. The repair may be minor or major. The equipment may be out of commission for a few hours or a few days. Emergency repair does not mean that the equipment will be thoroughly cleaned and checked out, as is the case with a preventive maintenance checkup. The repair will be made and the equipment placed on line as soon as possible. The equipment is not operational, so no further delay will be tolerated. In this case, a PM should be scheduled in the future.

SCHEDULING PREVENTIVE MAINTENANCE

A good preventive maintenance program must be designed to fit the individual equipment. Each type of system has certain critical functions that must be maintained in good working order. For most systems, a program should be arranged for at least two inspections per year—one for air conditioning and one for heating. Some of the PM activities overlap for air conditioning and heating. For example, a technician checks the filter, blower, and blower motor whether conducting a PM on the cooling or heating mode.

MAINTENANCE CONTRACTS

Many contractors offer maintenance contracts for providing periodic customer service to keep the equipment in good working order and to prevent major breakdowns. These contracts usually contain provisions for some of the following items:

1. Period of time covered by the contract
2. Equipment covered by the agreement
3. Work to be performed by the contractor, such as
   a. Preparing the system for seasonal operation
   b. Performing routine maintenance
4. Materials to be provided by the contractor included in the contract price and those materials for which the customer will be charged extra.

5. Emergency service
6. Labor to be provided by the contractor included in the contract price and labor for which the customer will be charged extra.
7. Items that the customer agrees to do, such as
   a. Operating the equipment in accordance with instructions provided
   b. Notifying the contractor of any unusual operating conditions
   c. Permitting only the contractor's personnel to work on the equipment
8. Exclusions of items not covered by the agreement, such as
   a. Plumbing services beyond the equipment proper
   b. Electrical service beyond the disconnect switch
   c. Repairs due to flood, freezing, fires, vandalism, lightening strikes, and other causes beyond the contractor's control
   d. Expenses caused by obsolescence or insurance requirements
9. Warranties on parts and labor
10. Prices for work covered by the agreement

TYPES OF CONTRACTS

Basically there are three types of maintenance contracts. Each carries a different price range.

1. *Inspection and lubrication.* This is a minimum-cost contract. It includes specified materials, such as filters. Materials and labor needed for repairs are extra. Some contracts offer a percent reduction on unscheduled repairs.
2. *Full labor.* This is similar to the first contract, except that labor is included and materials are extra. Sometimes labor above a certain limit is extra. This is not a common agreement.

PRACTICING PROFESSIONAL SERVICES

A service technician should be the number one salesperson for after-market air-conditioning accessories. A technician should evaluate a system while conducting service or preventive maintenance and make one or two upgrade recommendations to the owner. Suggesting too many options will confuse the owner, and it will seem as if the technician is just trying to sell something. Make only suggestions that bring real value to the customer. The customer may need a better filtration system, compressor-time delay, ducts cleaned or replaced, or additional ceiling insulation. Record the suggestions on the preventive maintenance report or service receipt for future reference. The customer may not accept the idea at the time; continue to suggest a valuable upgrade at every visit. A future suggestion may be made at the right time for the customer and the technician. Many of these upgrades can be scheduled during a slow work period.

3. *Full maintenance*. This contract includes inspection and labor and materials for repair. To restrict the liability to the contractor, limits are often placed on the amount or types of repair. For example, compressors and old equipment due for replacement are excluded.

Any combination of these contracts can be created to meet the needs of the customer.

◢ MEASUREMENT AND ANALYSIS

An important part of performing preventive maintenance is to make sufficient measurements to determine the operating condition of the equipment. An accurate comparison needs to be made between the proper performance and the actual performance after a definite period of operation.

TEST INSTRUMENTS

To make these measurements a technician must be equipped to use the necessary test instruments. Some of the essential instruments are described here.

VOM (VOLT, OHM, MILLIAMP) METER
These instruments have typical ranges as follows:

DC volts:	0.25 to 1000
AC volts:	10 to 1000
Ohms:	R × 1, R × 10, R × 100, R × 10,000
DC milliamps:	0.05 to 250
AC amps (with transducer):	0 to 100

These instruments can be supplied in either digital or analog type. The analog type is useful for observing changes and variations in readings. The digital type is accurate and rugged: most digital models will withstand a 5-ft drop on a concrete floor.

CLAMP-ON AC AMMETERS (AMPROBE)
These meters permit the technician to read line currents to fan motors, compressors, and electric heaters without contacting or disconnecting the wiring. Typical scales available include:

Amps:	0.5 to 1000
Volts:	0.1 to 1000
Ohms:	0.1 to 800

Terminal leads are used in measuring volts and ohms. A digital clamp-on meter with "peak hold" is the best recommendation.

INSULATION RESISTANCE TESTER (MEGGER)
These instruments are used to test motor windings using a potential of 500 to 1000 V. They are much more reliable for checking hermetic compressor windings for grounds than an ohmmeter powered by only a few volts of DC. They may be battery-powered, AC-powered, or operated from an attached hand-crank generator. They will read resistances of 100 MΩ and higher. Normally, a compressor-winding resistance to ground of under 1 MΩ is a suspect for a defective compressor. The megger can be analog or digital.

LEAK DETECTORS
The use of a quality electronic leak detector is recommended. These detectors are sensitive and reliable and can be secured for servicing all halocarbon refrigerants (HFC, HCFC, and CFCs). Purchasing inexpensive leak detectors is a waste of money. They give many false signals and fail to detect refrigerant leaks.

GAUGE MANIFOLD SET
This is an essential piece of instrument, including both high-pressure and compound gauges. Analog and digital gauges are available. The four-valve models have increased versatility compared to the two-valve models.

MANOMETERS
Manometers are useful in diagnosing problems in the airflow on air-conditioning units. Plastic tube manometers with magnetic clips are available for reading static pressures up to 10 in. WC and below 0.1 in. WC. When used with a Pitot tube they can reliably read air velocities in duct work. Used alone they can read static pressure differentials across filters, coils, and fans. They can also read air-discharge and return static pressures. These tests can indicate problems such as plugged coils, dirty filters, fans running backwards, and closed fire dampers.

For a dry-type differential-pressure gauge, the diaphragm-type magnetic-helix coupled (Magnehelic) models are recommended. These gauges have a simple frictionless movement and resist shock, vibration, and over pressure.

FLOW HOOD
A flow hood is used to measure airflow in cfm. The total airflow can be measured at the return air or by measuring and combining the supply airflow of all the supply outlets. The airflow that goes into a system as return air is supplied to the space as supply air; 400 cfm per ton is the expected outcome if a system is designed correctly and there are no restrictions.

TEMPERATURE TESTER
Two temperature testers are needed. A temperature tester that measures wet-bulb and dry-bulb is important when measuring outdoor and indoor

temperatures. A digital meter with multiple probes is recommended. Several dry-bulb probes are needed at the outdoor unit. A wet-bulb and dry-bulb sling psychrometer is needed for indoor measurements.

TOOLS, SUPPLIES, AND EQUIPMENT

In this section, we have reached the point where the components of the air-conditioning system have been identified; the system has been operating; vital measurements have been made; and a list of preventive maintenance tasks have been set up that need to be performed.

Some of the special tools, supplies, materials and servicing equipment that the technician will find useful, beyond the usual assortment of combination open/box wrenches, screwdrivers, and pliers, are the following:

1. Spray units for cleaning coils, either hand-operated or pressure-pump types. A sprayer unit handles cleaning detergents, which dissolve accumulations on the surface and between the fins of evaporator coils, heating coils, and condenser coils. Specialized cleaning agents should be used for the condenser and evaporator coils. In some cases, a mild soap solution is all that is needed for the job. A separate sprayer may be required for each type of cleaner. Carry a sprayer-repair kit, because the sprayer will need maintenance. Repair kits may not be available on older units. Purchase the repair kit when purchasing a new sprayer unit.

2. Refrigerant recovery and recycling units (Figure 18–1), which are essential in removing refrigerant from the system to permit the repair and replacement of components. Recovery tanks are required for servicing of the various types of refrigerants. An empty recovery tank is always valuable to have on the service vehicle.

3. Refrigerant-measuring devices, such as weight-charging cylinders or a digital scale, will be needed to accurately charge package units and monitor the refrigerant used for replacement.

4. Nitrogen or carbon dioxide pressure regulator, for use with a tank of nitrogen or carbon dioxide to purge tubing being brazed; for use in pressurizing a system for leak testing; to break the vacuum in a triple-evacuation procedure; or to blow out a plugged condensate line. Some technicians adapt a tire fill-up valve to allow them to pressurize tires. Nitrogen is a bigger molecule than oxygen and leaks more slowly through the tire rubber. This can be used to pressurize a tire in an emergency.

5. Torque wrench for working on open and semihermetic compressors. These wrenches ensure tightening bolts and nuts to the manufacturer's requirements.

6. One-quarter-inch socket-wrench set and driver bits, for adjusting valves.

7. Nut drivers, to save time in opening enclosures.

8. Pinch-off tools, to close compressor pinch-off tubes used in servicing hermetic systems.

PREVENTIVE MAINTENANCE ROUTINES

REFRIGERATION/AIR-CONDITIONING MAINTENANCE

Every maintenance contractor should perform a series of routine maintenance operations at the time of each inspection. The following list of maintenance items used by one contractor is an example:

1. Clean and flush condensate drain pans and drain lines. Clean, lubricate, and test condensate pumps.

Figure 18–1 Piping arrangement for push-pull liquid recovery (left) and for vapor recovery (right).

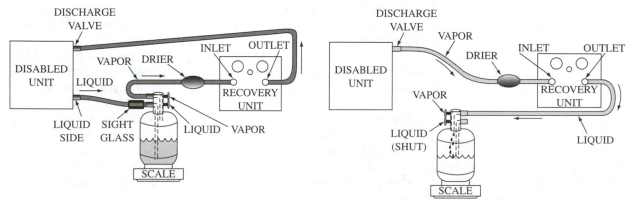

2. Check fan belts. Replace worn or cracked belts with new ones. Adjust belt tension.

3. Lubricate bearings. For oil-type bearings, use non-detergent oil (not automotive engine oil) of weight (viscosity) specified by bearing or motor manufacturers. Technicians find that plastic containers with extended spouts are useful for reaching out-of-the-way oil cups.

   For grease-type bearings, use grade and type of grease specified. Do not over-grease. Forcing excessive grease into bearings can damage the grease seals and lead to eventual failure. Where a drain or vent plug is provided, follow instructions and allow excess lubricant to escape.

4. Clean up the equipment. Wipe up any oil or grease spills. Clean equipment makes it easy to spot oil leaks, which sometimes indicate refrigerant leaks and bearing problems. Be sure to tighten packing glands on refrigerant valves and replace valve caps.

5. Check crankcase oil heaters. Follow the instructions for turning on the heater for 12 to 24 hrs before attempting to start the compressor.

6. Check and tighten all electrical connections. Loose connections will overheat, damaging wire insulation and could cause motor failures.

7. Examine starter contacts for burning or excessive pitting. Check for free operation of the armature without binding. Rebuild or replace defective units.

8. Examine the air filters. Replace dirty disposable filters or replace with washable filters. Ordinary disposable air filters only have an efficiency rating of 5 to 10%. Often it is possible to install a better filter with 30% cleaning efficiency (same size), at only a modest increase in cost. Most owners appreciate an improvement in indoor air quality.

9. Examine the water pumps. Mechanical seals do not require service. Packed stuffing boxes usually need repacking periodically. Do not tighten the packing gland so tight that no dripping occurs. This can score the shaft or burn the packing rings. Some dripping is desired and normal.

10. Check the refrigeration circuit. Attach a gauge manifold and record operating pressures. Convert refrigerant pressures to their corresponding refrigerant temperatures. Check the refrigerant sight glass (if provided) to see that the charge is adequate. Check compressor oil sight glass (if provided) to verify that the oil charge is adequate.

11. Check the cooling towers (Figure 18–2) or evaporative condensers. Clean the sumps before filling. Scrape and paint rusty spots. Make sure all spray nozzles or openings are clear and functioning properly.

12. Check the operating temperatures of the system. Many experienced technicians can make a preliminary check by feeling various parts of the system to determine any abnormal temperatures. It is best to use a multiprobe temperature tester and record these readings on a permanent record.

13. Check the condition of the evaporator and air-cooled condenser coils. They must be clean for the system to operate properly. Clean the condenser coil even if it appears clean. Dirt may be buried in the coil. Flush the coil with water in the opposite direction of the airflow through the coil.

ON THE JOB

The following is a partial list of items, forms, and records needed for a professional service technician:

Customer's Service Information
- Name
- Address
- Phone number
- Service complaint
- Dispatch time
- Arrival time
- Completed time

Forms
- Refrigerant use log
- Accidental refrigerant venting report
- Refrigerant removal form
- Material safety data sheets
- Truck parts inventory
- Truck tool inventory
- Maintenance agreement forms
- Time sheets
- Service tickets
- Sales lead forms
- Reference manuals

(a)

(b)

Figure 18–2 Two types of cooling powers: (a) mechanical draft and (b) atmospheric draft. *(Courtesy of Marley Towers)*

14. Always shut off the power before servicing equipment. Technicians tend to become complacent and forget the hazard of electric shock. Play it safe; always shut off the power before servicing. Cover motors, terminals, and components before applying detergents or washing with water.

15. Cleanup. A dirty, littered workplace can be hazardous and certainly makes a poor impression on the owner or manager. Old belts, dirty filters, broken controls, empty refrigerant drums and oil containers should be removed from the premises or placed in a trash receptacle. A well-cleaned-up space lets people know a technician is a professional.

16. Perhaps one of the most important aspects of a PM job is to be certain that the equipment is operating prior to beginning the job. A repair may be necessary before starting the checkout. Hook up the manifold gauge set and the Amprobe to the common of the compressor. Start equipment and allow it to run for 10 to 15 min before starting work. This is an opportunity to unload equipment and supplies needed for the job. If you begin the PM without checking the general equipment operation, you will be blamed if something needs repair after the work begins. Let the customer know if there is a problem prior to beginning the PM activities.

ULTIMATE SPRING TUNE-UP

The following items are listed as the ultimate spring tune-up on a residential air-conditioning system. In the ideal world, all technicians would complete the following items. The list is not in order of priority:

1. Clean evaporator.
2. Clean condenser coil.
3. Check controls.
4. Change filter.

ON THE JOB

Complete all forms, service tickets, and warranty claims at the service-call or job site. It is very important to document refrigerant use. The EPA requires refrigerant-use records. This is the best time to recall what was done, parts that were used, and time spent on the job. Remember that this is part of the time that should be charged to the job. You should not do your paperwork at home or in the shop, or give it to the dispatcher or service manager. Incomplete forms or forms that are lost or filled out incorrectly or in a messy manner cost you and your service organization money. Take pride in your work.

5. Lubricate all moving parts.
6. Clean drains and install pan tablets.
7. Conduct a *complete* charging checklist.
8. Level the evaporator coil.
9. Clean the blower wheel.
10. Replace missing screws.
11. Check the fan blade.
12. Check the capacitor.
13. Check the time delay.
14. Check high- and low-pressure switches.
15. Check attic ventilation.
16. Inspect return-air sealing.
17. Tighten electrical connections.
18. Check the amperage draw of motors
19. Wax the unit.
20. Determine if a float switch is needed.
21. Calibrate the thermostat.
22. Repair pipe insulation.
23. Place ant killer around the unit.
24. Level the condenser.
25. Check for vibrations.
26. Check the crankcase heater.
27. Check compression ratios.
28. Check attic insulation.
29. Review the PM with customer.
30. Check belt tension (if used).

PREVENTIVE MAINTENANCE REPORT

It is important to document the work conducted on a preventive maintenance activity. Figure 18–3 is a sample that can be used in many standard residential or commercial air-conditioning PM work orders. A preventive maintenance job as outlined in Figure 18–3 will take the experienced technician 2 h to complete. Some air-conditioning technicians do a PM in an hour or less. This type of hurried check is nothing more than a survey of the equipment. The dirtiest items will be cleaned and the rest left to become dirtier before cleaning is attempted. Unloading and loading the necessary items to conduct a professional and thorough inspection takes 20 to 30 min by itself. An inexpensive preventive maintenance job is merely a ploy to get a technician in the door in hopes of finding other problems. In PM jobs, customers get what they pay for.

The line items are numbered on Figure 18–3 to aid in explaining the report. Each company should make up its own report that represents the needs of its customers. The customer should be left with a copy of the report and the service company should maintain a copy for future use. Whenever a service call or PM is scheduled for that address, the previous PM reports should be taken to the job site and used as a resource for comparison.

At the completion of this explanation, optional PM items are discussed. Review this PM report and the optional information prior to customizing your report.

EXPLANATION OF A PREVENTIVE MAINTENANCE REPORT

Line 1. The complete address is important, along with the owner's name and phone number. Different owners have different operating habits, which may affect the way a system performs. The time of day may give clues, should problems arise with the equipment. The date tells a technician how soon the system becomes dirty.

Line 2. The name of the technician allows supervisors the opportunity to spot-check the completed work. Sometimes a technician fails to make sure the equipment is placed into full operation before leaving a job. Such information builds a track record that can merit promotion or lead to demotion to work in the warehouse. When a technician is familiar with a location, sending that technician back may speed up diagnosis of a problem. The service company name will remind the customer or future customer of a well-performed service call. For better visibility, the company name, address and telephone number should be at the top of the PM report. A permanent sticker with the company's name should also be left on the condenser and evaporator coil. Some customers will allow an attractive sticker on the thermostat. Ask permission before applying this sticker. Many states require contractors to place their license number on any document associated with servicework.

Line 3: The outdoor temperature at the condensing unit helps the technician determine if the condensing temperature and pressure are within range. Take the air temperature entering the condensing unit. Do not allow the temperature probe to touch the coil surface. Touching the coil will give a higher-than-normal temperature reading.

The return-air dry-bulb and wet-bulb temperatures are used on some superheat charging charts to obtain the temperature difference across the evaporator. Measure these indoor temperatures at the filter grille or air entering the evaporator coil.

Line 4. It is important to record the condition of the air filter. Sometimes customers forget to change their air filters as often as necessary. Noting the filter size on the report is helpful to the customer.

PREVENTIVE MAINTENANCE REPORT

❶ **Address**_____ **Time** _____ **Date** _____

❷ **Service Technician** _____ **Company** _____

❸ **Outdoor Temperature** _____ **Indoor**_____

❹ **Filter Condition:** _____ **OK** _____ **Dirty**

Corrective Action _____

❺ **Evaporator Condition:**_____ **OK**_____ **Dirty**_____ **Clogged**

Corrective Action_____

❻ **Condensate Drain Line:** _____ **OK** _____ **Drains Slowly** _____ **Blocked**

Corrective Action _____

❼ **Condenser Condition:** _____ **OK** _____ **Dirty** _____ **Clogged**

Corrective Action _____

❽ **Suction Pressure:** _____ **Discharge Pressure** _____

❾ **Discharge-Line Temperature:** _____ **°F** **Discharge Superheat** _____ **°F**

❿ **Suction Superheat:** _____ **°F** **Subcooling** _____ **°F**

⓫ **Crankcase Heater Check:** _____ **OK** _____ **N/A**

⓬ **Indoor-Fan-Motor Check:** _____ **Oiled** _____ **Sealed Bearing**_____ **Noisy Bearings**

⓭ **Outdoor-Fan-Motor Check:** _____ **Oiled** _____ **Sealed Bearing**_____ **Noisy Bearings**

⓮ **Compressor-Amperage Check:**_____ L_1_____ L_2

⓯ **Contactor-Condition Check:** _____ **OK** _____ **Sticking**_____ **Pitted**

⓰ **Contactor-Voltage Check:** _____ **Volts**

⓱ **Voltage-Under-Load Check:** _____ **Volts**

⓲ **Thermosat Calibration** _____ **OK** _____ **Recalibrated**

⓳ **Megohmmeter Check:**_____ **Ohms**

⓴ **Wire Connections:** _____ **OK** _____ **Tightened Terminals**

㉑ **Ductwork Connections:** _____ **OK** _____ **Repaired**

㉒ **Additional Comments:** _____

Figure 18–3   PM report.

The next time a technician returns, the service truck should be stocked with several of these filters. The customer may want to purchase filters from the technician instead of waiting to purchase them in the future. If the filter is dirty and a replacement is not readily available, then the technician can check the system without a filter. This should be noted and highlighted on the PM report. A dirty filter is more efficient for filtering the air, but it significantly blocks airflow through the evaporator.

Electronic air-filtration screens and cells must be cleaned periodically with detergent and hot water (Figure 18–4). Some filter designs include an automatic washing arrangement with the filter screens and cells in place. Other designs require the removal of the screens and cells for cleaning. The frequency of cleaning depends upon the application. After initial start-up, the filters should be inspected frequently. Once the rate of buildup has been observed, the appropriate frequency for regular cleaning is easily determined.

The filter-washing procedure for units without an automatic built-in washer is as follows:

- Turn the electronic filter, blower, and furnace power off.
- Remove lint screens and ionizing collecting cells.
- Clean the filter using hot soapy water, and rinse it thoroughly.
- Replace lint screens and ionizing collecting cells after they have dried thoroughly.

A properly operating unit turns the water black when the cell is washed. If cheesecloth placed over the air-outlet grille becomes discolored, the electronic air filter is not working properly.

Line 5. Checking the condition of the evaporator is important. If the customer changes the filter on a regular basis, then the evaporator will probably be clean. The condition of the evaporator needs to be checked. The most common access to the evaporator is by removing a piece of duct transition between the furnace and the evaporator. Sometimes it is easier to cut an access opening in this ductwork. The opening can be resealed for future use. The best way to reseal this opening is to take a 12-in. round or square piece of insulated sheet metal and screw it over the new access opening. Sheet metal shops that fabricate duct systems have 10- and 12-in. insulated sheet metal. Many times these pieces are scrap and can be obtained at little or no cost. Screw the sheet metal cover over the new access opening. At this time, it is also important to record the type of metering device found in or near the evaporator. Knowing the type of metering device is necessary when determining the correct system superheat. Record the type of metering device on the outside of the evaporator and inside the condenser near the charging instructions.

If the evaporator needs to be cleaned, use the following instructions. Use a brush to sweep off the face or inlet (Figure 18–5) of the evaporator. Take care not to allow the gummy dirt to drop into the evaporator condensate drain pan. This debris may plug up the condensate drain line. Spray hot water

Figure 18–4   Cleaning an electronic filter (left) and wall filter located behind a grille (right).

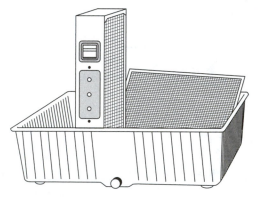
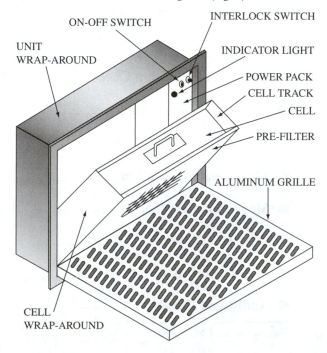

INSPECT FROM THE BOTTOM

Figure 18–5 Inspect "A" coil from the bottom side or air inlet side. *(Courtesy Bard Manufacturing)*

or coil cleaner across the face of the coil, taking care not to force any dirt into the coil fins. Clean out the condensate drain pan with a wet-dry vacuum cleaner. Clear out all the debris remaining in the drain pan.

In summary, the evaporator-coil drain pan, vent lines, condensate pumps, and humidifiers are a source of potential indoor-air-quality problems if they are not checked when doing PM. Scale, rust, and algae growth can prohibit proper drainage and water flow if parts are not cleaned regularly. Excess algae or mold growth can also contribute to poor indoor-air quality and be detrimental to your customer's health. Whenever working on this type of equipment, a technician should use protective gloves and respirator-protection devices to prevent potential self-contamination.

Line 6. Now is the time to check the flow of the condensate pan. Slowly pour a gallon of water into the drain pan. The water should drain from the pan in less than a minute. If the drain is plugged or sluggish, it needs to be blown out with nitrogen or carbon dioxide. Use a fish tape with cloth fastened to the tip to push out the slime and algae plug. Pour bleach in the drain pan and drain line to kill algae and slime growth. This is only a temporary fix. Place drain-pan tablets in the condensate pan to impede future growth of algae and slime. Care must be taken to capture bleach if it drains onto the lawn or flowerbed. Using chemicals indoors and around the evaporator may require the use of a portable fan to

prevent high concentrations of chemical fumes that might bother the technician or the building's occupants. Check with the customer before using bleach in the drain pan. Some customers may object to the temporary odor.

Line 7. The condenser should be cleaned each time a preventive maintenance is scheduled. Air passing through the condenser coil is unfiltered. Airborne dirt passes by the surface of the coil and imbeds itself between the fins of the coil. The dirt, fuzz, pet hair, and grass clippings reduce airflow through the condenser coils, thus reducing heat transfer from the refrigerant to the air.

The coil should be cleaned with a detergent or a condenser-coil cleaner. The coil should be flushed with water in the reverse direction that the air flows through the coil (Figure 18–6). It is important to use water to flush all cleaning agents from the coil surface and from the base that the condenser coil sits in. Some cleaning agents cause deterioration of the aluminum fins or copper tubing. Care must be taken to protect exposed electrical parts. Water-logged components may short out when electricity is applied.

Some condenser coils are double split coils. Two coils are placed one behind the other. This type is also called the back-to-back coil. There can be a problem cleaning this type of coil. The dirt gets trapped between the two coils and does not flush out using coil cleaners and water. In this instance, the refrigerant needs to be recovered from the coil and the coil cut and separated for cleaning. The double condenser coil that appears clean but has low subcooling and higher-than-normal condensing temperatures/pressures may have dirt trapped between the coils. Dirt may not be a good description of what is found when separating the coils. What is found is more like a thin layer of carpet or felt that cannot be flushed out with a normal cleaning

Figure 18–6 Water is flushed through the coil in the reverse direction of airflow.

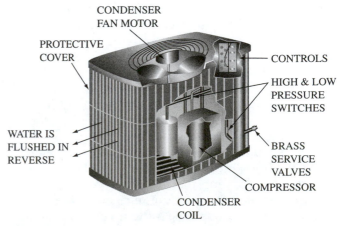

CONDENSER FAN MOTOR

PROTECTIVE COVER

CONTROLS

HIGH & LOW PRESSURE SWITCHES

WATER IS FLUSHED IN REVERSE

BRASS SERVICE VALVES

COMPRESSOR

CONDENSER COIL

process. In most instances, condensers with this construction do not need to be separated annually. Maybe once or twice in the life of the condenser is sufficient.

It may be difficult to determine that a condenser has a double-split coil. Many coils have a two- or three-row design but are not split. It is easier to clean this type of coil.

Another type of fin used is the spine fin, or Christmas garland, style. The spine fin must be flushed at about a 60° angle to prevent the water pressure from pushing over the spines and totally blocking the airflow. This style of fin is more difficult to clean and in some cases is impossible to clean once the coil becomes deeply impacted with dirt, fuzz, hair, and grass clippings.

Dirty coils show a reduction in heat transfer of 25% or greater. Even cleaned coils show loss in heat-transfer capabilities. New coils have the highest heat-transfer rates. It is impossible to get an old coil as clean as a new coil. Electrolysis causes the contact between the copper tubing and aluminum to reduce heat transfer. Copper fins are available from custom coil manufacturers. There is less electrolysis between copper fins and a copper coil. These copper fins can withstand harsh (such as offshore and chemical) environments.

Line 8. The suction and discharge pressures are needed in order to check the charge on the system. The corresponding suction and condensing temperatures should be recorded in this space. These temperatures are needed to check superheat and subcooling. With this information a technician may be able to determine if the system has a leak the next time a PM is conducted on the system. This is baseline-charging information.

Most window units and many package units do not have access valves for checking pressures. A temporary line-tap valve may be installed on these units to gain access to pressure or temperatures. The temporary valves need to be removed or replaced with permanent valves. This is not part of a normal PM program. Since this adds additional time and cost to the job, it must be discussed with the customer.

A way to check the general operation of the refrigeration system in a window or package unit is to block the airflow through the evaporator while the unit is operating. A piece of plastic sucked up against the evaporator will do. The suction line will begin to freeze or frost after the evaporator is blocked. A frost line (Figure 18–7) up to the compressor should develop in less than 10 min. Sometimes the frost line travels onto the shell of the compressor. A frost line that does not come near the compressor indicates a lack of refrigerant due to several possible reasons. The system could be undercharged or the compressor could be experiencing a

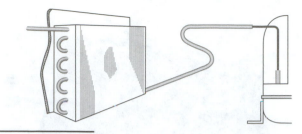

Figure 18–7 Determining the refrigerant change from the position of the frostline. The airflow through the evaporator is blocked to create this frostback condition.

lack of capacity. A liquid-line restriction can create the same situation. A frost line that covers a large portion of the compressor shell may indicate an overcharged condition.

Do not operate the compressor too long with the evaporator blocked. There is a chance that liquid refrigerant droplets will return to the crankcase of the compressor.

Ask the customer if the unit is cooling properly. If the unit is cooling adequately, a frost line at or near the compressor indicates that the system is probably functioning correctly (good compressor, refrigerant level, etc.), and refrigeration access valves are not necessary. Many customers pressure the technician to add refrigerant to a system when it is not needed. Explain to the customer that refrigerant is not always necessary unless it has leaked out of the system. Do not be pressured into adding access valves or refrigerant unless necessary.

Line 9. The discharge-line temperature is used to determine discharge superheat. Due to the cooling of this hot line, the discharge-line temperature is measured as close to the compressor as possible (Figure 18–8). Discharge superheat is the difference between the discharge-line temperature and the condensing temperature, as found on the gauges or a pressure-temperature chart. The normal discharge superheat is within a range of 50° to 125° F. The discharge line temperature should not exceed 250° F.

Line 10. This line records the suction superheat and condenser subcooling (Figure 18–8). The suction superheat depends on the type of metering device found in the system. If the system uses a capillary-tube or fixed-orifice device, it is important to use the superheat charging chart designed for these metering devices. If the system uses a thermostatic valve (TXV), it is important to charge to the manufacturer's specifications or a range of 15° to 20°F. The recommended superheat is discussed in the chapter on charging procedures for superheat methods.

Line 11. The crankcase heater (Figure 18–9) is needed to keep the oil warm enough so that refrigerant does not condense in the compressor shell, diluting the oil. A warm crankcase keeps the refrigerant

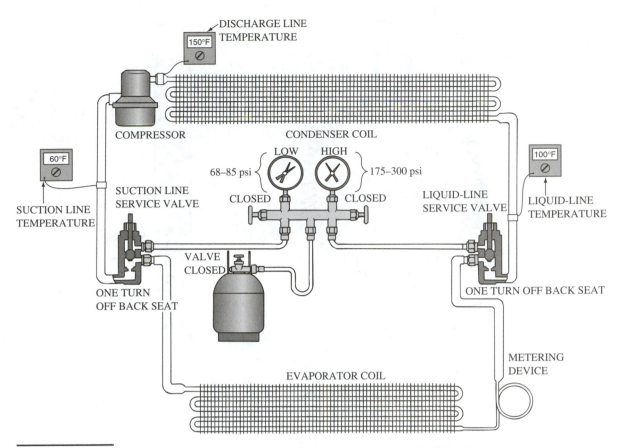

Figure 18–8 Normal range of operating pressures for an R–22 system. Checking the subcooling, discharge superheat and suction superheat.

pressure up in the compressor, so refrigerant migration to this area is less likely.

The crankcase heater should be warm or hot to the touch. Some crankcase heaters are operated continuously and some are designed to cut off when the compressor is operating. A crankcase heater is not needed when the compressor is operating. There are a few crankcase heaters that have a variable-temperature operation. The warmer it is outside, the less warmth is provided by the heater. Crankcase heaters are difficult to check in warm weather because they may seem to be defective and not producing heat. They may need to be removed and chilled and then checked for operation in a chilled condition. Some crankcase heaters are connected in series with a thermostat that opens at 70°F. The thermostat needs to be bypassed when it is checked in warm weather. Crankcase heaters that cut off as the temperature warms will save 40 to 80 W of energy per hour. An 80-W crankcase heater, operating continually, will use about $6 worth of electricity per month if the utility rate is 10¢ per kilowatt-hour.

Line 12. The indoor-fan motor should be oiled with nondetergent, or turbine, oil. The squirrel-cage blower should be checked and cleaned if necessary (Figure 18–10). The best way to clean a squirrel-cage blower is to remove it from the motor and cage assembly and take it outside for a good scrubbing. It is difficult to clean the blower wheel thoroughly while it is mounted in the blower housing. The inside portion of the blower housing may need to be cleaned also. Be sure not to remove or disturb the balancing weights used on the squirrel-cage blower. Cleaning the blower is a time-consuming job and is not a favorite of a technician. Loose dirt that is not removed by cleaning will be blown into the furnace or evaporator when the unit is started. Notice how the blower wheel is mounted in the blower assembly and on the motor shaft. It will be important to mount the blower wheel in the same position to prevent dragging of the wheel across the blower housing. If the blower is belt-driven, inspect belts for cracks or frays.

Line 13. The condenser-fan motor should be oiled with the recommended oil. Many motor manufacturers recommend a few drops of 10- or 20-weight nondetergent oil. Note the position of the fan blade in the throat of the fan housing. For maximum airflow, it is important to install the fan blade in the exact position designed by the manufacturer. Moving the fan blade up or down a couple of inches will have an impact on the airflow through the coil. It is important that the position of the fan blade on the motor shaft

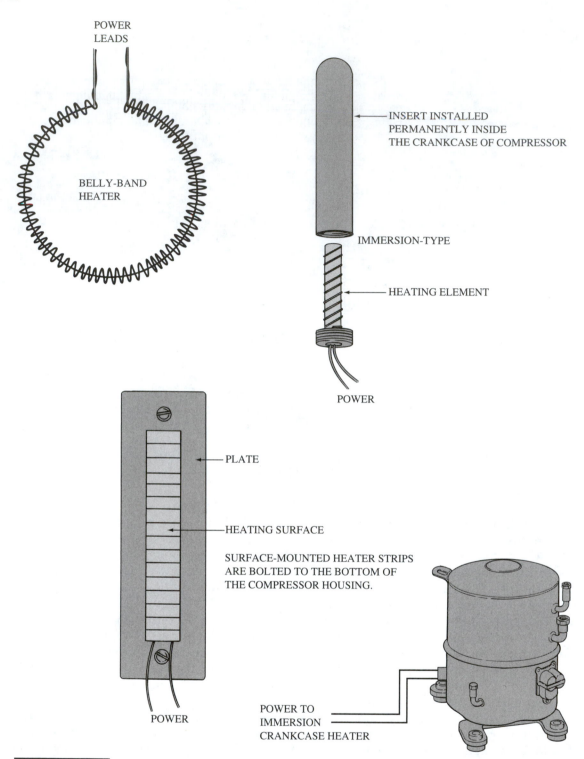

POWER
LEADS

BELLY-BAND
HEATER

INSERT INSTALLED
PERMANENTLY INSIDE
THE CRANKCASE OF COMPRESSOR

IMMERSION-TYPE

HEATING ELEMENT

POWER

PLATE

HEATING SURFACE

SURFACE-MOUNTED HEATER STRIPS
ARE BOLTED TO THE BOTTOM OF
THE COMPRESSOR HOUSING.

POWER

POWER TO
IMMERSION
CRANKCASE HEATER

Figure 18–9   Types of crankcase heaters.

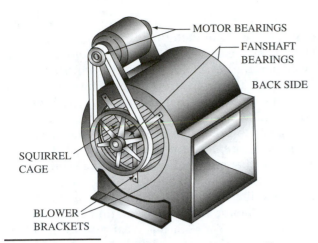

Figure 18–10   Belt-driven blower with two pulley wheels for motor and blower. Remove the pulley and blower brackets to access the squirrel cage.

be measured, or marked prior to removal. In most cases the blade will not need adjustment.

Inspect the fan blade for fatigue cracks and remove dirt that will reduce airflow. If fan blade is replaced, clean the shaft of dirt and corrosion prior to removal from the shaft. Grease the joint between the fan hub and shaft for quick removal in the future.

Line 14. The amperage draw of the common winding should be measured and documented. It is useful to record the amperage draw of the run and start winding. The amperage draw should not exceed rated load amps of the compressor in all but a few cases.

On a three-phase compressor, the amperage of each leg should be recorded. The amperage readings between the pairs of the three legs should be very close to each other. A deviation in this reading may indicate a problem. See the electrical troubleshooting chapter for information on calculating amperage and voltage imbalance.

Line 15. The condition of the contactor contacts should be inspected. Pitted and burned contacts should be replaced. The contactor should move freely in and out. Pitted and burned contacts can create higher amperage draw or cause the contact surfaces to weld together. One way to measure the condition of the contacts is to measure the voltage drop across the closed set of contacts operating under the compressor load (Figure 18–11). The voltage drop across a new set of contacts is 0 V. As the contacts open and close, they generate carbon deposits that interfere with the current transfer through contactor points. This carbon creates resistance to current flow and increases the current flow to the compressor and fan motor. If the voltage drop across a closed set of contacts under a load is 5 V or more, the contactor should be replaced.

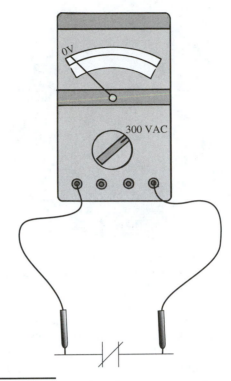

Figure 18–11   Testing closed contactor or relay contacts underload. No voltage should be shown under an operating load. Up to 5 V can be tolerated before the contactor/relay should be replaced.

Lines 16 and 17. Check the voltage on the line side of the contactor before energizing the compressor and after energizing the compressor. The voltage difference should be a few volts or less. Expect a small voltage drop after the condenser is energized. A large voltage drop between the stopped and running condition means that there is a bad connection or undersized wire feeding the condenser. Check the connections on both sides of the contactor, at the disconnect, at any junction boxes, and at the circuit breaker. These connections should be checked as noted in Line 20.

On three-phase equipment the voltage should be nearly the same between each of the phases, L_1 to L_2, L_2 to L_3, and L_1 to L_3. The calculation for an acceptable voltage deviation is found in the electrical troubleshooting chapter. The voltage readings should be measured before and after the equipment is energized. The voltage-deviation calculation should be measured when the equipment is operating.

Line 18. There are two temperatures to check on the thermostat. Check the thermometer on the cover or face of the thermostat (Figure 18–12). Use one of the your calibrated temperature sources to check and adjust the thermostat thermometer.

Next, on mechanical thermostats, blow or lightly brush away dust from the moving mechanical parts. Now check the temperature at which the thermostat opens and closes the cooling circuit. To

Figure 18–12   Room thermostat. Remove cover and calibrate room temperature thermometer on the lower part of this unit *(Courtesy of Honeywell Corporation)*

prevent short-cycling, it is best to turn off the condensing unit's disconnect while doing this exercise. A technician can use the indoor blower as an indication to determine at which temperature the thermostat is opening and closing. A technician must determine that the indoor blower is not using a time delay. A blower with an on-off time delay cannot be used with this test.

Calibrate a thermostat only it the reading is several degrees away from the true reading or the customer is not satisfied with operation of this device.

 Caution: Calibrating a thermostat can create more problems than it solves. A technician may devote too much time on this activity with little benefit. In some cases, the thermostat may be damaged and needs to be replaced at cost to the company.

Line 19. Using a megohmmeter or insulation tester has value if the customer schedules a regular PM or if the customer wants to predict compressor or motor failure. A megohmmeter (Figure 18–13) is used to track deterioration of motor-winding insulation. The megohmmeter checks the resistance from the compressor motor winding to the compressor case ground. The reading should be in millions of ohms; thus the term megohms. The rule of thumb is that the resistance from the motor winding to ground should be at least 1000 Ω/V.

WORKING EXAMPLE

A 240-V compressor motor should have a reading of at least 240,000 Ω (240 $\times$ 1000) between the motor winding and ground. Most readings will be in the millions of ohms and not this low. ∎

Just one megohmmeter reading is probably not significant. Megohmmeter readings must be tracked over several years. A technician should expect a slow deterioration in motor-winding insulation to ground. It is the rapid drop in resistance to ground that is an indication that the motor windings are ready to fail. Sometimes this failure is caused by moisture in the refrigerant or contaminated oil. To avert this failure, a technician can change the compressor oil, change the filter drier, add a suction-line drier, and recycle or change the refrigerant. Changing the oil, filters, and refrigerant will not benefit the compressor motor if the deterioration is caused by motor-bearing wear. Bearing wear allows the compressor rotor to slowly grind into the stator winding and short it out.

In the case of a commercial or industrial client, it may be important to predict motor failure. The motor can be replaced prior to failure. At a minimum, the motor can be purchased and available on site when the failure occurs, thus reducing downtime and loss of production.

Line 20. Wire connections are mentioned in other parts of this PM report. Wire connections are easily overlooked, and if a technician is pressed for time, are skipped completely. The most important wire connections to check are the connections that carry 120 V or more. Tighten every connection, including the one in the breaker or fuse panel. Connections become loose over time due to the heat developed in the wire and the expansion and contraction found at the connection. Loose connections create higher amperage draw and tend to overheat or burn up the wire.

A connection or wire that is discolored from overheating should be replaced. The whole piece of wire and connector should be replaced. A burned connector is the manifestation of too much current draw at that resistive point in the wire. There may be other undetected, overheated points in the wire that are not obvious. When locating a burned-out or discolored terminal connection, replace the total length of wire and connector. The wire may be burnt inside the insulation jacket without visible signs.

The low-voltage, 24-V circuit does not usually have burned-out wires. Check the connections for tightness. Do not forget the connections on the thermostat subbase.

Line 21. When checking for ductwork connections (Figure 18–14) a technician should inspect air leaks in the supply- and return-air duct systems. It is best to have the air handler operating when conducting this inspection. Feel for air leaks at all duct joints, starting collars, and duct boxes and between the transition and duct plenum. Return-air leaks are difficult to detect. A thorough inspection of the

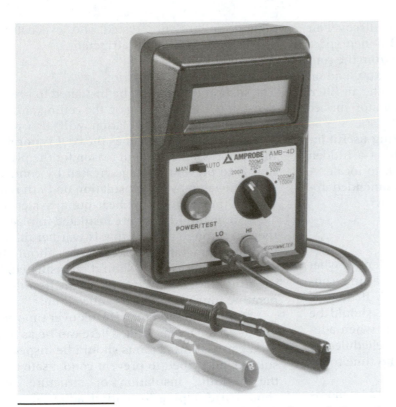

Figure 18–13 Digital megohmmeters. *(Courtesy of Amprobe Promax)*

Figure 18–14 Typical attic application of a horizontal gas furnace. *(Courtesy of Rheem/Rudd Corporation)*

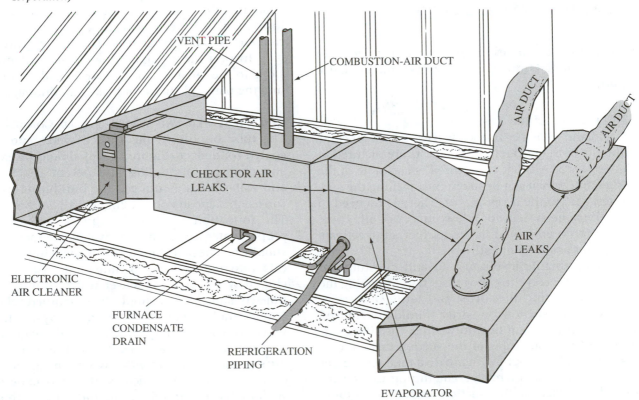

return-air plenum may help a technician discover air leaks. Return-air leaks from the conditioned space should also be sealed. Return-air leaks from the conditioned space do not contribute to an extra load on the evaporator, but they pass air around the air filter. Unfiltered air contributes to the evaporator-surface loading.

Line 22. This is a section for stating useful information about the equipment, such as a hidden disconnect or fuse that may be difficult to locate. A technician can also list one or two recommended upgrade items discussed with the customer.

SUMMARY

This PM report is a thorough reminder of what needs to be done at the job site. PM is a time-consuming task and should not be hurried. If there is not enough time to do it completely, it should be done at some other time. Reschedule it when adequate time is available. PM should be scheduled in slow periods when a technician has the time to do the job properly.

This report is a guideline and should be modified to meet the needs of a company and its customer base. A residential PM report may be different from a commercial PM report.

The next section discusses some options in regard to what could be added to a preventive maintenance report.

PREVENTIVE MAINTENANCE OPTIONS

This section discusses options that may need to be incorporated into the preventive maintenance report.

ACID AND MOISTURE TEST KIT

An acid and moisture test kit is used to test for contaminants in the refrigerant. Such kits are pictured in Figure 18–15. All that is required is a sample of the refrigerant, which can be taken by attaching the special test vial to the access valve. There is no need to break into the system for a few ounces of oil. There is enough oil entrained in the refrigerant to activate the test for acid and moisture.

Unacceptable levels of acid and moisture can be reduced by recycling the refrigerant, changing the oil and liquid-line filter-drier, and adding a suction-line drier. The suction-line drier should be removed after a short period of operating time. Installing cut-off valves on both sides of the suction drier allows easy removal without having to do a total refrigerant recovery. A piece of refrigerant tubing can be installed in place of the suction drier. This tubing

should be pressure tested and evacuated before charging the unit with refrigerant.

PIPE INSULATION INSPECTION

The suction line should be insulated to prevent condensation from dripping on the ceiling and wall surfaces. Suction-line insulation will also reduce the rise in suction superheat as the cool vapor returns to the compressor. The main reason for the pipe insulation is to prevent damage created by condensation. A technician may find insulation on both refrigerant lines in the case of some heat-pump systems.

Some liquid lines are insulated in places where they go through an area that is warmer than the liquid-line temperature. For example, a liquid line that passes through an attic should be insulated, because the attic temperature rises above the liquid-line temperature, thus reducing its subcooling.

Pipe insulation may shrink over time and separate at each circular joint. There can be as much as a 4-in. gap. All these joints should be inspected, reinsulated, and taped to prevent condensate damage to the building insulation or structure. Installers should use a pipe-insulation adhesive at each of these circular joints. The pipe insulation should be installed compressed, or slightly wrinkled, to allow for normal insulation shrinkage.

Inspect the pipe insulation for lateral splits. Deteriorated insulation should be replaced. The condensate drain line should be insulated at least the first 10 ft from the evaporator. Cold condensate can make this line sweat and create water damage. Pipe insulation is a mechanical-code requirement when condensate can damage the building structure.

Inspect the auxiliary drain pan for debris and slope to the outside. Pour a couple gallons of water into the pan and check for leaks and drainage.

CHILLED WATER AND WATER-COOLED CONDENSERS

The tube bundle in the chilled water and water-cooled condenser (Figure 18–16) should be exposed and checked for fouling and potential refrigerant and water leaks. Fouling is the buildup of water or corrosion deposits on the inner walls of the tubing. The formation of deposits can be reduced with proper water treatment. The water-treatment system should be checked for correct dispensing of chemicals into the water. Opening and inspecting the chilled-water and condenser-water tubing is very time consuming and should be scheduled when the equipment is not needed for several days. In the case where repairs are required, the equipment will be off-line for additional time. It is best to schedule this during the winter months when the demand for cooling is less. Technicians that are experienced with industrial equipment should conduct this PM.

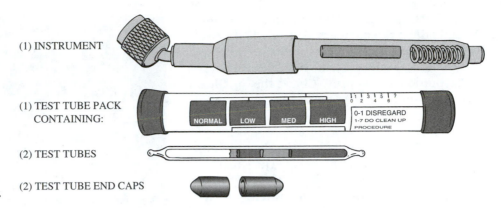

(1) INSTRUMENT

(1) TEST TUBE PACK CONTAINING:

(2) TEST TUBES

(2) TEST TUBE END CAPS

Figure 18–15 Oil-acid and refrigerant-moisture test kit.

DAMPERS AND LINKAGES

Specialty residential systems and larger commercial and industrial systems have dampers that control the airflow to various parts of the building. These dampers and their controlling linkages should be cleaned and lubricated. Their proper operation should be checked. Some dampers open for cooling and some dampers open for heating. In some installations the cooling and heating dampers are both open, mixing the air to the desired temperature. The mixing-damper operation is important for comfort cooling, heating, and dehumidification. Dirt collect-ing on the damper blades and linkages prevents a damper from sealing or causes sluggish operation.

OUTSIDE AIR DAMPERS

Outside air dampers are used in commercial air-conditioning systems. They are used to introduce outside air into the return-air side of the duct system. The dampers should be adjusted to admit the required CFM of fresh air. Too little fresh air prevents proper dilution of indoor-air pollution. This is becoming an indoor-air quality issue, sometimes abbreviated IAQ. Too much outdoor air increases the

Figure 18–16 (a) Shell-and-tube condenser. (b) Shell-and-tube chiller. *(Courtesy of York International Corporation)*

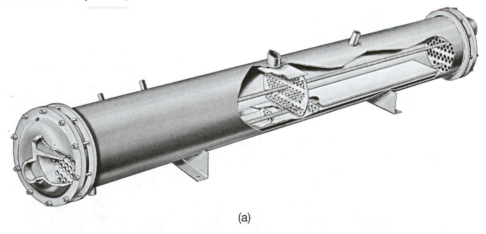

(a)

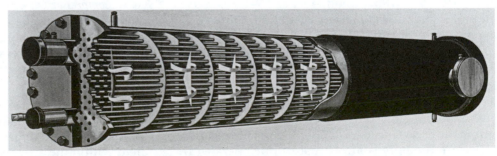

(b)

cooling and heating load of the building. Measure the CFM and adjust the dampers as needed.

The outside-damper screen should be checked for tears and holes. Many systems have filtering systems in the outdoor-air supply. Check the air filter that is used to filter the outside air.

DUCT INSULATION

Duct insulation is important when the duct system travels through unconditioned spaces. Duct insulation prevents heat gain and prevents the cold duct from sweating and damaging the structure.

The best way to seal duct insulation is with duct mastic such as the HARDCAST® system. Many technicians use duct tape. The duct-tape adhesive will eventually lose its strength and allow the insulation to open and expose the duct. If duct tape is used, it should be covered by duct mastic that overlaps the taped area. Duct mastic should be used in repairing duct leaks and in new-duct installations.

Some ducts and plenums have internal insulation. Insulation that has become dislodged inside the duct will be evident with sweating ductwork or restricted airflow in part or all of the duct system.

MISCELLANEOUS ITEMS

The list of PM checks can go on indefinitely. What is listed on the previous pages will cover most residential and commercial air-conditioning systems. Each job will have a customized PM report that varies with the equipment, types of controls, ductwork, and operating habits.

The following areas may need to be incorporated into the PM report.

1. *Air Balancing.* Air balancing is usually done in a large commercial building at the commissioning of the air-conditioning system. Air balancing is not scheduled on a regular basis but is done when one space has too much conditioned air and another space has too little conditioned air. If all spaces are conditioned properly, air balancing may not be done for many years of the building's occupancy. Air balancing is required when major remodeling disturbs the airflow pattern as established by the original floor plan. An airflow hood (Figure 18–17) is a useful tool in balancing duct systems and troubleshooting airflow problems.

2. *Energy-Management Systems.* Energy-management systems (EMS) should be cycled for programmed operation. Does the EMS shut off equipment at prescribed times? Is the exhaust system turned off during unoccupied times? Is the chilled-water temperature reset as the outdoor temperature varies? These are the only controls that can be operated by

Figure 18–17 Airflow hood. *(Courtesy Alnor Instruments)*

the EMS. The PM schedule should be arranged with the installer of the energy-saving device.

3. *Pressure-Switch Operation:* Some air-conditioning systems have low-, high-, and oil-pressure switches. These switches should be checked for correct operation. System pressures needs to be artificially raised or lowered to conduct tests on the operation of these devices.

4. *Cooling Tower.* The cooling tower should be cleaned at least once a year. An automatic chemical feed should be used to reduce corrosion and growth of algae and bacteria. Water quality should be checked on a daily or weekly basis.

5. *Compressor-Oil Level.* A semihermetic compressor with an oil sight glass is checked with the compressor operating. The right oil level should not be assumed. Some compressors require an oil level of an eighth of a sight glass, whereas others require sight-glass levels of ½ or ⅞. Too much or too little lubrication can cause damage to the compressor.

6. *Controls.* Pneumatic or direct digital controls should be checked for calibration. This is a specialty field within the air-conditioning field.

ON THE JOB

Public and commercial buildings are required to maintain a certain amount of outside air for ventilation purposes. Normally, this outside air is brought into the building through the economizer ductwork.

When on the job, to verify that an air-conditioning system is working properly, inspect the outside air dampers, protective screening, and damper actuators. Test the economizer circuitry. Normally the economizer control circuit modulates the dampers as needed to maintain a predetermined minimum of fresh air into the building. To verify operation, adjust the set point of the mixed-air controller from fully open to fully closed. The outside air dampers should never close down completely during normal daytime running conditions. If they do close down, readjust the minimum outside air to at least 10% fresh air or whatever is prescribed in the local building code. The outside dampers may be shut during times when the building is not occupied.

REVIEW

- Preventive maintenance is a necessary routine to keep an air-conditioning system operating at peak performance, to lower operating cost, and to reduce equipment run times, which should translate into longer equipment life. Preventive maintenance is one of the least-liked jobs by an air-conditioning technician. Its routine nature makes is more susceptible to cutting corners and falsifying PM report documents. Supervisors must occasionally check a PM job for quality conformance. The customer must receive what they are paying for; the only way to assure this is to double-check the PM work.

- PM reports should be customized for the type of building under consideration. Most residential structures can use one standard PM report. Commercial buildings vary widely in application and use; therefore, it is important to tailor a plan to meet the needs of each building.

PROBLEMS AND QUESTIONS

1. What is the difference between a preventive maintenance checkup and an emergency service repair?

2. Describe the difference between the three types of service contracts.

3. List five areas that would be checked on a commercial PM report but not on a residential report.

4. What are the benefits of a preventive maintenance report?

5. Select the 10 most important parts of an ultimate tune-up program.

6. List the various supplies needed to conduct a PM job.

7. What meter and instruments would the technician need on a PM work order?

8. Why is it stated that a technician is the best salesperson of air-conditioning-related options?

9. Describe the PM routine.

10. What are the benefits of recording the PM information?

11. A technician works on a PM where the air filter is extremely dirty. The customer states that he will replace it as soon as the technician leaves. What should the technician do?

12. The technician has an early-spring PM service call on a 10-year-old system. The technician cleans everything. The condensing temperature is above the 10% normal range and the subcooling is 5°F. What the likely problem?

13. Why is it important that a technician operate the equipment before beginning the PM process?

14. What type of bearing lubrication is generally required by a motor manufacturer?

15. What is the purpose of using a megohmmeter? What is its long-term value?

16. How can a technician service a window unit without access fittings? How can the technician determine if the refrigerant side is working properly without tapping into the system?

17. Why is a crankcase heater important?

18. How does the technician check a contactor?

19. What is the purpose of an acid-moisture test kit? How it is used?

20. Why should the outside air damper be adjusted to pull in the right amount of fresh air?

CHAPTER NINETEEN

HVAC/R
INDUSTRY CERTIFICATION

AFTER STUDYING THIS CHAPTER, THE STUDENT WILL BE ABLE TO:

- List the various industry certifications.
- State the reasons for industry certification.
- Describe the different types of industry certification.

- Complete practice exams on industry certification.
- Discuss why industry certification is important.

INDUSTRY CERTIFICATION

The HVAC/R industry is moving into the new millennium with a focus on improving itself and its public image through the process of certification testing. One of the first certification programs began in 1987 with the Industry Competency Exam (ICE). The ICE exam was developed to test the entry-level skills of the graduating HVAC/R student. The ICE program was developed by the Air-Conditioning Refrigeration Institute (ARI), Gas Appliance Manufacturers Association (GAMA), and a host of other air-conditioning-related organizations. The ICE exams were developed from input from all sectors of the air-conditioning industry.

In 1996–97 the HVAC/R industry took a significant step toward certification of experienced technicians. During this time period, the North American Technicians Excellence (NATE) and the Air Conditioning Excellence (ACE) certification programs emerged. These certifying groups cater to technicians with more than 1 year of experience and some formal or informal training. Certification programs are being developed for senior technicians with substantial field and supervisory experience.

Refrigeration Service Engineers Society (RSES) has been training and certifying air-conditioning technicians for many decades. Their training and study materials are excellent. After passing the rigorous exam, an RSES member may become a certified member, known as a CM, or a certified member specialist, known as a CMS.

In mid-1999, NATE, ACCA, and RSES merged to form one industry-certification program with one standard of excellence. The goal of this merger is to draft the best test of knowledge by all involved. Their goal will be to maintain the highest standards of knowledge testing while developing an exam that is meaningful for today's experienced technician. ACCA will administer the new industry exams through its chapter network and will provide training materials and products. RSES will administer the industry exam through its chapter network as part of its newly launched Technical Institute, a certification and training program for RSES members. It will continue to offer its special member-only certification programs, CM and CMS. NATE will offer exams to other technicians not affiliated with ACCA or RSES.

Another certification program was developed for technicians in the small-appliance-repair industry. The National Appliance Service Technician Certification (NASTeC) program is for technicians that service and repair household appliances such as refrigerators, freezers, window units, dehumidifiers, washers, dryers, and most kitchen appliances.

This chapter does not discuss EPA certification, which is covered in many other sources.

INDUSTRY COMPETENCY EXAM FOR STUDENTS

The purpose of the Industry Competency Exam (ICE) is to allow a student that is completing a training program the opportunity to exhibit his or her knowledge on a nationwide test of competencies. Students that have completed 75% of their program may attempt the exam. The ICE program has been very successful. Many training programs are now requiring that graduating students pass one of the ICE exams in order to receive their certificate of completion. Students passing the ICE exam exhibit that they have learned basic knowledge to compete as an entry-level technician. Training programs with high student-pass rates can be confident that their curricula meet the needs of the air-conditioning industry.

The ICE program offers three different certification exams. The exams are divided into the follow categories:

- Residential air-conditioning and heating certification
- Light commercial air-conditioning and heating certification
- Commercial refrigeration certification

A student needs to obtain a 60% grade in order to obtain certification. Each multiple-choice exam has 100 questions. The residential and commercial certification exams cover air conditioning, gas heating, oil heating, and heat pumps.

The commercial refrigeration certification exam covers the basic installation, service, and troubleshooting of commercial refrigeration. This is the only exam that does not test heating systems.

Subjects included on the exams are

- Fundamental refrigeration principles
- Refrigeration cycle
- Trade-related tools and materials
- Instruments and meters
- Properties of air and air distribution
- Theory of heat

- Safety practices
- Electricity, motors, and controls
- System installation

Students passing the ICE exam should attempt advanced certification exams after 1 year of practical experience. A student passing an exam will receive his or her score report, a certificate of competency, wallet card, and his or her name and school published in the trade press and on the internet HVAC job on-line. ICE certification is the best way to prove a new technician's skill level. Only students with a good level of HVAC/R knowledge pass these certification exams.

The following practice questions are provided to give the reader a flavor of the ICE exam. Memorizing these questions and answers will not help you pass the exam. Use the practice exam as an indication of your knowledge. The ICE exam is a broad examination of the student's skills and knowledge. The exam questions are designed to test long-term knowledge found in course and lab work. It would be difficult to study for a short period of time and succeed on the ICE exam. A person taking a few courses will not pass the exam.

Not every answer to the practice exam will found in this textbook. The practice exam should be attempted closed book, without access to the answers that can be found in the various sources used in your education. Passing this practice exam will not guarantee success in the ICE official examination; however, doing poorly on this practice exam is an indication that further review is necessary.

The following questions are courtesy of Prentice-Hall Publishing Company.

ICE PRACTICE QUESTIONS

RESIDENTIAL HEATING AND AIR-CONDITIONING EXAMINATION

1. After installing a new natural gas furnace, the technician notes the following conditions: manifold pressure is 3.5 in. of water column; return-air temperature is 70°F: and the supply-air temperature is 150°F. The rating plate on the furnace calls for a 35°F–65°F temperature rise. The technician's most appropriate action would be to

   A. Increase the blower speed
   B. Change the burner orifices to a larger size
   C. Decrease the volume of return air
   D. Increase the gas pressure

2. For proper burner operation, the manifold pressure on an LP gas–fired furnace should be approximately

   A. 3 in. of water column

   B. 4 in. of water column

   C. 9 in. of water column

   D. 11 in. of water column

3. When you are checking the power to the heat pump, you should determine the power level at the high-voltage terminals. A fault may be present in the power service if the voltage here is below what percentage of the design voltage?

   A. 3.5%

   B. 5%

   C. 10%

   D. 15%

4. A single-stage thermostat controls an electric furnace. On a call for heat, all connected circuits function properly; however, the heat rise across the furnace is much lower than design requirements. Which one of the following actions should the technician take to correct this problem?

   A. Install a two-stage thermostat.

   B. Reduce the air volume through the furnace.

   C. Increase the air volume through the furnace.

   D. Replace the dirty filter in the return air.

5. When you are using a manifold gauge set, the compound gauge must

   A. Never be connected to the low side of the system

   B. Always be on the right

   C. Always be on the left

   D. Never be connected to the high side of the system

6. A compressor has been off all winter. Upon checking, you find that it is warm, *most likely* because

   A. The crankcase heater is on.

   B. The furnace is radiating heat.

   C. It has low power.

   D. It has been short cycling.

7. On initial start-up of an air-conditioning system, you notice a temperature drop of 26°F across the evaporator coil. This drop *most likely* indicates a(n)

   A. Excessive airflow

   B. Lack of airflow

   C. Lack of refrigerant charge

   D. Refrigerant overcharge

LIGHT COMMERCIAL HEATING AND AIR-CONDITIONING EXAMINATION

1. A three-way valve that has two inlets and one outlet is classified as a

   A. Diverting valve

   B. Mixing valve

   C. Pilot-operated valve

   D. Two-position valve

2. The purpose of taking a pressure-drop reading across the suction line filter/dryer is to determine the

   A. Contaminant loading of the filter/dryer

   B. Proper size and charge

   C. Superheat reading

   D. Temperature and pressure of the gas

3. Most natural gas furnaces use an operating manifold pressure of

   A. 3.5 in. of water column

   B. 4.5 in. of water column

   C. 7 in. of water column

   D. 11 in. of water column

4. To determine the Btu/h output of a resistance heater at rated voltage, the wattage should be multiplied by

   A. 1.08

   B. 3.1415

   C. 3.413

   D. 4.31

5. On a liquid line, insulation may be necessary when

   A. Discharge temperature are low.

   B. Low suction pressures are required.

   C. Line routes are short.

   D. Subcooling is not available.

6. Temperature rise is correctly defined as the

   A. Temperature of supply air plus the temperature of the return air

   B. Temperature of outdoor air plus the temperature of the condenser

   C. Temperature of supply air minus the temperature of the return air

   D. Indoor temperature minus the outdoor temperature

7. The temperature rise of an electric furnace should be approximately

   A. 20°F

   B. 50°F

   C. 90°F

   D. 160°F

COMMERCIAL REFRIGERATION EXAMINATION

1. If an evaporator is located on a higher level than the compressor, the possibility is high for compressor damage due to
   A. Oil migrating to the compressor during the running cycle
   B. Liquid refrigerant draining to the compressor during the OFF cycle
   C. Excessive modulation of the expansion valve occurring during shutdown
   D. Excessive increase in suction-gas pressure during the OFF cycle

2. The *principal* reason double risers are used in the suction line is that they
   A. Assist in oil return to the compressor
   B. Reduce refrigerant migration to the compressor
   C. Reduce the amount of current drawn by the compressor
   D. Increase the system's total capacity

3. An overload protector mounted on the compressor shell by a removable bracket is usually
   A. Wired in parallel with the compressor
   B. Wired in series with the compressor
   C. Operated only by heat generated by the compressor shell
   D. Operated only by current generated in the heater wire with the protector

4. The result of reversing two leads on a three-phase motor would be
   A. A change in the direction of rotation
   B. A loss of power
   C. The creation of a direct short
   D. An increase in power

5. On an inside receiver with a safety-pressure relief device, the discharge of the device should be piped to the
   A. Inside compressor
   B. Outside condenser
   C. Outside atmosphere
   D. Inside evaporator

6. If a compressor must be located in a hot equipment room a long distance from the evaporator, the suction line must be insulated to reduce
   A. Pressure drop across the expansion valve
   B. Oil loss from the compressor
   C. Pressure drop in the suction line
   D. Superheat to the compressor

7. In a system using a thermal expansion valve, increasing the spring pressure against the expansion valve diaphragm will
   A. Increase the capacity of the evaporator coil
   B. Cause the compressor motor to become overloaded
   C. Cause the refrigerant to flood back to the compressor
   D. Increase the superheat of the refrigerant

EXPERIENCED TECHNICIAN CERTIFICATION PROGRAM

The consolidated **North American Technicians Excellence**, the **Air Conditioning Contractors of America**, and the **Refrigeration Service Engineers Society** certification programs are designed to test the comprehensive knowledge and skills of the experienced technician. The new exams are divided into various sections, or modules. Some of the available sector exams are Air Conditioning; Gas Heating; Oil Heating; Heat Pumps; and Air Distribution. These exams use the term *Sector* to identify these specific areas of the trade, such as air conditioning, gas heating, etc. For example, a technician may wish to begin by taking the *Air Conditioning Sector* followed by a *Gas Heating Sector* on another test date.

GOALS OF TECHNICIAN CERTIFICATION

The goal of technician certification is to

1. Recognize today's professional technician.
2. Develop a mechanism to promote and reward technicians.
3. Improve the public image of the air-conditioning industry.
4. Develop a recognized national industry standard.
5. Show technicians areas that need improvement.
6. Facilitate screening of new employees.
7. Develop a sense of pride for the technician.
8. Improve the HVAC/R industry.

ELEMENTS OF THE EXAMINATION

In order to succeed, the technician must have comprehensive experience and knowledge in several areas.

CORE ELEMENTS

The core elements cover a technician's communication skills. Communication skills include reading, writing, and English comprehension. This section also tests basic grammar, punctuation, spelling, and oral communication skills.

The core elements include questions using math-based computations and basic science knowledge. The core elements are not difficult. Most of the information is practical knowledge learned in secondary school. Non-air-conditioning personnel should be able to pass this section.

INTERPERSONAL RELATIONS

The module on interpersonal relations is another general test category. This module asks questions that relate to personal ethics and conduct and interpersonal relations with customers. Again, anyone should be able to pass this module.

TOOLS AND MEASUREMENTS

This modules tests information on the uses of fabrication tools and refrigerant-related instruments such as manifold gauges and recovery units. The exam may ask questions about the use of electrical meters, airflow devices, and temperature/humidity measuring instruments. Experienced technicians should have little difficulty with this section of the exam.

ELECTRICAL

This module is more of a challenge to the technician as compared to the previous three modules. This module tests skills in basic electricity, A/C motors, electrical diagrams, and electrical troubleshooting. Because the majority of field problems are electrical, a technician with strong electrical troubleshooting experience will master this section.

BASIC SYSTEMS COMPONENTS

Basic systems and components form a general category of the area being tested. It includes questions on system operations, duct systems, wiring layouts, and specific components. For example, if the technician is testing on heat-pump systems, the questions from this module would include questions about the refrigeration cycle, duct design, high- and low-voltage wiring, and components found in heat-pump systems. Some of the questions would be heat-pump specific and some of the questions would be general questions that relate to other air-conditioning or heating applications.

INSTALLATION PROCEDURES

The installation procedures section includes fabrication of copper tubing; installation of outdoor, indoor, and package units; installation of duct materials; installation of accessories; field wiring; and startup and checkout procedures. Some of the questions are specific to equipment being tested and

some of the questions are generic and could be found on any of the various exams.

SERVICE AND REPAIR PROCEDURES

This module examines preventive maintenance, diagnostics, and retrofitting equipment. For example, it may ask questions about conducting preventive maintenance on the equipment that is being tested. Other questions may relate to repair and replacement. Troubleshooting questions are an important part of this section.

CONTROLS

The questions in this module test the technician's knowledge for most types of controls. The questions are related to thermostats; contactors, and relays; flow and pressure switches; and general questions on electronic controls. The section does not ask questions relating to pneumatic or direct digital controls.

REGULATIONS, CODES, AND SAFETY

The technician can expect questions drawn from a wide variety of code sources. Questions will be derived from EPA regulations, the Uniform Mechanical Code, the National Electrical Code, the National Fire Code, and other state and national safety codes. Most of the questions can be answered from the general knowledge the technician has obtained on the job. Some of the answers can be reduced to two possible choices. Reviewing the Uniform Mechanical Code for equipment access, refrigerant-line routing, sizing, and installation of condensate line would be beneficial.

DESIGN

The final module examines system design. The technician can expect questions on bid proposals and design considerations for residential and commercial structures. Expect questions on diffusers, registers, grilles, temperature, humidity, estimating equipment installation, and indoor-air quality.

TEST QUESTIONS

The test participant can expect to take an exam that has as few as 200 questions or as many as 400 questions. The number of exam questions depends on the type of certification attempted by the technician. The **Installer** exam has 200 questions. The **Service Technician** exam has 300 questions. All modules or sections must be passed with a grade of 70% or higher in order to receive certification in that area or sector. In some instances, modules or sections passed can be carried to other sector examinations. Modules that are passed are officially recorded even if the technician does not obtain certification on that examination. On the next test date, the technician can retest on the module(s) needed for certification. Certification is an ongoing process. The technician may need to review missed concepts prior to attempting a retest of the sector that was not passed.

The exams are given in half-day sessions. Technicians that pass will receive a certification card and certification sleeve patch. A chevron patch is awarded to signify the sector, or area, of certification. Technicians can be certified in several sectors and may proudly display several certification chevrons.

PRACTICE QUESTIONS

The following practice questions are provided to give the experienced technician a sample test on industry certification. The questions are provided to promote HVAC/R training, testing, and certification. Use the practice exam as an indication of your knowledge and experience. The exams are used to test the technician's training, experience, and skills. Doing well on the practice exams does not mean that you will pass the comprehensive examination; however, doing poorly on the practice exams is an indicator that you will need to study prior to taking any of these challenging examinations. It is expected that industry certification will generate new training activities, which will in turn improve the performance and stature of the professional technician. HVAC/R certification is being promoted to the air-conditioning technician and the general public. Once aware of this certification, the general public will begin requesting certified technicians to do work on their equipment.

The answers to most of the practice exams are not found in the material in this textbook. This textbook provides the reader with some information that will aid in passing the exam. The practice exams should be attempted as closed-book examinations without access to answers that can be found in various technical sources, such as this textbook and the third edition of the *ARI* textbook *Refrigeration and Air Conditioning*.

The practice exams are divided into general questions that the technician might find on any certification examination. The general practice questions for all sectors are

- Core
- Interpersonal relationships
- Electrical
- Tools and measurements
- Codes, regulations, and safety

Specific practice questions are provided for the following technician sector exams:

- Air conditioning
- Gas furnaces
- Oil furnaces
- Heat pumps
- Air distribution

GENERAL PRACTICE QUESTIONS (COURTESY OF NORTH AMERICAN TECHNICIAN EXCELLENCE, INC.)

CORE MODULE

1. All substances contain a measurable amount of
   A. Air
   B. Heat
   C. Water
   D. Carbon

2. If you bought six cases of duct tape for $64.14, what was the price of each case?
   A. $10.59
   B. $10.96
   C. $10.60
   D. $10.69

3. The change caused by one chemical acting upon another is called a chemical
   A. Imbalance
   B. Reaction
   C. Transformation
   D. Formulation

4. A BTU is defined as the amount of heat that must be added to 1 _____ of water to raise its temperature 1°F.
   A. Pound
   B. Ounce
   C. Quart
   D. Gallon

5. In the following sentence, which of the following choices would best simplify and still express the meaning of the underlined words?

   In my opinion, the new thermostat design is a change for the better.

   A. An improvement
   B. A step in the right direction
   C. A help
   D. A badly needed modification

6. If you bought four thermostats for $91.12, what was the price of each thermostat?
   A. $22.98
   B. $23.78
   C. $22.87
   D. $22.78

7. The type of heat that must be added or removed to cause a change in the physical state of a substance is
   A. Specific heat
   B. Superheat
   C. Latent heat
   D. Sensible heat

8. Which of the triangles in Figure 19–1 is a 30-60-90 right triangle?
   A. a
   B. b
   C. c
   D. d

Figure 19–1

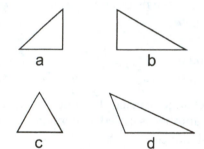

Interpersonal Relationships Module

1. A person who is persistent is one who
   A. Sticks by their friends no matter what they do
   B. Keeps to a course of action no matter how difficult
   C. Always disagrees with what is said by another person
   D. Always agrees with everything said by another person

2. An understanding of an HVAC service company's warranty policies and the ability to discuss them with a customer is the responsibility of the
   A. Technician
   B. Company sales manager only
   C. Manufacturer's sales representative only
   D. Equipment distributor

3. Carefully choosing questions that will give you a customer's true feelings is
   A. A technique to be avoided when asking questions of a customer who has a complaint
   B. A primary technique used when asking questions of a customer who has a complaint
   C. Less important than asking as many questions as possible of a customer who has a complaint
   D. A technique used only when asking questions of satisfied customers

4. A customer asks that you do something that your company rules specifically prohibit. When responding to such a request, you should
   A. Tell the customer you cannot comply with the request and leave it at that.
   B. Explore with the customer the circumstances under which you might do it.
   C. Tell the customer you cannot comply with the request and offer a brief explanation why you cannot.
   D. Whenever possible, prevent a conflict and go ahead and do it.

5. Why must a contract include one person's commitment in exchange for a second person's commitment?
   A. Only to let both people know their responsibilities
   B. Because it is a federal law
   C. To make the contract legally binding
   D. To make the warranty of the contract unlimited

6. It is essential that a technician service a gas furnace within a set work schedule in order to be efficient. The technician's ability to do the job correctly and keep on a set schedule must typically involve some
   A. Planning
   B. Missing steps
   C. Hastiness
   D. Gambling

Tools and Measurements Module

1. A core-removal tool is helpful when installing a heat pump to
   A. Reduce dehydration time
   B. Increase dehydration time
   C. Reduce needed charge
   D. Increase capacity in cooling

2. A recycling machine
   A. Can clean only recovered refrigerant and test it to ARI standards
   B. Can clean any refrigerant and test it to ARI standards
   C. Can clean any refrigerant, but cannot test it to ARI standards
   D. Can clean any refrigerant and can test it to EPA standards

3. To measure the difference in air pressure between two locations in a duct system, a technician uses a
   A. Diaphragm-type pressure differential gauge
   B. Digital anemometer
   C. Inclined manometer with one end open to the atmosphere
   D. Digital velometer

4. To reduce the end diameter of a section of round duct, a technician would typically use a
   A. Snap-lock punch
   B. Hand seamer
   C. Notching tool
   D. Hand crimper

5. An electrical meter that displays volts, ohms, or milliamperes on an analog scale is called an
   A. Analog continuity tester
   B. Analog megohmmeter
   C. Analog voltmeter
   D. Analog VOM

6. Static pressure is the
   A. Volume of air through a duct section in a given amount of time
   B. Velocity of air through a duct section
   C. Pressure that is exerted equally on all interior walls of a duct section
   D. Maximum speed at which air can travel through a duct section without generating static electricity

7. An airflow hood is used to measure
   A. CFM
   B. Velocity
   C. FPM
   D. Air quality

8. To calculate the cfm of airflow in a section of ductwork, you need the
   A. Static pressure and duct length
   B. Duct cross section and length
   C. Duct cross sectional area and air velocity
   D. Duct length and air velocity

9. Which of the following thermometers would you typically use to obtain a temperature measurement of an inaccessible area?
   A. A pocket test thermometer
   B. An infrared thermometer
   C. A surface probe thermometer
   D. A pressure probe thermometer

10. After cutting copper tubing, the tube edge should be cleaned with a
    A. Wire brush
    B. Hacksaw blade
    C. Round file
    D. Reamer

11. Under which of the following conditions would you read 0 psig on the low-side and high-side gauges of a properly calibrated manifold set?
    A. Whenever the high and low side hoses are connected to a charged, nonoperating system and both manifold hand valves are closed
    B. When both manifold hand valves are closed and all hoses are open and disconnected
    C. High and low sides connected to a charged and operating system, center hose connected to refrigerant cylinder, and both manifold hand valves closed
    D. High and low sides connected to a charged and operating system, center hose connected to refrigerant cylinder, and both manifold hand valves open

12. To check a deenergized, normally open relay for fused contacts, a VOM would be set to check
    A. Current
    B. Ohms
    C. Amps
    D. Voltage

13. Which of the following equipment may be used to remove refrigerant from a system without a requirement to test the refrigerant's purity?
    A. Refrigerant recycling machines only
    B. Refrigerant reclaim machines only
    C. Refrigerant recovery machines only
    D. Any recovery, recycling, or reclaim machine

14. The following readings of a conditioned space were obtained with a sling psychrometer:

    Wet bulb = 53°F
    Dry bulb = 73°F

Using the psychrometric chart in Figure 19–2, what is the relative humidity?

A. 25%

B. 15%

C. 35%

D. 45%

15. Which of the following instruments are typically used to measure static and total pressure in a duct system?

A. Psychrometer

B. Diaphragm-type differential pressure gauge

C. Portable air hood

D. Anemometer

16. An 80% efficient gas furnace has a lower operating cost than a 60% efficient gas furnace with the same BTU input. Why?

A. The 80% furnace produces more heat.

B. The 80% furnace produces the same heat with 20% less gas.

C. The 80% furnace produces the same heat with 20% less electricity.

D. The 80% furnace produces the same heat with 20% less gas and 20% less electricity.

17. A typical liquefied-petroleum (LP) gas furnace has its manifold pressure adjusted to between

A. 1 to 2 in. of water column

B. 3 to 4 in. of water column

C. 5 to 8 in. of water column

D. 9 to 11 in. of water column

18. To detect leaks in the flue pipe of a natural gas furnace, an electronic leak detector must be capable of sensing

A. Carbon dioxide

B. Carbon monoxide

C. Natural gas

D. Refrigerants

ELECTRICAL MODULE

1. In a simple resistive circuit, the voltage is 24 V and the current is 8 A. According to Ohm's law, what is the resistance in the circuit?

A. 16 Ω

B. 3 Ω

C. 0.5 Ω

D. 192 Ω

2. According to Ohm's law, as voltage is increased in a circuit, current

A. Increases

B. Decreases

C. Stays the same

D. Can either increase or decrease

Figure 19–2

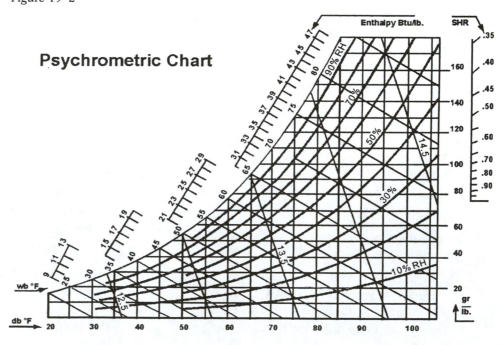

Psychrometric Chart

3. As the number of poles in a PSC motor are decreased from 4 to 2, what happens to the rpm?

   A. The rpm increases.

   B. The rpm decreases.

   C. The rpm stays the same.

   D. The number of poles in a motor has no effect on the rpm.

4. Which of the following would a technician use to check the operating sequence of a heat pump?

   A. Pictorial diagram

   B. Schematic

   C. Isometric drawing

   D. Illustration

5. What reason would a technician have for tracing unit's electrical circuit on a sketchpad during troubleshooting procedures?

   A. The technician is unable to read wiring diagrams.

   B. The technician has no prior experience with the unit.

   C. The unit wiring diagram is missing from the access panel.

   D. A technician should use a sketch to confirm all components are in place.

6. Using Figure 19–3, what designation is given to the dual capacitor used for the compressor and fan motor:

   A. RC

   B. R

   C. 2TH

   D. ISC

7. 3412 Btus of heat is equal to how many watts of electrical power?

   A. 1 W

   B. 3.4 W

   C. 1 kW

   D. 3.4 kW

8. A packaged cooling system has a 240-V blower motor with an open internal overload device. Which of the following VOM readings would typically indicate this condition?

   A. 244 V measured across the motor with a normal amp draw

   B. 120 V measured across the motor with high amperage draw

   C. 0 V measured across the motor and a 0-A draw

   D. 244 V measured across the motor with a 0-A draw

9. Which of the following will generate a voltage in a coil of copper wire?

   A. Freezing the coil

   B. Moving the coil through a magnetic field

   C. Grounding each end of the coil

   D. Heating the coil

10. Which of the following actions would cause a motor to reverse rotation?

    A. Removing one of the three leads at the terminal connections to a three-phase motor

    B. Reversing the power connections to a 220-V, single-phase motor

    C. Reversing the terminal connections to the run capacitor of a 220-V, single-phase motor

    D. Reversing two of the three leads at the terminal connections to a three-phase motor

11. Which of the following would a technician use to check the operating sequence of a gas furnace?

    A. Pictorial diagram

    B. Schematic

    C. Isometric drawing

    D. Illustration

12. Which of the following is a typical test available when checking an electronic ignition control board?

    A. Input voltage check

    B. Microprocessor bias check

    C. Solder joints resistance check

    D. Rectifier current check

REGULATIONS, CODES, AND SAFETY MODULE

1. A technician's understanding of proper response procedures to accidents involving refrigerants is a requirement of

   A. All refrigerant manufacturers

   B. All HVAC equipment manufacturers

   C. The National Fire Protection Association

   D. The federal government

2. EPA regulations require that recovered refrigerant can be used only in

   A. Any equipment using a similar refrigerant

   B. Any equipment with the same evaporator pressure

   C. Equipment with the same owner as the original equipment

   D. Any equipment as long as it is recycled

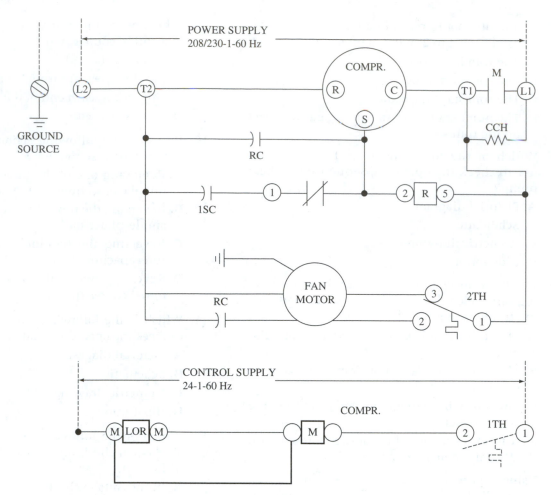

POWER SUPPLY
208/230-1-60 Hz

GROUND
SOURCE

COMPR.

CONTROL SUPPLY
24-1-60 Hz

COMPR.

LEGEND

M	CONTACTOR
RC	CAPACITOR, RUN (C-HERM, COMPR.)
	(C-FAN, FAN MOTOR)
1TH	THERMOSTAT, COOLING
2TH	THERMOSTAT, FAN MOTOR
	(HI SPEED CLOSE @ 80°F)
	(LOW SPEED OPEN @ 70° F)
CCH	CRANKCASE HEATER (IF USED)
1SC	CAPACITOR, COMPR. START
R	RELAY, START DEVICE
LOR	LOCK-OUT RELAY
LPS	LOW PRESSURE SWITCH
	FACTORY FIELD WIRING AND DEVICE

NOTES:

1.   ALL FIELD WIRING PER.
    (a)   NATIONAL ELEC. CODES (NEC)
    (b)   LOCAL OR CITY CODES.
2.   DRAFTING PRACTICES & SYMBOLS PER ARI
    GRAPHIC ELECTRICAL STANDARDS.
3.   REPLACE ANY ORIGINAL WIRE WITH
    EQUIVALENT.
4.   MOTORS ARE INTERNALLY PROTECTED.
5.   CONNECTORS SUITABLE FOR COPPER
    CONDUCTORS ONLY.

Figure 19–3

3. Rated-load current is defined as
   A. The current that results when a hermetic compressor is operated at rated load, voltage, and frequency
   B. The current that results when a hermetic compressor is operated under any conditions
   C. The current that results during initial start-up of a hermetic compressor
   D. The maximum allowable current for a particular hermetic compressor

4. Which of the following is typically covered by local codes?
   A. Maximum CFM per diffuser
   B. Wire sizing
   C. Refrigerant line sizing
   D. Location of condensate drains

5. Which of the following is covered by the National Fire Protection Code?
   A. Wire sizing
   B. Use of return-air sensors

C. Location of service platforms

D. Clearances around condensing units

6. Which of the following jobs most requires the wearing of safety glasses or goggles?

A. Changing the oil in a vacuum pump

B. Cutting a cap tube

C. Brazing copper tubing

D. Demonstrating programming on a thermostat

7. The efficiency of a heat pump operating in the cooling mode is expressed in which of the following standards?

A. AFUE rating

B. SEER rating

C. COP rating

D. LRA rating

8. The regulations concerning termination locations of secondary condensate drains for attic-mounted horizontal blower coils are found in the

A. Uniform Mechanical Code

B. Uniform Building Code

C. National Fuel Gas Code

D. National Electric Code

SPECIFIC SECTOR PRACTICE EXAMINATIONS

AIR CONDITIONING SECTOR EXAMINATION

1. A technician on a preventive maintenance call notices that a lawn trimmer has frayed the low-voltage wiring to an outdoor unit. What should be done?

A. Replace and bury the new wire.

B. Tape the frayed wire and bury it.

C. Relocate the wire and cover it with tape.

D. Replace and relocate the wire on the unit.

2. Excessive liquid-line length or lift can cause which of the following conditions?

A. Flash gas in the liquid line

B. Low superheat

C. Ice formation on the condenser coil

D. High liquid-line pressures

3. To determine the amount of sensible heat added to the refrigerant above the saturation point in an evaporator coil, the installer would

A. Perform a subcooling check on the liquid line.

B. Perform a superheat check on the evaporator line.

C. Perform a cfm check across the evaporator coil.

D. Perform a delta-T check across the evaporator coil.

4. A check of a residential HVAC system reveals a supply-air static pressure of 0.20 and a return-air static pressure of −0.15. What do these readings tell you?

A. The difference between supply static and return static is unacceptably low.

B. The system is operating normally.

C. Supply static is unacceptably low.

D. Return-air static is unacceptably high.

5. A customer had a system installed more than a year ago. In the past few months, the air conditioner has slowly stopped cooling as well as it once did. Air volume to all rooms is very low. The problem is most likely a

A. Clogged filter

B. Bad fan blade

C. Blower motor that is going bad

D. Refrigerant circuit that is leaking

6. A technician finds that an R-22 unit with a freeze-protection low-pressure control occasionally trips on startup, but it operates normally when it runs. The technician determines the unit shuts down when the low-side pressure drops below 56 psig. Which of the following is the most likely problem?

A. The low-pressure control has drifted to an excessively high cutout.

B. The low-pressure control has drifted to an excessively low cutout.

C. The internal overload is opening.

D. The unit is overcharged.

7. A customer's compressor has failed, and the customer wishes to replace the outdoor unit only. To properly replace the unit, it must be sized to match

A. The cooling load of the space

B. The capacity of the evaporator

C. The size of the slab

D. The available clearances around the unit

8. A tripped internal protector on a hermetic compressor should be reset by

A. Resetting the wall thermostat

B. Manually resetting it

C. Waiting for it to cool off

D. Replacing the protector

9. A vapor-charged temperature control uses what in its power element?
   A. Oil
   B. Water
   C. Antifreeze
   D. Refrigerant

10. A simple time-activated switch is an example of a control that does not use
    A. Feedback
    B. Electronics
    C. Calibration
    D. Manual adjustment

11. Any moisture remaining in a refrigerant circuit after evacuation is removed from the refrigerant by
    A. A receiver
    B. An accumulator
    C. A filter drier
    D. The moisture indicator

12. A thermistor is a component used in an electronic wall thermostat to measure
    A. Temperature
    B. Humidity
    C. Atmospheric pressure
    D. Barometric pressure

13. A rejected bid that was accompanied by a specification sheet for proposed work by an HVAC contractor becomes the property of
    A. The customer
    B. The contractor
    C. Any competitor
    D. Anyone who has a copy

14. What legal condition exists after a verifiable verbal agreement on a bid is reached by both the customer and a contracting company?.
    A. The contracting company has the legal right to change its bid within 7 days.
    B. A legal condition does not exist unless the attorneys for both parties sign an agreement.
    C. The contracting company could be legally liable for the bid.
    D. A legal condition does not exist until a written contract is signed by both parties.

15. Which of the following can reduce indoor air quality?
    A. Oversized supply diffusers
    B. Undersized supply diffusers
    C. Low humidity in the supply air
    D. Excessive humidity in the supply air

16. The slopes of the condensate lines in an attic are often set by
    A. Underwriter's Laboratory
    B. AGA
    C. ARI
    D. Uniform Mechanical Code

17. The number of air changes recommended for a commercial building can be found in the specifications of
    A. ARI
    B. ASHRAE
    C. OSHA
    D. Uniform Mechanical Code

18. The sizing of an expansion valve is critical because
    A. A valve too large will be erratic and a valve too small introduces excessive pressure drop.
    B. A valve too small will be erratic and a valve too large introduces excessive pressure drop.
    C. The larger the valve the larger its superheat setting.
    D. The larger the valve the smaller its superheat setting.

GAS FURNACE SECTOR EXAMINATION

1. The flue connection for a natural draft gas furnace is typically located on the
   A. Heat exchanger
   B. Draft diverter
   C. Burner manifold
   D. Bottom of the furnace

2. The products of combustion produced when natural gas is completely burned are
   A. Fluorine, carbon dioxide, and water
   B. Fluorine, carbon monoxide, and water
   C. Nitrogen, carbon monoxide, and water
   D. Nitrogen, carbon dioxide, and water

3. The vent pressure of a natural draft gas furnace immediately after the heat exchanger is
   A. Positive
   B. Negative
   C. Atmospheric
   D. The same as the blower outlet

4. All residential gas furnaces use some type of
   A. Inducer blower
   B. Standing pilot
   C. Heat exchanger
   D. Electronic flame prover

5. Which of the following components is energized first after an intermittent ignition gas furnace receives a call for heat?

   A. Induced draft motor

   B. Pilot valve

   C. Blower motor

   D. Gas valve

6. Which of the following statements is true concerning in-shot burners?

   A. In-shot burners are typically quieter than ribbon-style burners.

   B. In-shot burners typically generate more heat than ribbon-style burners.

   C. In-shot burners typically generate less heat than ribbon-type burners.

   D. In-shot burners typically generate greater amounts of carbon monoxide than ribbon-type burners.

7. Which of the following is an advantage of a hot-surface ignition system over a standing-pilot system?

   A. They do not require low voltage transformers.

   B. They are less affected by air drafts.

   C. They are unaffected by supply voltage polarity.

   D. They eliminate delayed ignition.

8. Which of the following components is typically used only on induced-draft condensing-gas furnaces?

   A. Electronic fan control

   B. Hot-surface ignition

   C. Induced-draft blower motor

   D. Secondary heat exchanger

9. Condensate piping must be properly trapped to prevent

   A. Excessive air noise in the piping

   B. Ice formation in colder temperatures

   C. Condensate running off the evaporator coil

   D. Condensate overflowing the drain pan

10. A belt-driven blower with a three-phase motor is generally found in

    A. All residential HVAC systems

    B. All commercial HVAC systems

    C. HVAC systems with cooling capacities under 5 tons

    D. HVAC systems with cooling capacities of 10 tons and above

11. Color-coded wires are typically used in which circuits of HVAC equipment?

    A. High-voltage AC circuits only

    B. High-voltage DC circuits only

    C. Control or low-voltage AC wiring

    D. Low-voltage DC circuits only

12. Which of the following examples requires an additional combustion-air opening capable of supplying 100% of the air requirements of a nondirect-vent gas furnace?

    A. Furnace located in an interior close (confined space) of a residence that obtains combustion air through a louvered door

    B. Furnace located in an exterior area (unconfined area) of a residence that obtains combustion air from surrounding sources

    C. Furnace located in an equipment room (confined area) of a commercial building having two combustion-air openings

    D. Furnace located in an interior closet (confined area) of a residence having two combustion-air openings

13. To reduce customer complaints about an attic-mounted horizontal furnace producing excessive noise, a technician should

    A. Secure the furnace to the ceiling joists using wood screws.

    B. Set the furnace on concrete blocks.

    C. Suspend the furnace from the roof trusses.

    D. Locate the return-air grille as close as possible to the return-air plenum.

14. The first event you should look for after the thermostat is set for a call for heat on a hot-surface ignition, noncondensing gas furnace is the:

    A. Gas valve opens

    B. Induced-draft motor begins to run

    C. Ignitor glows

    D. Supply blower begins to run

15. During a preventive maintenance check, light deposits of soot and rust are discovered in the bottom of a heat exchanger with ribbon-style natural gas burners. In addition to tests for proper burner-flame patterns, what other actions, if any, should be taken prior to returning the furnace to operation?

    A. Check for unsafe levels of carbon monoxide in the residence.

    B. The gas valve should be changed whenever sooting occurs.

    C. Check for proper ignition sequence of all burners.

    D. No extra steps are needed.

16. A customer calls in to report insufficient heat from a gas furnace. Which of the following could be causing the customer complaint?

    A. Inadequate flow rate at the gas supply

    B. Cracked heat exchanger

    C. Excessive combustion air

    D. A too-high manifold pressure setting

17. A technician is called to check the performance of a gas furnace. On arrival, you wish to compare the temperature rise across the furnace to the nameplate rating. What measurement would you take to accomplish this task?

    A. Temperature of the combustion air compared to the temperature of the supply air

    B. Temperature of the combustion air compared to the temperature of the return air

    C. Temperature of the exhaust gas compared to the temperature of the combustion air

    D. Temperature of the supply air compared to the temperature of the return air

18. A homeowner calls complaining of headaches and nausea during operation of his natural gas furnace. A service technician observes that all burners appears to have the same flame pattern on start-up, but the burners begin to "roll out" after the furnace blower is energized. This symptom is typically an indication of

    A. Insufficient combustion air

    B. Incorrect gas pressure

    C. Blocked return air

    D. Incorrect blower setting

19. Which of the following conditions would typically create a "lazy" burner flame pattern on a natural gas furnace using ribbon-style burners?

    A. Gas manifold pressure set too high

    B. Supply-gas pressure set too high

    C. Excessive vent pipe height

    D. Blocked heat exchanger

20. On start-up, a natural gas heating system has cold air blowing from the registers. What could be the cause of this?

    A. Burner manifold pressure is too high.

    B. Supply duct runs are too short.

    C. Blower-ON delay is set too short.

    D. Blower-ON delay is set too long.

21. The benefit of using a two-stage gas valve on a residential furnace is that it

    A. Modulates gas flow to a constant temperature rise

    B. Extends the life of the gas valve

    C. Permits gas flow to the main burner with a stepped opening time for safe ignition

    D. Permits gas flow to the main burner in several stages controlled by the thermostat

22. Thermostatically controlled gravity-vent-wall furnaces that do not require external electrical power typically rely on which of the following ignition-control systems?

    A. A standing pilot and hot-surface ignition

    B. A standing pilot and powerpile generator

    C. A standing pilot and thermocouple

    D. A standing pilot only

23. What accessory would you quote to a customer with allergies who wants to improve her indoor-air quality?

    A. An electronic air cleaner

    B. An economizer

    C. A dehumidifier

    D. A carbon monoxide detector

24. In dry climates during the winter months, what effect would increasing the humidity have on the thermostat settings in the heated space?

    A. Increased comfort at increased temperature settings

    B. Increase comfort at decreased temperature settings

    C. Increased comfort regardless of temperature settings

    D. Increasing the humidity would have no effect on setting for comfort

25. Selection of a residential gas furnace is typically based on structural heat loss, physical location of the equipment, fuel efficiency rating, and

    A. Number of conditioned rooms

    B. Average indoor ceiling height

    C. Type of fuel available

    D. Total height of the vent pipe

26. Which of the following conditions would typically require locating supply registers in the floors of both the upper and lower levels of a two-story residence?

    A. In predominately colder climates when the furnace is located in the basement

    B. In predominately colder climates when the furnace is located in the attic

    C. In predominately warmer climates when the furnace is located in the basement

    D. In predominately warmer climates when the furnace is located in the attic

OIL FURNACE SECTOR EXAMINATION

1. A fuel-line heater would typically be used with an oil furnace to
   A. Decrease oil viscosity
   B. Increase oil viscosity
   C. Increase nozzle fuel-flow rate
   D. Increase oil pressure

2. The height of the vent pipe will have what effect on draft intensity?
   A. The lower the vent, the greater is the draft.
   B. The higher the vent, the greater is the draft.
   C. Draft doubles as vent height doubles.
   D. The height of the vent has no effect on draft.

3. Modern homes and buildings are being constructed tighter to help them become more energy efficient. The significant negative effect of this is that
   A. Supply blowers operate with a heavier load because of a positive pressure in the space.
   B. Air infiltration for ventilation is reduced.
   C. The system run times are longer, adding wear to all heating equipment.
   D. Any door opening quickly drops indoor temperature in the heating season.

4. Oil is sometimes the preferred heating fuel in some regions of the United States. Why?
   A. Electricity and gas are higher in cost.
   B. Electricity and gas are not available.
   C. Electricity and gas equipment are not available.
   D. The comfort of oil heat is preferred over gas and electric.

5. Which of the following locations should be avoided when locating a supply register?
   A. The floor of a residence having high side-wall returns
   B. An interior wall of a residence next to a return grille
   C. A ceiling of a residence having low sidewall returns
   D. A residential kitchen

6. In HVAC equipment, color-coded wire is typically used in which of the following applications?
   A. In high-voltage AC circuits only
   B. In high-voltage DC circuits only
   C. In control or low-voltage AC wiring
   D. In low-voltage DC circuits only

7. Which of the following residential oil-furnace locations requires an additional opening capable of supplying 100% of its combustion air requirements?
   A. A small closet (confined space) drawing combustion air through a louvered door
   B. A garage (unconfined area) drawing combustion air from surrounding sources
   C. A large attic (unconfined area) drawing combustion air from surrounding sources
   D. A cellar (confined area) having two openings, each capable of supplying 100% of the combustion air

8. Which of the following oil-furnace components should be replaced annually?
   A. Burner head
   B. Fuel filter
   C. Flame prover
   D. Combustion chamber

9. Two-stage-heat, single-stage-cool thermostats are typically used on dual-fuel heat-pump systems using an oil furnace for second-stage heating. How would the heat pump operate if the second stage of the thermostat malfunctioned to an open circuit on an extremely cold day? Assume the emergency heat switch is not turned on.
   A. The heat pump will operate on calls for heat without energizing the oil furnace.
   B. The oil furnace will operate on calls for heat without energizing the heat pump.
   C. Neither the heat pump nor the oil furnace will operate in heating modes.
   D. The oil furnace will operate in emergency-heat mode only.

10. A customer calls in to report insufficient heat from his oil furnace. Which of the following could be causing the customer complaint?
    A. Regulator set too high
    B. Inadequate flow rate at the oil supply
    C. Cracked heat exchanger
    D. Cracked combustion chamber

11. A technician has dismantled an oil furnace during a check for excessive wear. The nozzles that were removed from the oil burner should *not* be reused unless
    A. They are soaked overnight in nonflammable solvent.
    B. They are blown out with high-pressure air.
    C. They are cleaned and polished.
    D. No other replacement is available.

12. An oil furnace installed with an undersized fuel nozzle could result in a customer complaint of
    A. Excessive heat
    B. Insufficient heat
    C. High return-air noise
    D. Tripped breakers

13. Voltage to an oil furnace may be
    A. No more than 10% over nameplate voltage but not under nameplate voltage
    B. No more than 15% over nameplate voltage but not under nameplate voltage
    C. No more than 10% under or over nameplate voltage
    D. No more than 15% under or over nameplate voltage

14. What is the difference between the wiring of an electromechanical thermostat and an electronic programmable thermostat?
    A. Both the hot leg and the common of a 120-V supply must be run to an electronic programmable thermostat.
    B. The wire size used on an electronic thermostat is of a heavier gauge than that of an electromechanical thermostat.
    C. Both the hot leg and the common of the 24-V supply must be run to an electronic programmable thermostat.
    D. Typically there is no difference; they are designed to be interchangeable.

HEAT PUMP SECTOR EXAMINATION

1. Which of the following occurs when a heat pump terminates the defrost cycle?
    A. The outdoor fan is shut off.
    B. The reversing valve switches to the heating mode.
    C. The reversing valve switches to the cooling mode.
    D. All auxiliary heat is immediately energized.

2. A single-trunk duct system that extends in one or two directions from the unit and has multiple branch ducts is called a
    A. Perimeter duct system
    B. Extended-plenum duct system
    C. Radial-duct system
    D. Reducing trunk system

3. In HVAC equipment, color-coded wire of red to the hot side of the transformer, yellow to the contactor, green to the fan relay, and white to the heat relay is typically used in which of the following applications?
    A. In high-voltage AC circuits only
    B. In high-voltage DC circuits only
    C. In control or low-voltage AC wiring
    D. In low-voltage DC circuits only

4. Which of the following is the most likely cause of low indoor-coil pressure in a properly charged heat pump running in the cooling mode?
    A. An undersized metering piston
    B. An oversized metering piston
    C. Excessive blower speed
    D. Oversized suction line drier

5. Condensate piping must be properly trapped
    A. When the condensate drain pan is in at a negative pressure
    B. When excessive air noise occurs in the piping
    C. When ice formation occurs in colder temperatures
    D. When condensate runs off the evaporator coil

6. Which of the following is used to automatically bypass refrigerant flow around a metering device in a heat-pump circuit?
    A. A king valve
    B. A check valve
    C. A globe valve
    D. A relief valve

7. To maintain maximum comfort in a heat-pump installation while operating in the heating mode, the duct system must be sized to supply sufficient CFM to match heat loss at
    A. The lowest blower speed
    B. The highest blower speed
    C. At a blower speed to maximize discharge temperature
    D. At a blower speed to minimize discharge temperature

8. Which of the following joints, when properly done, will best join a copper tube to a copper fitting?
    A. Compression
    B. Flare
    C. Sweat
    D. Interference

9. A condensing unit should be mounted on a slab that is level to
   A. Allow any accumulated water to evenly run out of the cabinet
   B. Increase airflow through the fin area
   C. Prevent the unit from tipping over
   D. Make certain any top discharge air blows straight up

10. Packaged units should not be installed directly on
   A. A foundation slab
   B. A concrete slab
   C. Wooden timbers
   D. The ground

11. Piston-type metering devices can be installed on heat pumps at
   A. The outdoor unit only
   B. The indoor unit only
   C. Both the outdoor and indoor units
   D. Either the indoor unit or outdoor unit but not both

12. A heat pump is supplied from the factory with charge for a 15-ft line set. The dataplate shows the required system charge to be 5 lb 2 oz. The line set uses a ⅜-in. liquid line and is 35 ft long. The manufacturer's instruction sheet specifies 0.8 oz of refrigerant to be added for each added foot of refrigerant line length. What is the additional refrigerant charge required for the system?
   A. 6 lb 2 oz
   B. 1 lb 4 oz
   C. 16 oz
   D. 0.8 oz

13. Which of the following connections should typically be done using sheet-metal screws only?
   A. Attaching plenums to coils
   B. Attaching start collars to plenums
   C. Connecting sections of rectangular metal duct
   D. Mounting diffusers to boots

14. A heat pump operating in the heating mode with inadequate airflow across the indoor coil will typically exhibit which of the following symptoms?
   A. A higher-than-normal temperature rise across the outdoor coil
   B. A lower-than-normal temperature rise across the outdoor coil
   C. A higher-than-normal temperature rise across the indoor coil
   D. A lower-than-normal temperature rise across the indoor coil

15. Excessive liquid-line length or lift in a heat pump can cause which of the following conditions?
   A. Flash gas in the liquid line
   B. Low superheat
   C. Ice formation on the condenser coil
   D. High liquid-line pressure

16. Pressure readings on an operating heat pump indicate low head pressure and high suction pressure. Which of the following could be a cause of these readings?
   A. A clogged filter drier
   B. A restricted metering device
   C. A reversing valve with an internal leak under the slide
   D. Operation in the heating mode during very cold outdoor temperatures

17. Which of the following symptoms would a TEV-equipped heat pump typically exhibit if the check valve on the outdoor unit were stuck in the closed position?
   A. The compressor pumps down in both heating and cooling modes.
   B. The system performs properly in cooling but the compressor pumps down in heating.
   C. The system performs properly in heating, but the compressor pumps down in cooling.
   D. The system performs properly in cooling, but it has extremely high suction pressures in heating.

18. A technician would require the outdoor-coil pressure and liquid-line temperature of a heat pump operating in the cooling mode to determine the
   A. Amount of subcooling of the refrigerant at the liquid line
   B. Total system capacity
   C. Amount of refrigerant superheat
   D. Indoor-coil temperature

19. A customer had a system installed over a year ago. In the past few months, the heat pump has slowly stopped cooling as well as it once did. Air volume to all rooms is very low. The problem is most likely a
   A. Bad fan blade
   B. Clogged filter or dirty evaporator coil
   C. Blower motor that is going bad
   D. Refrigerant circuit that is leaking

20. A typical heat pump uses a high-pressure control to shut off the compressor if the condensing pressures become excessive. This control uses a switch that breaks the compressor contactor on

    A. Rise at the compressor discharge
    B. Fall at the compressor discharge
    C. Difference between the high and low side
    D. Combined high and low side

21. During replacement of a heat-pump outdoor unit, the technician used one ⅞-in. and one ½-in. short radius elbow and the existing copper lines. Using Figure 19–4, what was the equivalent length added to the refrigerant lines?

    A. 1.2 ft
    B. 1.8 ft
    C. 2.6 ft
    D. 3.0 ft

22. Slightly undersizing a heat pump will typically reduce the number of customer complaints in locations with

    A. Exceptionally low temperatures in the winter
    B. Exceptionally high temperatures in the summer
    C. Low humidity levels in the summer
    D. High humidity levels in the summer

23. What term describes the outdoor conditions at which the cost of heating using gas or oil is the same as when using the heat pump alone?

    A. Comfort-heat index
    B. Outdoor-temperature index
    C. The economic balance point
    D. The thermal balance point

Figure 19–4

DESIGN

EQUIVALENT LENGTHS OF ELBOWS		
LINE SIZE INCHES (O.D.)	90° SHORT RADIUS ELBOW (FT)	90° LONG RADIUS ELBOW (FT)
¼	0.7	0.6
5/16	0.8	0.7
⅜	0.9	0.8
½	1.2	1.0
⅝	1.5	1.3
¾	1.6	1.4
⅞	1.8	1.6
1⅛	2.4	2.0

AIR DISTRIBUTION SECTOR EXAMINATION

1. Customer complaints of "cold air blowing" from the registers of a properly charged heat pump system are typically reduced if

    A. Air ducts and supply registers are slightly undersized to increase the air velocity.
    B. Air ducts and supply registers are properly sized to maintain the correct velocity.
    C. Only floor registers are used in a residence.
    D. Only sidewall air registers are used in a residence.

2. A perimeter-loop supply-duct system using floor or low sidewall outlets is preferred in which of the following types of residential construction?

    A. Pier and beam construction in a warm-weather climate using a closet-installed up-flow gas furnace
    B. Pier and beam construction in a cold-weather climate using a closet-installed up-flow gas furnace
    C. Slab construction in a cold-weather climate using a closet-installed counterflow gas furnace
    D. Slab construction in a warm-weather climate using an attic-mounted horizontal gas furnace

3. Which of the following adds significantly to the pressure drop of a duct system?

    A. Fittings
    B. External insulation
    C. Oversized diffusers
    D. Oversized returns

4. Large metal duct that is suspended should be supported using

    A. Plumber's strap screwed to the side wall
    B. Metal strap formed into a saddle
    C. Cloth strap formed into a saddle
    D. U-channel and rods

5. After connecting two sections of metal ductwork using drive cleats, you should

    A. Bend the ends of the drives over the duct
    B. Bend the ends of the drives away from the duct
    C. Cut off the ends of the drives
    D. Crimp the ends of the drives and cut them off

6. If it is necessary to convert a section of a flexible duct trunk line to rectangular metal duct, the connection point should be fitted with a

    A. Flexible collar
    B. Ductboard transition

C. Sheet-metal transition

D. Plywood transition

7. Metal duct that has hung for several years must be checked for sags and misalignments because

A. They can place excessive stress on joints and seals.

B. They increase static pressure in the duct.

C. They decrease static pressure in the duct.

D. They increase CFM at the diffusers.

8. Blower belt noise is caused by which of the following?

A. Overvoltage to motor

B. Worn pulley

C. Undervoltage to motor

D. Wrong blower wheel

9. Which of the following could result from an undersized supply-duct system?

A. Increased capacity of the system in cooling

B. Poor dehumidification in cooling

C. Compressor short cycling

D. Wide temperature swings in the space

10. An undersized supply-duct system could lead to what condition in heating?

A. Preignition

B. Cut out on high limit

C. Flame rollout

D. Condensation in the vent

11. A properly balanced air-distribution system delivers

A. The same total velocity of supply air to each space controlled by the system

B. A proportional total velocity of supply air to each space controlled by the system

C. The same quantity of supply air to each space controlled by the system

D. A proportional quantity of supply air to each space controlled by the system

12. The next test a technician would typically perform after adjusting dampers to a conditioned area of a residence would be to

A. Measure CFM and air-velocity levels at supply-air grilles.

B. Measure CFM levels on all other ductwork connected to the same system.

C. Leak-check all duct joints.

D. Check refrigerant pressures.

13. Humidity control is an important contributor to total comfort in a cooling system. A properly sized duct system is an important part of humidity control because it

A. Ensures the maximum amount of air is moving over the people in the space

B. Ensures the proper amount of air is moving over the evaporator

C. Maintains a constant static pressure in the supply plenum

D. Maintains a constant static pressure in the return plenum

14. In which of the following installations would the supply and return ductwork need to be insulated and weatherproofed?

A. A curb-mounted, packaged A/C unit supplying a totally enclosed interior-duct system

B. A split-system A/C unit supplying a totally enclosed interior-duct system

C. A ground-level packaged system requiring ductwork to be totally enclosed in a crawl-space

D. A ground-level packaged system requiring ductwork to run up an outside wall and into an attic

15. When selecting grilles and diffusers, which of the following should a technician avoid to reduce noise problems?

A. Undersizing supply-air diffusers

B. Oversizing supply-air diffusers

C. Oversizing return-air grilles

D. Using fixed-vane diffusers

APPLIANCE CERTIFICATION

The final certification under discussion is the small-appliance certification program sponsored by National Appliance Service Technician Certification (NASTeC). This not-for-profit corporation was created to promote skilled and professional serving of major home appliances. The Association of Home Appliance Manufacturers (AHAM) provides NASTeC with administrative, consulting, clerical, and legal services.

The goals of NASTeC are in line with the goals of the other industry programs. Technicians passing the extensive exam receive a certificate, wallet card, patch, and bar designating the area of certification. The exam is divided into the following two categories and three specialties:

- Basic Skills Examination
- Specialty Examinations
  - Refrigerators, freezers, window units, and dehumidifiers

- Washers, dryers, dishwashers, disposals, and trash compactors
- Ranges, ovens, and microwave ovens

The Basic Skills Examination has 80 multiple-choice questions. The basic exam and specialty exam must be passed in order to receive technician certification. Technicians passing the basic exam and all three specialty exams receive Universal Appliance Certification. A technician failing the Basic Skills Examination and passing one of the other specialty exams during the same test day is not considered certified. The technician must retake the Basic Skills Examination and the specialty exams.

The Basic Skills Examination is very similar in content to the other industry exams. The technician can be expect to see questions on

- Interpersonal relationships
- Personal behavior and work habits
- General information
- Safety
- Communication skills
- Basic mechanical skills
- Basic electrical skills
- Basic gas skills

Because this is an air-conditioning textbook we emphasize only the refrigeration specialty. The technician will find questions that relate to

- General questions on heat transfer, pressure-temperature charts, recovery procedures, and safety practices
- Refrigerant questions covering types of refrigerants, refrigerant oils, and troubleshooting refrigerant-side problems
- Refrigerant cycle operation
- Specific questions on refrigerators, freezers, window units that heat and cool, and dehumidifiers

REVIEW

- HVAC/R certification is making itself felt throughout the air-conditioning industry. Something has to be done to improve the image of the professional technician. Certification is one of the stepping stones to this goal. Let us not kid ourselves into thinking that industry certification is a sure bet. Everyone needs to promote quality industry certification. Air-conditioning and refrigeration companies that hire technicians have to hire and promote on the basis of certification. Educators need to improve their curricula. Customers need to demand that only certified technicians service their equipment. Builders and construction groups should contract with HVAC/R companies that have certified installers and technicians. Suppliers and manufacturers may offer incentives to companies with certified technicians because the job will be done right the first time without the need for warranty replacement because the equipment or part was damaged during installation. Industry certification will benefit everyone, and it is long overdue. There are many certification exam options available to the technician. Certifications associated with the Air-Conditioning Refrigeration Institute (ICE), NATE, RSES, ACCA and NASTeC are the only quality exams that test the skills and knowledge of those attempting to exhibit their professionalism and experience in the field.

USEFUL FORMULAS, CONVERSIONS, AND COMMON ABBREVIATIONS

FORMULAS

- Volts = amps × ohms ($V = I \times R$)

- Watts = volts × amps ($P = V \times I$)

- Power factor = $\dfrac{\text{watts}}{\text{volts} \times \text{amps}}$

- Resistance in series = $R_1 + R_2$

- Two resistors in parallel = $\dfrac{R_1 \times R_2}{R_1 + R_2}$

- Multiple resistors in parallel =

  $\dfrac{1}{R_T} = \dfrac{1}{R_1 + R_2 + R_3 + \cdots}$

- Capacitance in parallel = $C_1 + C_2$

- Capacitance in series = $\dfrac{C_1 \times C_2}{C_1 + C_2}$

WATER HEATING, COOLING AND HEAT RECLAIM COILS, WATER CHILLERS, CONDENSERS, ETC.

$Q = 500 \times \text{gpm} \times \Delta T = \text{BTU/h}$

$\Delta T = \dfrac{Q}{500 \times \text{gpm}}$

For brines, Q = 500 × gpm × ΔT × (sp. ht. × sp. gr. of brine) = BTU/h

AIR COILS

Q sensible = 1.08 × cfm × ΔT = Btu/h

Q latent = 0.68 × cfm × ΔSH = Btu/h

Q total = 4.5 × cfm × ΔH = Btu/h

Lb/h condensate = $\dfrac{4.5 \times \text{cfm} \times \Delta \text{ SH grains}}{7000 \text{ grains/lb}}$

SHR (sensible heat ratio) = $\dfrac{Q \text{ sensible}}{Q \text{ total}}$

HEAT TRANSMISSION

Q total = U × A surface × ΔT = Btu/h

PRODUCT

All conversion factors used in standard calculations must be corrected for other than standard properties

Sensible heat in Btu/h = lb/h × sp. ht. × ΔT

Latent heat in Btu/h = lb/h × lt. ht. in Btu/lb

Heat of resp. in Btu/h = lb × heat of respiration in Btu/lb/h

- Pulley-speed ratio

  $\dfrac{\text{Drive pulley (dia.)}}{\text{Driven pulley (dia.)}} \times \dfrac{\text{driven pulley (RPM)}}{\text{drive pulley (RPM)}}$

- Air mixture

  $\dfrac{\substack{\text{Return air} \\ \text{(CFM × temp)}} + \substack{\text{outside air} \\ \text{(CFM × temp)}}}{\text{RA CFM + OA CFM}} =$

  Final mixture, air temperature

- Sensible heat formula

  CFM $\times$ 1.08 $\times$ Δt = Btuh

- Converting Celsius to Fahrenheit

  $(C \times 1.8) + 32 = F$

  Converting Fahrenheit to Celsius:

  $$\frac{F - 32}{1.8} = C$$

- Static water pressure

  $$\frac{\text{Height (l.f.)}}{2.31} = \text{psig} \qquad \text{(l.f. = linear feet)}$$

- Specific volume of air to pounds of air per hour (minute)

  $$\frac{Q \times 60}{\text{SP. V.}} = \text{lb/h}$$

  $$\frac{Q}{\text{SP. V}} = \text{lb/min}$$

  Q = Quantity (CFM)

  60 = 60 minutes in one hour

  Sp. V. = specific volume

- Btu output of electric motor

  H.P. $\times$ 746 $\times$ 3.416 = Btuh

- Sensible heat factor (SHF)

  $$\frac{\text{Sensible heat}}{\text{Total heat}} = \text{SHF}$$

- U factor for heat load transmission through walls, etc.

  $A \times \Delta t \times U$ = Btuh

  A = Area of wall in square feet

  t = Temperature difference inside to outside

  U = U factor (given)

- Area of a circle

  $R^2 \times \pi$ = area

  R = radius of circle

  π = 3.1415926

- To find the area of a circle, multiply the square of the diameter by 0.7854.

  $A = d_2 \times 0.7854$

- To find the circumference of a circle, multiply the diameter by 3.1416.

  $C = d \times 3.1416$

- To find the volume of a ball (sphere), multiply the cube of the diameter by 0.5236.

  $V = d^3 \times 0.5236$

- Doubling the diameter of a pipe increases its capacity four times.

- To find the pressure in pounds per square inch at the base of a column of water, multiply the height of the column in feet by 0.433.

- CFM by duct size

  $A \times V = Q$

  A = area of duct in square feet

  V = velocity of air in fpm

  Q = quantity of air in cmf

- To change square feet to square inches:

  $$\frac{\text{in.}^2}{144} = \text{ft}^2$$

  To change square inches to square feet:

  $\text{ft}^2 \times 144 = \text{in.}^2$

- Total heat formula:

  CFM $\times$ 4.5 $\times$ Δt = Btuh

- Sides of square duct from area:

  Side = $\sqrt{\text{area}}$

- Water velocity:

  $$\frac{\text{gpm} \times 8.35}{62.4} = \text{ft}^3\text{/min}$$

- Air velocity

  $$\frac{\text{CFM}}{\dfrac{\pi(D^1/12)^2}{4}} = \text{velocity 1}$$

  $$\frac{\text{CFM}}{\dfrac{\pi(D^2/12)^2}{4}} = \text{velocity 2}$$

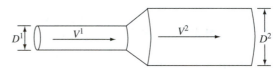

- U-factor from R, C, K, or U

  $$\frac{1}{U} = R$$

  $$\frac{1}{R} = U$$

  $$\frac{1}{C} = U$$

  $$\frac{K}{\text{thickness in inches}} = C$$

U-factor formula

$$U = \frac{1}{R + R + R + R} + \cdots$$

Step 1   Convert C, U, and K to R.

Step 2   Put these R values into the U-factor formula.

Step 3   Add the R values.

Step 4   Divide 1 by the total of the R values.

Step 5   The dividend is the new U value.

- Hydronic coil process formula

$$4.5 \times 500 \times \Delta t = \text{Btuh}$$

For example, the difference in degrees Fahrenheit of the water temperature entering and water temperature leaving is Δt (delta *t*).

Plug this Δt into the formula, and the answer will be the amount of heat lost or gained by the *water* passing through the coil (or heat exchanger).

Heat of rejection (condenser) = Heat of compression + heat of absorption (evaporator)

Net refrigerating effect = Heat content of vapor leaving evaporator Btu/lb − heat content of liquid entering evaporator, Btu/lb

Net refrigerating effect, Btu/lb = Latent heat of vaporization, Btu/lb − change in heat content of liquid from condensing to evaporating temperature, Btu/lb

Net refrigerating effect, Btu/lb =

$$\frac{\text{capacity, Btu/min}}{\text{refrigerant circulated, lb/min}}$$

Refrigerant circulated, lb/min =

$$\frac{\text{load or capacity, Btu/min}}{\text{net refrigerating effect, Btu/lb}}$$

Compressor displacement, ft³/min = Refrigerant circulated lb/min × volume of gas entering compressor, ft³/lb

Compressor displacement, ft³/min =

$$\frac{\text{capacity, Btu/min} \times \text{volume of gas entering compressor, ft}^3/\text{lb}}{\text{net refrigerating effect, Btu/lb}}$$

Heat of compression, Btu/lb = Heat content of vapor leaving compressor, Btu/lb − heat content of vapor entering compressor, Btu/lb

Heat of compression, Btu/lb =

$$\frac{42{,}418 \text{ Btu/min} \times \text{compression horsepower}}{\text{refrigerant circulated, lb/min}}$$

Compression work, Btu/min = Heat of compression, Btu/lb − refrigerant circulated, lb/min

Compression, horsepower =

$$\frac{\text{compression work, Btu/lb}}{\text{conversion factor, 42,418 Btu/min}}$$

Compression, horsepower =

$$\frac{\text{heat of compression, Btu/lb} \times \text{capacity, Btu/min}}{42{,}418 \text{ Btu/min} \times \text{net refrigerating effect, Btu/lb}}$$

Compression, horsepower =

$$\frac{\text{capacity, Btu/lb}}{42{,}418 \text{ Btu/min} \times \text{coefficient of performance}}$$

Compression, hp/ton =

$$\frac{4.715}{\text{coefficient of performance}}$$

Power, W = compression hp/ton × 745.7

Coefficient of performance =

$$\frac{\text{net refrigerating effect, Btu/lb}}{\text{heat of compression, Btu/lb}}$$

Capacity, Btu/min = refrigerant circulated, lb/min × net refrigerating effect, Btu/lb

Capacity, Btu/min =

$$\frac{\text{compressor displacement, ft}^3/\text{min} \times \text{net refrigerating effects, Btu/lb}}{\text{volume of gas entering compressor, ft}^3/\text{lb}}$$

Capacity, Btu/min =

$$\frac{\text{compression horsepower} \times 42{,}418 \text{ Btu/min} \times \text{net refrigerating effect, Btu/lb}}{\text{heat of compression, Btu/lb}}$$

OHM'S LAW WHEEL

This chart shows four ways to figure each value; Amps (I), Volts (E), Ohms (R) or Watts (W).

Example: A 4800-W electric heat element is connected to a 240-V circuit. How many amps (I) does it draw? Solution: Locate Amps section of chart—

$$\frac{\text{Watts (W)}}{\text{Volts (E)}} = \text{Amps (I)}$$

Thus 4800 ÷ 240 = 20 A. Carried further, what is the resistance?

$$\frac{\text{Volts}^2 \ (E^2)}{\text{Watts (W)}} = \Omega \ (R) \quad 240 \times 240 \div 4800 = 12 \ \Omega$$

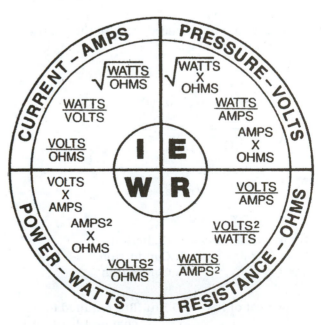

SINGLE-PHASE LOADS

OHM'S LAW FOR DIRECT CURRENT

W = watts
I = current (amperes)
E = electromotive force (volts)
R = resistance (ohms)

To obtain any value in the center circle, for direct or alternating current, perform the operation indicated in one segment of the adjacent outer circle.

3-PHASE DELTA LOADS

3ϕ balanced loads = $P_1 + P_2 + P_3$

Total line current = total power (balanced load)

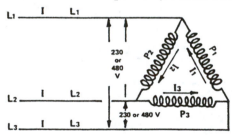

If the phases are unbalanced, each of the phases will differ from the others:

Formulas:

$$IL_1 = \sqrt{I_2^2 + I_1^2 + (I_1 \times I_2)}$$

$$IL_2 = \sqrt{I_2^2 + I_3^2 + (I_2 \times I_3)}$$

$$IL_3 = \sqrt{I_3^2 + I_1^2 + (I_1 \times I_3)}$$

Ways to transfer heat: 1. Radiation
                               2. Conduction
                               3. Convection

Specific heat of ice = 0.5 Btu (Btu required to change 1 lb of ice 1°F)

3-phase voltage and current deviation formulas:

Percentage voltage deviation =
$$\frac{\text{Maximum deviation from average voltage}}{\text{average voltage}} \times 100$$
The maximum voltage deviation is 2%

Percentage current deviation =
$$\frac{\text{maximum deviation from average current}}{\text{average current}} \times 100$$
The maximum current deviation is 10%

ABBREVIATIONS

1 Btu = British thermal unit = energy required to raise the temperature of 1 lb H_2O by 1°F (from 59° to 60°F)
= 350 calories
= 1 wooden kitchen match burned end to end

K = thermal conductivity: measure of heat flow through 1 ft², 1 in. thick, in 1 h, 1°F diff. between two faces of homogeneous material

C = thermal conductance: measure of heat flow through actual thickness of material used
$$C \text{ value} = \frac{K}{\text{thickness}}$$

R = thermal resistance: measure of ability to retard heat flow
$$= \text{reciprocal of } C = \frac{1}{C}$$

R is additive: R5 + R11 = R16

Higher R = higher insulating value

U = overall coefficient of heat-transmission rating of a total building section
= no. Btu/ft² of wall (etc.)/1 h/1°F
$$= \text{reciprocal of } R_{\text{total}}; U = \frac{1}{R_{\text{total}}}$$

Lower U = higher insulating value

Degree days: For estimating *heating* consumption:
For any one day, when the mean (average) temperature is less than 65°F, there exist as many degree days as there are °F difference between the mean temperature for the day and 65°F. Such degree days total for a heating season.
Varies with location.

Cooling hours: For estimating *cooling* consumption:
The number of hours in the cooling season when the temperature is over 80°F. (NAHB)
Varies with location.

EER: *Energy efficiency ratio:*
The ratio of net cooling capacity in Btu per hour to total rate of electrical input in watts under designated operating conditions. EER is an instantaneous rating that varies with the outdoor temperature.

$$EER = \frac{Btu}{Watts}$$

SEER = 10 minimum per model code
    = 12 to 15 is optimal

SEER: *Seasonal Energy Efficiency Ratio:*
The ratio of the cooling capacity in Btus for an entire cooling season divided by the total rate of electrical input in watts for the same cooling season.

$$SEER = \frac{Btu\ per\ cooling\ season}{Watts\ per\ cooling\ season}$$

COP: *Coefficient of Performance:*
The ratio of the rate of net heat removal or output to the rate of total energy (kWh) input. See also: model code definitions

HSPF: *Heating Seasonal Performance Factor:*
Used in reference to the heating efficiency of *heat-pump* systems only. Definition is the same as COP but obtained by calculating actual performance of the system at 5° outdoor temperature increments
SPF = 8.0 = good (varies with location).

Q = heat flow in Btu/h

T = temperature in °F (ΔT = temp. diff.)

A = Area in ft^2

U = Coef. of heat transfer in BTU/h/ft^2/°F

H = total heat of air at wet-bulb temp. Btu/lb

ΔH = enthalpy difference between entering and leaving air

SH = specific humidity in grains of moisture/lb of dry air (ΔSH = specific humidity difference for entering and leaving air)

cfm = ft^3/min

gpm = Gal/min

ESP = external static pressure

ADP = apparatus dew point

N.P.S.H. = net positive suction head

Cu ft = cubic foot

Ft^3 = cubic foot

Cu in = cubic inch

in^3 = cubic inch

CONVERSIONS

CONVERSION FACTORS AND FORMULAS

TO CONVERT	MULTIPLY BY	TO OBTAIN
Atmospheres	14.7	Pounds/sq. in.
Atmospheres	760	Millimeters of mercury
Bars	14.5	Pounds/sq. in.
Btu/h	0.0003929	Horsepower
Btu/h	0.2931	Watts
Cubic centimeters	0.06102	Cubic inches
Cubic inches	16.39	Cubic centimeters
Cubic feet	1728	Cubic inches
Cubic feet	7.48052	Gallons (US liquid)
Cu. meters/h	0.2778	Liters/s
Cu. meters/h	16.67	Liters/min
Cu. meters/h	0.589	Cu. feet/min
Cu. feet/min	0.4719	Liters/s
Cu. feet/min	28.32	Liters/min
Feet	0.3048	Meters
Gallons	0.1337	Cubic feet
Gallons	3.785	Liters
Grams	0.002205	Pounds
Grams/sq. cm.	2.0481	Pounds/sq. ft
Horsepower	42.44	Btu/min
Horsepower	0.7457	Kilowatts
Inches	2.540	Centimeters
Inches of mercury	0.4912	Pounds/sq. in.
Kilowatts-h	3413	Btu
Liters	0.2642	Gallons
Liters/s	2.12	Cu. feet/min
Liters/s	3.60	Cu. meters/h
Liters/min	0.0353	Cu. feet/min
Liters/min	0.060	Cu. meters/h
Meters	0.0006214	Miles (statute)
Meters/min	3.281	Feet/min
Million gal/day	1.54723	Cu. ft/s
Ounces	28.349	Grams
Pounds/sq. in.	0.06804	Atmospheres
Pounds/sq. in.	2.036	Inches of mercury
Scruples	20	Grains

PROPERTIES OF WATER AT 39.2°F

Density of water = 62.4 lb/ft^3

Specific heat of water = 1 Btu/lb/°F

Latent heat of vaporization =
    970 BTU/lb at 212°F and 1 atm =
    1054.3 BTU/lb at 70°F

Specific heat of ice = 0.5 Btu/lb/°F

Latent heat of fusion = 144 Btu/lb

1 gal of water = 8.33 lb

1 gallon of water = 231 in^3

1 lb of water = 7000 grains

CONVERSION FACTORS (CONSTANTS)

WATER

$500 = 8.33$ lb/gal $\times$ 60 min (Converts gpm to lb/h)

AIR

$4.5 = \dfrac{60 \text{ min}}{13.35 \text{ ft}^3/\text{lb}}$ (Converts CFM to lb/h)

$1.08 = 4.5 \times 0.241$ BTU/lb/°F (lb/h $\times$ sp. ht. of air)

$0.68 = \dfrac{4.5 \times 1054.3 \text{ BTU/lb}}{7000 \text{ gr/lb}}$

(4.5 combined with heat of vaporization of water at 70°F and grains per pound of water)

COMMON CONVERSIONS

1 gal water $= 231$ in^3

1 gal water $= 8.35$ lb

1 ft^3 water $= 62.4$ lb

1 ft^3 of water $= 1,728$ in^3

1 W $= 3.416$ Btu

1 horsepower (hp) $= 746$ W

7000 grains of moisture per pound of water

1 atmosphere (atm) $= 14.7$ psia at sea level

1 atm $= 29.92$ in. of mercury

1 atm $= 33.9$ ft of water column

psia = pounds per square inch absolute

psig = pounds per square inch gauge

types of energy = kinetic and potential

144 Btu to change 1 lb of water to ice at 32°F (vice versa)

0.5 Btu to change 1 lb of ice 1°F

970 Btu to change 1 lb of water to steam

0.48 Btu to change 1 lb of steam 1°F

latent heat: hidden heat, cannot be measured with a thermometer

sensible heat: heat that can be felt and measured with a thermometer

specific heat: amount of heat in Btus to change a substance 1°F

CONVERTING PRESSURE UNITS USED IN HIGH VACUUM

- 1 in. Hg (mercury) = 25.4 mm Hg = 25,400 microns
- 1 torr = 1 mm Hg = 1000 microns
- 1 millibar = 0.75 mm Hg
- 1 bar = 750 mm Hg = 29.53 in. Hg

CONVERTING HORSEPOWER TO AMPS

HORSEPOWER—AMPERE TABLE

APPROXIMATE HORSEPOWER		120 VOLTS		240 VOLTS	
		FULL LOAD	LOCKED* ROTOR	FULL LOAD	LOCKED* ROTOR
1/6	AC	4.4	26.4	2.2	13.2
	DC	—	—	—	—
1/4	AC	5.8	34.8	2.9	17.4
	DC	2.9	29.0	1.5	15.0
1/3	AC	7.2	43.2	3.6	21.6
	DC	3.6	36.0	1.8	18.0
1/2	AC	9.8	58.8	4.9	29.9
	DC	5.2	52.0	2.6	26.0
3/4	AC	13.8	82.8	6.9	41.4
	DC	7.4	74.0	3.7	37.0
1	AC	16.0	96.0	8.0	48.0
	DC	9.4	94.0	4.7	47.0
1½	AC	20.0	129.0	10.0	60.0
	DC	13.2	132.0	6.6	66.0
2	AC	24.0	144.0	12.0	72.0
	DC	17.0	170.0	8.5	85.0
3	AC	34.0	204.0	17.0	102.0
	DC	25.0	250.0	12.2	122.0
5	AC	56.0	336.0	28.0	168.0
	DC	—	—	20.0	200.0
7 1/2	AC	80.0	480.0	40.0	240.0
	DC	—	—	29.0	290.0

* Locked rotor ratings shown are 6 times full load on AC and 10 times full load on DC. The above chart is offered as a guide only, as all motors do not necessarily came within the maximum ratings shown in chart.

GLOSSARY

Absolute Pressure Gauge pressure plus atmospheric pressure. Gauge pressure plus 14.7 at sea level.

Accumulator Storage tank which receives liquid refrigerant from the evaporator and prevents it from flowing into suction line.

Air Changes The amount of air leakage through a building in terms of the number of building volumes exchanged. Expressed in air changes per hour.

Air Conditioner Device used to control temperature, humidity, cleanliness, and movement of air in conditioned space.

Alternating Current Abbreviated ac. Current that reverses polarity or direction periodically. It rises from zero to maximum strength and returns to zero in one direction then goes to similar variation in the opposite direction. This is a cycle which is repeated at a fixed frequency. It can be single phase, two phase, three phase, and poly phase. Its advantage over direct or undirectional current is that its voltage can be stepped up by transformers to the high values which reduce transmission costs.

Ambient Temperature Temperature of fluid (usually air) which surrounds an object.

American Wire Gauge Abbreviated AWG. A system of numbers which designate cross-sectional area of wire. As the diameter gets smaller, the number gets larger, e.g., AWG #14 = 0.0641 in., AWG #12 = 0.0808 in.

Ammeter An electric meter used to measure current.

Ampere Unit of electric current equivalent to flow of one coulomb per second.

Armature The moving or rotating component of a motor, generator, relay or other electromagnetic device.

Atom Smallest particle of element.

Automatic Expansion Valve (AEV) Pressure controlled valve used as a metering device. Tries to maintain evaporator pressure.

Back Pressure Pressure in low side of refrigerating system; also called suction pressure or low side pressure.

Back Seat The position of a service valve that is all the way counter-clockwise or the stem is backed out.

Blower A centrifugal fan.

Brazing Soldering with a filler material whose melting temperature is higher than 800°F, but below the melting temperature of the metal brazed.

Break Electrical discontinuity in the circuit generally resulting from the operation of a switch or circuit breaker.

British Thermal Unit (Btu) Quantity of heat required to change temperature of one pound of water one degree F.

Burnout Accidental passage of high voltage through an electrical circuit or device which causes damage.

Capacitor Type of electrical storage device used in starting and/or running circuits on many electric motors.

Capillary Tube A type of refrigerant control consisting of several feet of tubing having small inside diameter. The refrigerant flow rate depends on the condenser pressure.

Cfm Cubic feet per minute.

Charge The amount of refrigerant in a system.

Charging Cylinder A device for charging a predetermined weight of refrigerant into a system.

Check Valve A one-direction valve. Allows flow one way.

Circuit A tubing, piping or electrical wire installation which permits flow from the energy source back to energy source.

Circuit Breaker A device that senses current flow, and opens when its rated current flow is exceeded.

Clearance Space in cylinder not occupied by piston at end of compression stroke.

Coefficience of Performance (COP) Usable energy output divided by the energy input.

Coil A wound conductor, which creates a strong magnetic field with current passage.

Cold Cold is the absence of heat.

Commercial Buildings Such buildings as stores, shops, restaurants, motels, and large apartment buildings.

Compound Gauge Pressure gauge that has scales both above and below atmospheric pressure.

Compression Ratio The absolute discharge pressure divided by the absolute suction pressure for a compressor.

Compressor The pump of a refrigerating mechanism which draws a vacuum or low pressure on cooling side of refrigerant cycle and squeezes or compresses the gas into the high pressure or condensing side of the cycle.

Compressor Displacement The volume discharged by a compressor in one rotation of the crankshaft.

Compressor Seal Leakproof seal between crankshaft and compressor body.

Condensible A gas which can be easily converted to liquid form.

Condensation Liquid which forms when a gas is cooled below its dew point.

Condenser The part of refrigeration mechanism which receives hot, high pressure refrigerant gas from compressor and cools it until it returns to liquid state.

Condenser Water Pump Forced water moving device used to move water through condenser.

Condensing Pressure The refrigerant pressure inside the condenser coil.

Condensing Temperature The temperature at which refrigerant condensation is taking place inside the condenser.

Condensing Unit Refrigeration unit consisting of a compressor, condenser, and controls.

Conductance (Thermal) "C" factor—The time rate of heat flow per unit area through a material.

Conduction (Thermal) Particle to particle transmission of heat.

Conductivity The ability of a substance to allow the flow of heat or electricity.

Conductor Material or substance which readily passes electricity.

Constrictor Tube or orifice used to restrict flow.

Contact The part of a switch or relay that carries current.

Contactor A device for making or breaking load-carrying contacts by a pilot circuit through a magnetic coil.

Contaminant A substance (dirt, moisture, etc.) foreign to refrigerant or refrigerant oil in system.

Controls Devices designed to regulate the gas, air, water or electricity.

Convection Transfer of heat by means of movement of a fluid. The warm fluid rises and the cold fluid falls.

Cooling Anticipator A resistor in a room thermostat that causes the cooling cycle to begin prematurely.

Cooling Tower Device which cools water by water evaporation in air. Water is cooled to the wet bulb temperature of air.

Coulomb An electrical unit of charge, one coulomb per second equals one ampere or 6.25×10^{18} electrons past a given point in one second.

Counter EMF Counter electromotive force; the EMF induced in a coil, which opposes applied voltage.

Counterflow Flow in opposite directions.

"Cracking" a Valve Opening valve a small amount.

Crankshaft Seal Leakproof joint between crankshaft and compressor body.

Current The flow of electrons through a conductor.

Cut-In The temperature or pressure at which an automatic control switch closes.

Cut-Out The temperature or pressure at which an automatic control switch opens.

Cylinder Head Part which encloses compression end of compressor cylinder.

Dead Band A range of temperature in a heating/ cooling system in which no heating or cooling is supplied.

Decibel Unit used for measuring relative loudness of sounds.

Degree-Day Unit that represents one degree of difference from given point in average outdoor temperature of one day.

Dehydrate To remove water.

Delta Connection The connection in a three-phase system in which terminal connections are triangular similar to the Greek letter delta.

Density Weight per unit volume.

Desiccant Substance used to collect and hold moisture.

Design Load The amount of heating or cooling required to maintain inside conditions when the outdoor conditions are at design temperature.

Dew Point Temperature at which vapor (at 100 percent humidity) begins to condense.

Differential As applied to refrigeration and heating: difference between "cut-in" and "cut-out" temperature or pressure.

Diffuser Air distribution outlet designed to direct airflow into a room.

Dilution Air Air which enters a draft hood and mixes with the flue gases.

Direct Current Abbreviated dc. Electric current that flows only in one direction.

Direct Expansion Evaporator An evaporator containing liquid and vapor refrigerant.

Disconnecting Switch A knife switch that opens a circuit.

Domestic Hot Water The heated water used for cooking, washing, etc.

Double-Pole Switch Simultaneously opens and closes two wires of a circuit.

Double Thickness Flare Tubing end which has been formed into two-wall thickness.

Drier A substance or device used to remove moisture from a refrigeration system.

Drift Entrained water carried from a cooling tower by wind.

Drip Pan Pan-shaped panel or trough used to collect condensate from evaporator coil.

Dry Bulb Temperature Air temperature as indicated by ordinary thermometer.

Dry Ice Solid carbon dioxide.

E Symbol for volts.

Eccentric A circle or disk mounted off center on a shaft.

Effective Area Actual net flow area.

Efficiency (electrical) A percentage value denoting the ratio of power output to power input.

Electric Field A magnetic region in space.

Electric Heating Heating system in which heat is produced from electrical resistance units.

Electrolyte A solution of a substance (liquid or paste) that is capable of conducting electricity.

Electromagnet A magnet created by the flow of electricity through a coil of wire.

Electromotive Force Abbreviated EMF. The difference in potential electrical energy between two points.

Electronics Field of science dealing with electron devices and their uses.

Electrostatic Filter Type of filter which gives particles of dust electric charge.

End Bell End structure of electric motor which usually holds motor bearings.

End Play Slight movement of shaft along center line.

Energy Efficiency Ratio (EER) Btu/hr of cooling produced per watt of electrical input.

Enthalpy Total amount of heat in one pound of a substance.

EPR (Evaporation Pressure Regulator) An automatic pressure regulating valve to maintain a predetermined pressure in the evaporator. Located in suction line.

Equalizer Tube Device used to maintain equal pressure or equal liquid levels between two containers.

Evaporation A term applied to the changing of a liquid to a gas.

Evaporative Condenser A device which uses a water spray to cool a condenser.

Evaporator Part of a refrigerating mechanism in which the refrigerant evaporizes and absorbs heat.

Evaporator, Dry Type An evaporator into which liquid refrigerant is fed from a pressure reducing device and boil off to a vapor. The vapor (dry) leaves the evaporator by way of the suction line.

Evaporator Fan Fan which moves air through the evaporator.

Evaporator, Flooded An evaporator containing liquid refrigerant at all times. Liquid leaves the evaporator.

Exfiltration Air flow outward through a room.

Exhaust Opening Any opening through which air is removed from a space.

Expansion Valve A device in refrigerating system which maintains a pressure difference between the high side and low side. Also known as thermostatic expansion valve (TXV).

Expendable Refrigerant System System which discards the refrigerant after it has evaporated.

Extended Surface Heat transfer surface, one side of which is increased in area by the use of fins.

Fahrenheit The common scale of temperature measurement in the English system of units.

Fan-Coil A terminal unit consisting of a finned-tube coil and a fan in a single enclosure.

Farad A large unit of measurement for capacitance.

Feedback The transfer of energy output back to input.

Female Thread A thread on the inside of a pipe or fitting.

Field The space involving the magnetic lines of force.

Filter Device for removing small particles from a fluid.

Fin Comb Comb-like device used to straighten the metal fins on coils.

Flare An angle formed at the end of a tube.

Flare Nut Fitting used to clamp tubing flare against another fitting.

Float Valve Type of valve which is operated by sphere which floats on liquid surface and controls level of liquid.

Floc Point The temperature at which the wax in oil will start to separate from the oil.

Flooded System Type of refrigerating system in which liquid refrigerant fills evaporator. Liquid leaves the evaporator.

Flow Switch An automatic switch that senses fluid flow.

Fluid Substance in a liquid or gaseous state.

Flux-Brazing, Soldering Substance applied to surfaces to be joined to free them from oxides.

Flux (Electrical) The electric or magnetic lines of force in a region.

Flux, Magnetic Lines of force of a magnet.

Foaming Formation of a foam in an oil-refrigerant mixture due to rapid evaporation of refrigerant dissolved in the oil that occurs in the compressor.

Forced Convection Transfer of heat resulting from forced movement of liquid or gas by means of fan or pump.

Free Area The total minimum area of the openings in a grille.

Freezing Change of state from liquid to solid.

Freezing Point The temperature at which a liquid will solidify upon removal of heat.

Freon Trade name for a family of synthetic chemical refrigerants manufactured by DuPont, Inc.

Frequency The number of complete cycles in a unit of time.

Friction Head In a hydronic system, the loss in pressure resulting from the flow of water in the piping system.

Front Seat The position of a service valve that is all the way clockwise or turned in blocking the flow of refrigerant.

Frost Back Condition in which liquid refrigerant flows from evaporator into suction line and causes the suction line to be covered with frost.

Full-load Amperage (FLA) The amount of amperage an inductive load (motor) will draw at its full design load.

Fuse Electrical safety device consisting of strip of fusible metal in circuit which melts when the circuit is overloaded.

Fusible Plug A plug or a mode with a metal of low melting temperature to release pressures in case of fire.

Gasket A resilient material used between mating surfaces to provide a leakproof seal.

Gas—Noncondensible A gas which will not form into a liquid under pressure-temperature conditions.

Generator A machine that converts mechanical energy into electrical energy.

Grille An ornamental or louvered opening placed at the end of an air passageway.

Ground Connection between an electrical circuit and earth.

Ground Wire An electrical wire which will safely conduct electricity from a structure into the ground.

Halide Refrigerants Family of refrigerants containing halogen chemicals.

Halide Torch Type of torch used to detect halogen refrigerant leaks.

Head As used in this course, head refers to a pressure difference. (See Pressure Head, Pump Head, Available Head, High side or discharge pressure)

Head (Total) In flowing fluid, the sum of the static and velocity pressures at the point of measurement.

Header A piping arrangement for inter-connecting two or more supply or return tappings of a boiler.

Head Pressure Pressure which exists in condensing side of refrigerating system. Also known as the high side or discharge pressure.

Head-Pressure Control A pressure-activated control that prevents the head pressure from falling too low.

Head, Static Pressure of fluid expressed in terms of height of column of the fluid, such as water or mercury.

Head, Velocity In flowing fluid, height of fluid equivalent to its velocity pressure.

Heat Form of energy the addition of which causes substances to rise in temperature; energy associated with random motion of molecules.

Heat Anticipator A resistor in a room thermostat that shuts the heating cycle off prematurely preventing overshooting of the room temperature.

Heat Exchanger Device used to transfer heat from a warm or hot surface to a cold or cooler surface.

Heat Load The amount of heat that must be removed from the air conditioned or refrigerated space in order to maintain the design temperature.

Heat Loss The rate of heat transfer from a heated building to the outdoors.

Heat of Compression The heat added to the refrigerant by the work being done by the compressor.

Heat of Fusion The heat released in changing a substance from a liquid state to a solid state.

Heat of Respiration When carbon dioxide and water are given off by foods in storage.

Heat Pump Air-conditioning system that is reversible so as to be able to remove heat from or add heat to a given space.

Hermetic Completely sealed.

Hermetic Motor Compressor drive motor sealed within same casing which contains compressor.

High Side Parts of a refrigerating system which are under condensing or high side pressure.

Horsepower A unit of power equal to 33,000 foot pounds of work per minute.

Hot Gas Defrost A defrosting system in which hot refrigerant gas from the high side is directed through evaporator.

Hot Gas Line The line that carries the hot discharge gas from the compressor to the condenser.

Humidifiers Device used to add humidity.

Humidistat An electrical control which is operated by changing humidity.

Humidity Moisture in air.

Impedence A type of electrical resistance that is only present in alternating current circuits.

Impeller Rotating part of a centrifugal pump.

Inductance The characteristic of an alternating current circuit to oppose a change in current flow.

Induction The act that produces induced voltage in an object by exposure to a magnetic field.

Induction Motor An AC motor which operates on principle of rotating magnetic field. Rotor has no electrical connection, but receives electrical energy by transformer action from field windings.

Inductive Load A device that uses electrical energy to produce motion.

Inductive Reactance Opposition, measured in ohms, to an alternating or pulsating current.

Infiltration The air that leaks into the refrigerated or air conditioned space.

IR Drop Voltage drop resulting from current flow through a resistor.

Joule A measure of heat in the metric system.

Journal, Crankshaft Part of shaft which contacts the bearing.

Junction Box Group of electrical terminals housed in protective box or container.

Kilopascal A measure of pressure in the metric system.

Kilowatt Unit of electrical power, equal to 1000 watts.

Kilowatt-Hour A unit for measuring electrical energy equal to 3,413 Btu.

King-Valve The shut off valve located at the outlet of the receiver.

Latent Heat Heat characterized by a change of state of the substance concerned.

Line Voltage Voltage existing at wall outlets or terminals of a power line system above 30 V.

Liquid Line The tube which carries liquid refrigerant from the condenser or liquid receiver to the pressure reducing device.

Liquid Nitrogen Nitrogen in liquid form.

Liquid Receiver Cylinder connected to condenser outlet for storage of liquid refrigerant in a system.

Load A device that converts electrical energy to another form of usable energy.

Locked Rotor Amperage (LRA) The amount of energy a motor will draw under stalled conditions.

Lock-Out Relay A control scheme that prevents the restarting of a compressor after a safety control opens, even after the safety control resets itself.

Louvers Sloping, overlapping boards or metal plates intended to permit ventilation and shed falling water.

Low Pressure Cut-Out Device used to keep low side evaporating pressure from dropping below certain pressure.

Low Side That portion of a refrigerating system which is under the lowest evaporating pressure.

Low Voltage Control Controls designed to operate at voltages of 20 to 30 V.

Magnetic Field The space in which a magnetic force exists.

Make-up Air The air which is supplied to a building to replace air that has been removed by an exhaust system.

Make-up Water The water required to replace the water lost from a cooling tower by evaporation, drift, and bleedoff.

Male Thread A thread on the outside of a pipe or fitting.

Manifold Gauge A device constructed to hold compound and high pressure gauges and valved to control flow of fluids through it.

Manometer Instrument for measuring pressure of gases and vapors.

Mass A quantity of matter cohering together to make one body which is usually of indefinite shape.

Mean Temperature Difference The average temperature between the temperature before process begins and the temperature after process is completed.

Mechanical Cycle Cycle which is a repetitive series of mechanical events.

Mega Prefix for one million; e.g., megohm—one million ohms.

Megohm 1,000,000 ohms of electrical resistance.

Melting Point Temperature at atmospheric pressure, at which a substance will melt.

Meter Metric unit of linear measurement.

MFD Abbreviation for microfarad or μfd

Micro A combining form denoting one millionth.

Microfarad The unit of measurement for capacitance, equal to a millionth of a farad.

Micron Unit of length in metric system; a thousandth part of one millimeter.

Micron Gauge Instrument for measuring vacuums very close to a perfect vacuum.

Milli A combining form denoting one thousandth; example, millivolt, one thousandth of a volt.

Modulating A type of device or control which tends to adjust by increments (minute changes) rather than by either full on or full off operation.

Moisture Indicator Instrument used to measure moisture content of a refrigerant.

Molecule The smallest portion of an element or compound which retains the identity and characteristics of the element or compound.

Motor Burnout Condition in which the insulation of electric motor has deteriorated by overheating.

Motor Control Device to start and/or stop a motor at certain temperature or pressure conditions.

Multimeter A volt-ohm meter.

National Electrical Code Abbreviated NEC. A code of electrical rules based on fire underwriters' requirements for interior electric wiring.

Natural Convection Circulation of a gas or liquid due to difference in density resulting from temperature differences.

Needle Valve A valve used to accurately control very low flow rates of fluids.

Nominal Size Tubing Tubing measurement which has an inside diameter the same as iron pipe of the same stated size.

Noncondensible Gas Gas which does not change into a liquid at operating temperatures and pressures.

Nonferrous Group of metals and metal alloys which contain no iron; also metals other than iron or steel. In heating systems the principal non-ferrous metals are copper and aluminum.

Normally Closed Switch contacts closed with the circuit deenergized.

Normally Open Switch contacts open with the circuit deenergized.

Off Cycle That part of a refrigeration cycle when the system is not operating.

Ohm A unit of resistance.

Oil Rings Expanding rings mounted in grooves and piston; designed to prevent oil from moving into compression chamber.

Oil Separator A device used to separate refrigerant oil from refrigerant gas and return the oil to crankcase of compressor. It is located in discharge line.

Open Compressor One with a separate motor or external drive.

Orifice Accurate size opening for controlling fluid flow.

Outside Air Air exterior to refrigerated or conditioned space.

Overload Load greater than load for which system or mechanism was intended.

Overload Protector A device which will stop operation of unit if dangerous conditions arise.

Packaged Unit A complete refrigeration or air conditioning system in which all the components are factory assembled into a single unit.

Parallel Circuit connected so current has two or more paths to follow.

Parallel Circuit An electrical circuit in which there is more than one different path across the power supply.

Partial Pressures Condition where two or more gases occupy a space and each one creates part of the total pressure.

Pinch-Off Tool Device used to press walls of a tubing together until fluid flow ceases.

Piston Close fitting part which moves up and down in a cylinder.

Piston Displacement Volume obtained by multiplying area of cylinder bore by length of piston stroke.

Pitch Pipe slope.

Potential The amount of voltage between points of a circuit.

Power Time rate at which work is done or energy emitted.

Power Element Sensitive element of a temperature operated control.

Power Factor The rate of actual power as measured by a wattmeter in an alternating circuit to the apparent power determined by multiplying amperes by volts.

Pressure Force per unit area.

Pressure Limiter Device which remains closed until a certain pressure is reached and then opens and releases fluid to another part of system.

Pressure Motor Control A device which opens and closes on electrical circuit as pressures change to desired pressures.

Pressure-Operated Altitude (POA) Valve Device which maintains a constant low side pressure independent of altitude of operation.

Pressure-Reducing Device The device used to produce a reduction in pressure and corresponding boiling point before the refrigerant is introduced into the evaporator. Also known as metering device or flow control.

Pressure Reducing Valve A diaphragm operated valve installed in the make-up water line of a hot water heating system.

Pressure Regulator A device for controlling a uniform outlet gas pressure.

Pressure Relief Valve A device for protecting a tank from excessive pressure by opening at a predetermined pressure.

Process Tube Length of tubing fastened to hermetic unit dome, used for servicing unit.

psi Pressure measured in pounds per square inch.

psia Pressure measured in pounds per square inch absolute.

psig A symbol or initials used to indicate pressure in pounds per square inch gauge. It is the same as psi.

Psychrometer or Wet Bulb Hygrometer An instrument for measuring dry-bulb and wet-bulb temperature.

Psychrometric Chart A chart that shows relationship between the temperature and moisture content of the air.

PTC Positive temperature coefficient; a thermister used as a start relay.

Pump A motor driven device used to mechanically circulate water in the system.

Pump Down A service procedure where the refrigerant is pumped and trapped into the receiver.

Pump Head The difference in pressure on the supply and intake sides of the pump created by the operation of the pump.

Purging Releasing compressed gas to atmosphere through some part or parts for the purpose of removing contaminants from that part or parts.

Quenching Submerging hot solid object in cooling fluid.

Quick Connect Coupling A device which permits fast connecting of two fluid lines.

Radiation Transfer of heat by heat rays.

Range Pressure or temperature settings of a control.

Reactance Opposition to alternating current by either inductance or capacitance or both.

Reciprocating Action in which the motion is back and forth in a straight line.

Recirculated Air Return air passed through the conditioner before being again supplied to the conditioned space.

Reducing Fitting A pipe fitting designed to change from one pipe size to another.

Refrigerant Substance used in refrigerating mechanism to absorb heat in evaporator coil and to release its heat in a condenser.

Refrigerant Charge Quantity of refrigerant in a system.

Refrigerating Effect The amount of heat in Btu/h the system is capable of transferring.

Refrigeration The process of transferring heat from one place to another.

Refrigeration Oil Specially prepared oil used in refrigerator mechanism. It circulates with referigerant.

Register Combination grille and damper assembly covering on an air opening or end of an air duct.

Relative Humidity A ratio of the weight of moisture that air actually contains at a certain temperature as compared to the amount that it could contain if it were saturated.

Relay Electrical mechanism which uses small current in control circuit to operate a valve switch in an operating circuit.

Relief Opening The opening in a draft hood to permit ready escape to the atmosphere of flue products.

Relief Valve Safety device designed to open before dangerous pressure is reached.

Resistance The opposition to current flow by a physical conductor.

Return Air Air returned from conditioned or refrigerated space.

Reverse Acting Control A switch controlled by temperature and designed to open on temperature drop and close on temperature rise.

Rotary Compressor Mechanism which pumps fluid by using rotating motion.

Rotor Rotating part of a mechanism.

Run Winding Electrical winding of motor which has current flowing through it during normal operation of motor.

Saddle Valve Valve body shaped so it may be silver brazed to refrigerant tubing surface. It is an access valve.

Safety Control Device which will stop the refrigerating unit if unsafe pressures and/or temperatures are reached.

Safety Plug Device which will release the contents of a container above normal pressure conditions and before rupture pressures are reached.

Saturated Vapor A vapor condition which will result in condensation into droplets of liquid as vapor temperature is reduced.

Saturation A condition existing when a substance contains maximum of another substance for that temperature and pressure.

Schrader Valve Spring loaded device which permits fluid flow when a center pin is depressed.

Scotch Yoke Mechanism used to change reciprocating motion into rotary motion or vice-versa.

SCR A solid state device (silicon controlled rectifier) used to modulate the capacity of an electric heating element.

Secondary Voltage The output, or load-supply voltage, of a transformer.

Second Law of Thermodynamics Heat will flow only from material at certain temperature to material at lower temperature.

Semihermetic Compressor A serviceable hermetic compressor.

Sensible Heat A term used in heating and cooling to indicate any portion of heat which changes only the temperature of the substances involved; also heat which changes the temperature of a substance without changing its form.

Series A circuit with one continuous path for current flow.

Serviceable Hermetic Hermetic unit housing containing motor and compressor assembled by use of bolts or threads.

Service Valve Device used by service technicians to check pressures and charge refrigerating units.

Shaded Pole Motor A small ac motor used for light start loads.

Shaft Seal A device used to prevent leakage between shaft and housing.

Short Circuit A low-resistance connection (usually accidental and undesirable) between two parts of an electrical circuit.

Short Cycling Refrigerating system that starts and stops more frequently than it should.

Sight Glass Glass tube or glass window in refrigerating mechanism which shows amount of refrigerant, or oil in system.

Silver Brazing Brazing process in which brazing alloy contains some silver as part of joining alloy.

Sling Psychrometer Humidity measuring device with wet and dry bulb thermometers.

Slugging A condition where liquid refrigerant is entering an operating compressor.

Solar Heat Heat from visible and invisible energy waves from the sun.

Soldering Joining two metals by adhesion of a low melting temperature metal (less than 800°F).

Solenoid A movable plunger activated by an electromagnetic coil.

Solenoid Valve Valve actuated by magnetic action by means of an electrically energized coil.

Specific Gravity For a liquid or solid, the ratio of its density compared to water. For a vapor, the ratio of its density to air.

Specific Heat The heat absorbed (or given up) by a unit mass of a substance when its temperature is increased (or decreased) by 1-degree common units: Btu per (pound) (Fahrenheit degree), calories per (gram) (centigrade degree). For gases, both specific heat at constant pressure (c_p) and specific heat at constant volume (c_v) are frequently used. In air-conditioning, c_p is usually used.

Specific Volume The volume of a substance per unit mass: the reciprocal of density units: cubic feet per pound, cubic centimeters per gram, etc.

Split-Phase Motor Motor with two stator windings.

Split System Refrigeration or air-conditioning installation which places condensing unit remote from evaporator.

Squirrel Cage Fan which has blades parallel to fan axis and moves air at right angles to fan axis.

Standard Conditions Temperature of 68°F, pressure of 29.92 in. of Hg and relative humidity of 30 percent.

Starting Relay An electrical device which connects and/or disconnects starting winding of electric motor.

Starting Winding Winding in electric motor used only during brief period when motor is starting.

Static Pressure The pressure exerted against the inside of a duct or pipe in all directions.

Stator Stationary part of electric motor. The stator has the motor windings.

Strainer A screen used to retain solid particles while liquid passes through.

Stratification Condition in which air lies in temperature layers.

Subcooling Cooling of liquid refrigerant below its condensing temperature.

Sublimation Condition where a substance changes from a solid to a gas without becoming a liquid.

Suction Line Tube or pipe used to carry refrigerant gas from evaporator to compressor.

Suction Service Valve A two-way manual-operated valve located at the inlet to compressor. The lowside hose can be connected to this valve.

Superheat Temperature of vapor above boiling temperature of its liquid at that pressure.

Swaging Enlarging one tube end so another tube of same size will fit within.

Sweating Condensation of moisture from air on cold surface.

Temperature Degree of hotness or coldness as measured by a thermometer.

Therm A unit of heat having a value of 100,000 Btu.

Thermal Conductivity The ability of a material to transmit heat. Known as c-value.

Thermal Resistance The resistance a material offers to the transmission of heat. Known as R-value.

Thermistor A semiconductor which has electrical resistance that varies with temperature.

Thermocouple Device which generates electricity, using the principle that if two dissimilar metals are welded together and junction is heated, a voltage will develop across open ends.

Thermostat Switch responsive to temperature.

Thermostatic Expansion Valve A valve operated by temperature and pressure within evaporator coil.

Thermostatic Valve Valve controlled by thermostatic elements.

Ton Refrigerating effect equal to the melting of one ton of ice in 24 hours. Commonly referred to as 12,000 Btuh.

Torque Turning or twisting force.

Transformer A device designed to change voltage.

"U" Factor Thermal conductivity.

Ultraviolet Invisible radiation waves with frequencies shorter than wave lengths of visible light and longer than X-Ray.

Underwriters Laboratories Abbreviated UL. Underwriters Laboratories, Inc. maintain and operate independent laboratories for the examination and testing of devices.

Vacuum Pressure below atmospheric pressure.

Vacuum Pump Special high efficiency compressor used for creating high vacuums.

Valve Device used for controlling fluid flow.

Valve Plate Part of compressor located between top of compressor body and head which contains compressor valves.

Variable Pitch Pulley Pulley which can be adjusted to provide different pulley ratios.

Velocity Pressure The pressure exerted in direction of flow.

Ventilation The introduction of outdoor air into a building by mechanical means.

Viscosity Measure of a fluid's ability to flow.

Volt The unit of electrical potential or pressure.

Voltage Relay One that functions at a predetermined voltage value. Also known as a potential relay.

Volumetric Efficiency Ratio of the actual performance of a compressor and calculated performance.

VOM A meter that measures voltage, resistance (ohms) and milliamps.

Water Column Abbreviated as WC. A unit used for expressing pressure.

Watt A unit of electrical power.

Zone That portion of a building whose temperature is controlled by a single thermostat.

INDEX